AF893421

Springer Geography

The Springer Geography series seeks to publish a broad portfolio of scientific books, aiming at researchers, students, and everyone interested in geographical research. The series includes peer-reviewed monographs, edited volumes, textbooks, and conference proceedings. It covers the major topics in geography and geographical sciences including, but not limited to; Economic Geography, Landscape and Urban Planning, Urban Geography, Physical Geography and Environmental Geography. Submit a proposal: Proposals for the series will be considered by the Series Advisory Board. A book proposal form can be obtained from the Publisher, Dr. Robert K. Doe (robert.doe@springer.com).

More information about this series at http://www.springer.com/series/10180

Ján Buček • Andrew Ryder
Editors

Governance in Transition

Springer

Editors
Ján Buček
Department of Human Geography and Demography
Comenius University
Bratislava, Slovakia

Andrew Ryder
Department of Geography
University of Portsmouth
Portsmouth, UK

ISSN 2194-315X ISSN 2194-3168 (electronic)
Springer Geography
ISBN 978-94-007-5502-4 ISBN 978-94-007-5503-1 (eBook)
DOI 10.1007/978-94-007-5503-1

Library of Congress Control Number: 2015936134

Springer Dordrecht Heidelberg New York London
© Springer Science+Business Media Dordrecht 2015
This work is subject to copyright. All rights are reserved by the Publisher, whether the whole or part of the material is concerned, specifically the rights of translation, reprinting, reuse of illustrations, recitation, broadcasting, reproduction on microfilms or in any other physical way, and transmission or information storage and retrieval, electronic adaptation, computer software, or by similar or dissimilar methodology now known or hereafter developed.
The use of general descriptive names, registered names, trademarks, service marks, etc. in this publication does not imply, even in the absence of a specific statement, that such names are exempt from the relevant protective laws and regulations and therefore free for general use.
The publisher, the authors and the editors are safe to assume that the advice and information in this book are believed to be true and accurate at the date of publication. Neither the publisher nor the authors or the editors give a warranty, express or implied, with respect to the material contained herein or for any errors or omissions that may have been made.

Printed on acid-free paper

Springer Science+Business Media B.V. Dordrecht is part of Springer Science+Business Media (www.springer.com)

Preface

This book explores some of the practical issues involved in trying to decentralise government, reshape territorial administrative systems, and improve intergovernmental cooperation. There is often a profound mismatch between policy aims and policy implantation, as this book shows. The chapters were prepared within the framework of the International Geographical Union Commission on Geography and Governance. Although the contributions reflect experience in governance and administration in countries undergoing economic and political transition, this includes not only countries from Central and Eastern Europe, but countries from Western Europe, and even the Republic of Korea. As is the tradition within Commission activities, this book provides an opportunity for young scientists and doctoral students to publish together with distinguished scholars. As well, the book brings together experts from a wide variety of fields, including economics, geography, political science, and town planning. Many of the authors have worked in local or regional government. They also include a range of people with experience as consultants for local and national governments and international bodies. The result is a collection of papers covering a range of technical and practical, as well as theoretical, issues in contemporary governance within different social, economic and institutional contexts. Some of them deal with the latest changes in modes of governance, and evaluate introduced reforms. Others analyse longer term development in the context of historical changes. The first part of the book provides seven chapters on governance and development in large cities, intergovernmental cooperation, and the construction of regional identities. It starts with *Tomasz Kaczmarek's* chapter on top-down and bottom-up metropolitan integration in Poland. He describes progress in metropolitan area management in Poland. The rapid spread of metropolitan settlements through suburbanisation has created problems in managing infrastructure and the environment, and complicated the creation of regional plans and development strategies. Consequently, governments at all levels have both competed for investment and development, and tried to develop strategies for cooperation and collaboration on managing regional problems and policies. Proposals for changes in governance come from national government (proposed legislative solutions) and from local governments (the establishment of partnerships, associations and joint metropolitan strategies). The next two chapters focus on various aspects of global cities' development. *Andrew Ryder* looks at the relationship between devolution and local development. Looking at London and Seoul, he discusses the impact of local government autonomy on development.

The United Kingdom and Korea are among the most centralised countries in the Organisation for Economic Cooperation and Development. However, even with limited powers, London has succeeded in establishing a strong local agenda and successfully lobbied for new investment in transport and housing. *Pat Fensham* evaluates Sydney's Metropolitan Strategy and its main principles. As in the case of Poland, he stresses the importance of intergovernmental relations in creating and implementing a strategy, including State and Federal government involvement, as well as balancing various scales of interests acting on the entire metropolitan region. The next two chapters discuss changes in larger post-socialist cities, city regions, and larger meso-regions – Poznan (Poland) and South Trans-Danubia, focussed on Pecs (Hungary). *Łukasz Mikuła* addresses issues of metropolitan reforms and administrative adjustments. After reviewing concepts of urban and metropolitan government and governance, he looks at how they might apply to post-socialist cities such as Poznan. In the absence of legislation initiated at the national level, he emphasises the importance of voluntary co-operation among local-governments in metropolitan regions. The last two chapters in the section look at problems involved in governance across larger regions – trans-regional governance, and at problems involved in creating new area-wide networks which bring together governmental and non-governmental organisations. *Ilona Pálné Kovács* focuses on region-building processes in Hungary, looking at efforts to encourage cooperation imposed from above, as well as the success of cooperation developed from below. She discusses the formation of networks of actors at the regional level in Hungary, including ways in which networks and decision making have adapted to embrace EU regional policy. This contribution also portrays the opinions of regional elites on regional reform. According to them, regionalisation in Hungary has failed due to a lack of preparation and consensus building. *Marek Furmankiewicz*, *Wojciech Knieć* and *Jane Atterton* examine the implementation of a new form of rural governance through territorial partnerships in Poland from 2004 to 2008 and their subsequent impact. They examine the institutional base of rural governance in Poland, and the creation of local government partnerships, initially under the European Union's Leader+programme, and its successor. They compare Polish experience with experience in the UK, noting that in the United Kingdom, there is a much stronger tradition of local action and cooperation. Their final recommendations focus on improving governance through cross-sectoral partnerships in the rural areas of post-communist states.

The last 20 years have witnessed the growth of cross-border and international co-operation, particularly within the European Union. Therefore, the second part of the book contains four chapters presenting various aspects of development in this field. *David Rylander* and *Pontus Tallberg* deal primarily with transnational cooperation between regions without joint borders in Europe. They compare the co-operation activities of regions in Sweden, Germany, Denmark and the Czech Republic. Various types of strategies, organisation, formalisation, and fields of co-operation are identified.

Péter Balogh examines the city of Szczecin located on the Polish-German border. He discusses its evolving spatial orientation, focussing on its current geopolitical position on the Polish-German border, within a short distance of Berlin. He also looks at formal and informal cross-border cooperation, and assesses the prospects for future development. Transformation in former 'Black Triangle', located in the tri-border region between Germany, Poland and the Czech Republic, is depicted by *Rick Glöckner and Isolde Roch*. Until recently, this region was known for its mining and industrial activities, and their deleterious impacts on the environment (especially high air pollution). The authors assess the different economic, social, and demographic trajectories of the three national component parts of the region after 1989. They found that economic and political conditions at the national level played a strong role in regional estructuring, leading to growth in areas on the Czech and Polish sides of the borders, whereas the part of the region within Germany has stagnated. *Stefan Obkircher* turned his attention to another cross-border region in Europe – the Alpine Rhine Valley between Austria and Switzerland. He focuses on the intensity of cross-border cooperation and interactions, regional identities and regional interests in this region. Despite quite rich informal co-operation, this region lacks formal joint institutions, and lacks the administrative and financial framework for the expansion of co-operation. However this framework is certain to move ahead, particularly as there is a widespread perception of the Rhine Valley as one living space, and efforts to develop this space in common are evident.

The third part of the book focuses on some of the technical aspects of devolution of powers from central to local government, changes in administrative systems, service delivery, and local finance. It provides a wide-ranging overview on various fields at the regional and local levels in Hungary, Portugal and Slovakia. The chapter written by *Carlos Nunes Silva* addresses recent changes in public administration in Portugal. It looks at central-local relations in terms of administration, functions, finance, and regulation. It discusses the modernisation of public administration, which includes the transfer of powers and resources to local governments. *Nunes Silva* concludes that local government reforms were less radical than expected by Portugal's municipalities, and argues that issues of local democracy and participatory rights of local citizens did not receive sufficient attention within local government reform. The chapter by *Gábor Kovács* looks at the use of municipal bonds in local finance in Hungary. Issuing municipal bonds has led to the rapid growth of local government debt. He calls for a more sophisticated regulatory framework to regulate local borrowing, tailored to different size categories of municipalities, and advocates improving access to fundraising to other local organisations. The development of municipal assets during the post-communist transition period after 1989 is addressed by *Ján Buček*. He documents how Slovak local self-governments have improved their financial situation. In addition, he discusses strategies for managing and owning municipal assets, referring to Bratislava City. He also covers earlier phase of financial and economic reform and restricting their impact on asset base of the city. *Pavel Šuška* analyses the power relations and position of local governments within the urban development process. He has been inspired by the unprecedented transformation of the built environment in Bratislava within the last decade. He documents the destruction of industrial heritage sites, arguing that such destruction reflects the lack of efficient local institutions, signalling weak local democracy and limited local citizen participation in urban development.

Perhaps the lesson which can be drawn from all these papers is that administrative reform, intergovernmental cooperation, and the devolution of powers are long-term projects. They must have tangible benefits rather than being just a paper exercise. Reform is a learning experience, requiring the accumulation of institutional knowledge as well as individual knowledge. Local governments, particularly relatively young local governments with little accumulated experience, can find themselves out of their depth when coping with new powers. Often, they must rely on consultants and external advisors to guide them, and their inexperience makes them susceptible to manipulation. In the case of London, powers have been gradually transferred from central government to the Greater London Authority, and this has given the Authority a chance to prepare for and adapt to change in an incremental manner.

Change must also be a resilient process. Although bottom-up cooperation can be initiated by just a few individuals, to succeed it must gain adherents and broader institutional support. Otherwise, if the main promoter leaves or moves on, the cooperation will die. Top down cooperation also requires institutional support to succeed, and often requires financial carrots as well as sticks.

We want to thank the Department of Human Geography and Demogeography at the Comenius University in Bratislava and International Geographical Union for their support. We are grateful to two reviewers for careful reading of all the manuscripts.

Bratislava, Slovakia
Portsmouth, UK
December 2013

Ján Buček
Andrew Ryder

Contents

List of Figures

List of Tables

About the Authors

Jane Atterton Ph.D. in Geography (2005) is a Policy Researcher in the Rural Policy Centre at SAC (Scottish Agricultural College), based in Edinburgh. She was previously a Lecturer in Rural Development in the Centre for Rural Economy (CRE) at Newcastle University, UK. Her research interests focus broadly on issues of rural and regional development, with a particular emphasis on rural businesses and their networks, the characteristics of rural communities, ageing in rural areas, the interrelationships between rural and urban areas, and the rural policy-making process. She has previously worked for the Countryside Agency (Cheltenham, UK), and as a researcher at the Arkleton Centre for Rural Development Research, University of Aberdeen. Her Ph.D. research focused on the geography of networks maintained by business owners in the Highlands and Islands of Scotland.

Péter Balogh is a Ph.D. student at the Centre for Baltic and East European Studies, Södertörn University (Sweden), where he is a member of a research team on the resurgence of Pomerania, and on Szczecin's potential as a cross-border regional centre. He is also affiliated with the Department of Human Geography at Stockholm University. He received his master's degree from Uppsala University. His research interests include political geography, border studies, city-hinterland relations, and cross-border residential mobility.

Ján Buček is Associate Professor and Head of the Department of Human Geography and Demogeography at the Comenius University in Bratislava (Slovakia). He received a Ph.D. in Economic Geography from the department in 1996. His research focuses on governance and development issues at the local and regional levels. Since 2008 he has been Chair of the International Geographical Union Commission on the Geography of Governance.

Patrick Fensham has a Bachelor of Town and Regional Planning (First Class Honours) from Melbourne University, and an M.Sc. (Urban Planning Studies) from

Oxford Polytechnic. He is currently a Principal and Partner of SGS Economics and Planning based in the firm's Sydney office. Pat has extensive consulting experience in strategic planning, economic analysis, retail studies and urban and regional economic development. Pat was seconded to the New South Wales Department of Planning as Director of Strategic Development for the Sydney Metropolitan Strategy from March 2004 until the release of the Strategy in 2005.

Marek Furmankiewicz Ph.D. in Geography (2003), is a Lecturer in the Department of Spatial Economy, Wrocław University of Environmental and Life Sciences, Poland. His main research interests are: local government regional and international cooperation, partnerships and governance, and the management of local resources. He has also worked in government for the Lower Silesian Regional Development Agency and for Lower Silesia (Dolnosląskie) Voivod Government Office, where he dealt with European Union support programmes for local development.

Rick Glöckner is currently a scientific assistant at Leibniz-Institute for Regional Geography (Leibniz-Institut für Länderkunde) in Leipzig. He graduated from the Technical University of Dresden in 2008. Previously, he worked as a research assistant at Leibniz-Institute of Ecological and Regional Development (Leibniz-Institut für ökologische Raumentwicklung) in Dresden. He works on regional development processes in Central and Eastern Europe. His professional interests include cross-border research, regional development, and settlement development, mostly focusing on Central and Eastern Europe.

Tomasz Kaczmarek is a Professor at the Adam Mickiewicz University (AMU), Poznań, Poland, in the Institute of Socio-Economic Geography. Currently he is Head of Settlement Systems and Territorial Organization Research Unit, and Director of the AMU Center for Metropolitan Research. His research focuses mostly on socio-economic geography, local and regional administration in Europe and the governance in metropolitan areas. He is author of over 90 publications, including 8 books, among them *Struktury terytorialno-administracyjne i ich reformy w krajach europejskich* (Territorial-administrative structures and their reforms in European states) and *Ustroje miast i obszarów metropolitalnych* (The territorial-administrative systems of metropolitan areas in Europe). He is also member of the editorial board of the journals *Europe Regional* (Leipzig) and *Geospace* (Usti-nad-Labem, Czech Republic), and author of a series of monographs on the Poznan agglomeration, part of the series Biblioteka Aglomeracji Poznańskiej.

Wojciech Knieć Ph.D. in Rural Sociology, is employed as Lecturer and senior researcher in the Rural Sociology Department, Institute of Sociology, Nicolaus Copernicus University, Toruń, Poland. His main areas of research are related to local partnerships in rural areas, including LEADER LAGs, sustainable rural development, especially with regard to sustainable agriculture, rural social and cultural revival processes, and sustainability. He is an active member of several non-governmental organisations dealing with rural development, including local action groups and rural revival associations. He works as an expert advisor to the EU

Committee of Regions, Ministry of Agriculture and Rural Development, regional governments and local associations.

Gábor Kovács Ph.D., is a Senior Lecturer at Department of Economics in the Kautz Gyula Faculty of Economics at the Széchenyi István University in Győr (Hungary). He received his Ph.D. degree in Management and Business Administration Sciences in 2007. His main research interest and activities are local government finance and sub-national capital markets.

Ilona Pálné Kovács is a scientific advisor and Head of Department at the Research Centre for Economic and Regional Studies of the Hungarian Academy of Science, and a full time Professor in the Department for Political Studies of University of Pecs. She graduated as a lawyer in 1978, and received a Ph.D. in 1988. She became a doctor of the Hungarian Academy of Sciences in 2000, and received her Habilitation Degree from the University of Pecs in 2003. Her research interests deal with the implementation of regional policy, regional governance, and local government. She is the leader of the doctoral programme for political studies at University of Pecs. Dr. Kovacs has led or participated in several domestic and international projects, including FP, ESF, INTERREG, TEMPUS, ESPON, UNESCO, EC DG Regio programmes. She has on several occasions been invited to be a member of Hungarian government think tanks, and worked with the Council of Europe and the Soros Foundation preparing legislation and other documents on local government, regional development.

Łukasz Mikuła earned M.A. degrees in Spatial Management (2005) and Law (2007) from the Adam Mickiewicz University of Poznań (AMU). His doctoral thesis (2009) was titled *Zarządzanie terytorialne w obszarach metropolitalnych w Polsce* (Territorial governance in metropolitan areas of Poland). Since 2009 he has been an Associate Professor at the Institute of Socio-Economic Geography and Spatial Management, AMU. He has specialised in metropolitan government problems and spatial planning policies in Poland and Europe. He is a co-author of the Poznan Metropolis 2020 strategy.

Stefan Obkircher (born in 1982) is a geographer who works in the planning department of the Federal State of Vorarlberg, Austria. Previously he was a research associate at the University of Applied Sciences, Rapperswil, Switzerland, and at the University of Innsbruck, Austria. He received a Master of Science from the University of Innsbruck and is currently finishing his doctoral thesis. His main research interests are regional governance, spatial development in cross-border regions, and participatory planning processes.

Isolde Roch recently retired from her post as the Deputy Director of the Leibniz Institute of Ecological, Urban, and Regional Development. She holds an Honorary Professorship at China University of Mining and Technology Xuzhou. She has a doctorate in agronomy. Her professional interests are development of settlements in urban spaces, landscape development and sustainable regional development.

Andrew Ryder is a Senior Lecturer in the Department of Geography at the University of Portsmouth. He received his D.Phil. from Oxford University. He has a background in town planning and economic development, and has worked as a consultant on regional policy for the Organisation for Economic Cooperation and Development in a dozen countries including many in Central-Eastern Europe, Canada, and the Republic of Korea. More recently, he has worked as a consultant in Tanzania with the Ministry of Finance and the Eastern and Southern African Universities Programme and the Tanzanian Revenue Authority on ways to improve tax performance and ways to encourage the transformation of the informal sector into the formal sector.

David Rylander is a consultant in regional development and Senior Lecturer at Sodertorn University (2011) and Stockholm University (2012). He earned his Ph.D. in Human Geography at the University of Goteborg (2004). David's research has in recent years centered round entrepreneurship, policy and transnational cooperation in local and regional development, carried out in collaboration with the National Institute of Working Life, Swedish Knowledge Foundation and Jonkoping International Business School as well as several regional and local partners.

Carlos Nunes Silva Ph.D., is Professor Auxiliar at the Institute of Geography and Spatial Planning, University of Lisbon, Portugal. He has a degree in Geography (University of Coimbra), a post-graduate degree in European Studies (University of Coimbra – Faculty of Law), a Master's degree in Human Geography, Regional, and Local Planning (University of Lisbon) and a Ph.D. in Geography, Regional and Local Planning (University of Lisbon). His research interests are mainly focused on local government policies, history and theory of urban planning, urban and metropolitan governance, urban planning ethics, urban planning in Africa, research methods, e-government and e-planning. Among his recent publications are the *Handbook of Research on E-Planning: ICT for Urban Development and Monitoring* (2010) and *Online Research Methods in Urban and Planning Studies: Design and Outcomes* (2012). He is the Editor-in-Chief of the *International Journal of E-Planning Research* (IJEPR).

Pavel Šuška is a researcher at the Institute of Geography of the Slovak Academy of Sciences. Bratislava. His research interests are related to issues of urban change and post-socialist transformations.

Pontus Tallberg is a senior analyst for the Skåne Region in Sweden. He works in the area of public administration, particularly on relations between municipalities and regions and relations of both with national level government. He previously worked for the Swedish Tax Authority, and therefore also has a strong interest in fiscal federalism and public finance. He has been editor of several anthologies about cities and regions in Sweden. His latest book *The Role of Regions?* was jointly edited by Tassilo Herrschel, Westminster University, London, and published in 2011.

Chapter 1
Introduction

Andrew Ryder

Abstract In recent years, there has been a revolution in governance. Policymakers have argued that decentralised governance enhances democracy, accountability and transparency. Echoing the phrase 'no taxation without representation', supporters of fiscal decentralisation have said that spending and tax rates should be set locally. Experience in Central and Eastern Europe after the communist rule ended in 1989 changed the views of governance. The reorganisation of administrative regions and the reallocation of powers, driven by economic and administrative necessity, may be one of the greatest public administration experiments ever undertaken. Simultaneously, Western European governments were decentralised to increase accessibility and accountability. The December 1995 Madrid European Council helped this transformation, requiring candidate European Union members to change their systems of governance to foster citizen participation in policy formation at all levels of decision-making. The idea of devolving powers appeared in Spain's 1978 constitution, which transferred authority to 17 autonomous regions. Each was allowed to negotiate its relationship with the national government. However, decentralisation poses difficulties. It can foster regional inequality. No one local government or collection of local governments can encompass all regional problems. Local government reorganisation causes a loss of institutional memory. An alternative is to improve intergovernmental relations among local and regional governments, among different agencies and between local and regional governments and national ministries, departments and agencies. Good governance involves cooperation, collaboration and institution building. It is about fostering dialogue, exchanging views and reaching agreement. As this book shows, governance is a learning experience, shared among institutions and people. New laws and administrative systems require training and experimentation. The rights and obligations of local governments depend not just on written law but on its interpretation and on what becomes accepted practice. Legislators and government employees at national, regional and local levels need time to understand and successfully manage new competences. Local governments are often unprepared to take on new responsibilities, leaving them at the mercy of sophisticated investors and developers who

A. Ryder (✉)
Formerly Department of Geography, University of Portsmouth, Portsmouth PO1 3HE, UK
e-mail: andrewrydernyc@gmail.com

© Springer Science+Business Media Dordrecht 2015
J. Buček, A. Ryder (eds.), *Governance in Transition*, Springer Geography,
DOI 10.1007/978-94-007-5503-1_1

can hire experts to negotiate their way through relatively young legal systems and inexperienced staff. As London shows, reform can be gradual, allowing time for local governments to acquire new skills, develop new relationships with existing governments and agencies and gain confidence and experience.

Keywords Administrative decentralisation • Fiscal decentralisation • Reform as a learning process • Collaboration • Consensus building • Intergovernmental relations

Governance in a Changing World

During the last 25 years, there has been a revolution in governance. Researchers and policymakers have argued that decentralised government enhances democracy, accountability and transparency and increases satisfaction with the government. Multiple governments encourage political pluralism. Calls for administrative decentralisation have been accompanied by calls for fiscal decentralisation. Echoing the phrase 'no taxation without representation', supporters of fiscal decentralisation have argued that spending and tax rates should be set locally. In addition, if a small local government makes a mistake, it is less harmful than a mistake made at the national level. Throughout Europe, new regional governments have been created, in France, Italy, Spain, Sweden and Denmark. Administrative systems have been reorganised in Poland, Slovakia, the Czech Republic, Hungary, eastern Germany and the Baltic States. Belgium has become a federal entity. Even centralised governments, such as those in Portugal, discussed in Chap. 12, and the United Kingdom, discussed in Chap. 6, have seen some devolution of powers to lower levels of the government. This impetus towards decentralisation has been seen beyond Europe, for example, in the Republic of Korea.

In most countries, local authorities have broad land-use planning powers and provide a wide range of local amenities. Some manage spending on education, health and social services. However, when new competences are added, additional resources have not always followed, causing, as Caulfield (2000) noted, a mismatch between the functional competencies and fiscal resources. At the same time, even in decentralised systems, local governments have become increasingly dependent on transfers from higher levels of government, as is the case in the United States, which has perhaps the world's most decentralised system of government.

To a considerable extent, new views of governance reflect changes which occurred in Central and Eastern Europe after the collapse of communist rule. The collapse of communism gave the debate about governance and the tasks of government a special urgency. After 1989, it became necessary to shrink the size of the state. National governments had to reduce spending and cut back transfers to local governments. Regional governments did not exist, and local governments were essentially a form of local enterprise responsible for managing housing and local transport (see, e.g. Ryder 1990). As a consequence, national governments reallocated competences among national and subnational governments. Faced with funding cutbacks as wage costs rose, local governments had less to spend on transport and had to spend more on trash collection.

The subsequent reorganisation of administrative regions and the reallocation of powers which occurred in the region since the early 1990s may be one of the greatest experiments ever undertaken in the realm of public administration. In some cases, administrative regions remained essentially unchanged. In others, the system was completely transformed. This process went furthest in Poland. Forty-nine administrative regions were replaced by 16 larger ones. A new mid-level government, the powiat (or county), between local and regional governments, was introduced. Hungary represents an alternative approach. While this may have led to a more accountable system of governance, in creating new levels and districts, it complicated collaboration among different local governments and between local and regional governments, as is discussed in Chaps. 2 and 4. For the most part, regions and local government boundaries remained unchanged. However, many local authorities, forcibly amalgamated under communism, declared their independence and re-established themselves. The re-establishment of local government autonomy was viewed as an important benefit of the collapse of communism, but resulted in a large number of micro-jurisdictions which are reluctant to cede powers. Where Hungary stands out is in passing legislation which gave all local governments, from village to county, from urban district to city government, the same legal standing. No one government is superior to any other. Counties have the same standing as villages, and local districts in Budapest have the same legal standing as the city government. As a result, as Chap. 5 shows, it has been difficult to merge or abolish local governments. From the mid-1990s, successive Hungarian governments have unsuccessfully attempted to remedy this problem.

Reform in Central and Eastern Europe was driven by economic and administrative demands. In Western Europe, governments were decentralised in a quest for greater transparency, accountability and accessibility. It is widely argued that local autonomy and greater local control over local decision-making enhance democracy and create stronger local participation and engagement. The Madrid European Council in December 1995 played a significant role in this transformation. It stated that to meet the accession criteria for European Union membership, each candidate country should create a participatory democracy which would allow citizens to participate in decision-making at all levels of government (European Commission 2012).

However, the idea of devolving powers to lower levels of the government goes back to the late 1970s. Spain's 1978 constitution devolved authority to 17 autonomous regions, allowing each to negotiate its relationship with the national government. In the United Kingdom, since 1997, successive national governments have supported decentralisation. This followed an earlier withdrawal of the government from the direct management of the economy through state ownership, using the argument that the economy was best 'managed' by market forces.

More recently, the concept of decentralisation has been broadened to include a shift from national to local wage agreements (see, e.g. Economist 2012c). In centralised states, tax levels are similar throughout the country, but living costs vary among regions and localities. In decentralised ones, tax rates and service charges vary widely. In the United Kingdom, in 2007 house prices ranged from an average of £134,524 in the northeast region and £152,479 in Scotland to £302,486 in London, a difference of almost 250 %. By contrast, wages were fairly uniform across the country, averaging £456.70 per week, but ranging from £401.90 in Northern Ireland to £402.90 in the northeast, £441.50 in Scotland and £580.90 in London (Unison 2007). The narrow range of wages – London was just 27.5 % higher than Northern Ireland and the northeast – reflects national wage agreements. It also reflects national pricing for goods and the fact that most taxes are set nationally. In the United States, wages, income and taxes vary widely. At the state level, in 2010, annual incomes ranged from $31,186 in the state of Mississippi to $56,001 in Connecticut and $71,044 in Washington D.C. Disposable income averaged $36,808, but ranged from $29,155 in Mississippi to $48,596 in Connecticut and $63,619 in Washington D.C. At a metropolitan scale, average annual incomes ranged from $59,993 in the San Francisco area to just $29,680 in the San Bernardino-Riverside area. Tax rates also varied. Rates on an annual income of $50,000 per year ranged from 4.4 % in Anchorage, Alaska, to 13.7 % in Philadelphia (US Bureau of the Census 2012a, tables 447, 681, 682, 683). As the United States shows, to attract investment and gain the full benefits of lower prices, wages and taxes should be lower than those in more expensive regions, reflecting lower costs of living.

The Spectrum of Subnational Governance

Subnational governance ranges from extreme centralisation to extreme autonomy. It can allow considerable administrative freedom, but few taxation powers, leaving lower levels of the government dependent on central government for funding (Caulfield 2000). Equally, it can allow considerable fiscal freedom. Greater London has only limited freedom to set its own transport, policing and housing policies and has virtually no powers to raise money through taxation. According to The Economist (2012b), only 7 % of the Greater London Authority's revenue comes from direct taxation. Central government provides most funds. London's mayor has little influence over regulation or policy. In Slovakia, local and regional governments receive the bulk of their funding through income taxes, which are set by the national government and distributed on the basis of population. In the United States, states and local governments set their taxes and fees and determine the quality and range of services they offer relatively freely. The United Kingdom offers a different approach. Distrust of the local government remains strong at the national level, but even there, the central government has sought to decentralise governance through other means. The term 'localism' has become popular, meaning a mixture of cooperation and collaboration among local governments within metropolitan areas with, at the same time, cooperation with private businesses. However, as London shows, administrative decentralisation can be successful without fiscal decentralisation. London has successfully implemented significant policy changes in transport and, more recently, housing.

The United States lies on the far end of the decentralised side. It may have one of the densest systems of local government in the world. Most local governments have less than 10,000 residents (16,573 out of 19,540 in 2010), although they account for less than 15 % of the total population (Bureau of the Census 2012a, Table 28). Many metropolitan areas consist of several hundreds or more local governments, and over the past several decades, local governments increased from 78,269 in 1972 to 89,004 in 2012. This reflects two opposing trends. On the one hand, local governments are consolidating services to reduce sending: school districts fell from 15,781 (and from 34,678 in 1962) to just 12,881 in 2012. On the other hand, a need to meet mandates from the national (federal) level of government has required the creation of special purpose districts. These 'special districts' rose from 29,885 to 37,203 during the same period (Bureau of the Census 2012b, Table 2). Although the average share of own-revenue local government funding has fallen in recent years, it is still high, 57.3 % in 2008 (Bureau of the Census 2012a, Table 455, pp. 292–3), and compares favourably with London's 7 % (Economist 2012b).

Since 2007, cutbacks at the federal and state level have forced small local governments to reduce spending. As is the case in the United Kingdom, joining together allows them to share services and reduce per capita spending on schools and teaching equipment, emergency equipment and technical services. This cooperation can be described as bottom-up. By contrast, the rise in special districts devoted to community development and natural resource management is often due to top-down pressures: the need to comply with national and state requirements to obtain federal and state funds.

The United States' large number of local governments has made necessary for economic, political and technical reasons, including waste and water management, environmental quality and transport planning. Since 1940, metropolitan statistics have been collected on a metropolitan basis. Bottom-up and top-down cooperation among governments has become more common. An example of the latter is the Metropolitan Planning Organisations (MPOs). These consist of representatives appointed by elected local governments as well as from a variety of interest groups, including local business chambers, local transport providers, local transport and road users and representatives from state and federal agencies. As described in Solof (1998), they have existed in some form since 1962, when they were mandated to ensure that local plans which relied on the use of federal money for roads and public transport did not duplicate or contradict themselves. Today (2013), there are 387 (Metropolitan area data base 2013). They are required for all urban areas with a population of 50,000 or more (Metropolitan Planning Organisation (MPO) Database 2015). Since 2005, MPOs have been designated by an agreement between a state governor and local governments representing at least 75 % of the affected population or in accordance with state or local laws. Once established, an MPO can be altered only by agreement between the state governor and units of local government which represent at least 75 % of the affected population, including the central city or cities. The MPO itself must include not only the currently urbanised area but the contiguous geographic area likely to become urbanised within the 20-year forecast period covered by the transportation plan (Texas Department of Transportation 2013).

Independence is an extreme form of devolution. In the 1990s, several countries dissolved, sometimes peacefully, as in the case of the Czech and Slovak Federal Republic, or with extreme hostility, as in the case of the former Yugoslavia. Slovakia offers an argument in favour. In 1989, the Slovak economy collapsed, due partly to federal policies. The break-up of the Czech and Slovak federation allowed significant changes in the Slovak economy and eventually led to substantial growth. Between 1997 and 2012, Slovakia's per head income rose from 51.3 % of the EU average to 76 %, equalling Portugal's (Eurostat 2014). During the same period, per head income in the Czech Republic, its former national partner, went from 77 to 81 % of the EU average. Today (2013), the country has a new industrial economy. Car manufacturing and associated industries, the manufacture of flat-screen television and monitors and a large white goods industry have all developed since 1993. Before 1993, the Czech Republic accounted for almost half of all trade. In 2012, it accounted for just 9.8 % of imports and 14.0 % of exports (Statistical Office of the Slovak Republic 2014).

In the absence of formal regional metropolitan structures, local governments in England have increasingly banded together to share expenses and to lobby central government (see, e.g. Economist 2012g, 2013). Perhaps the best example is Manchester, which is often held up as an example of bottom-up collaboration among local governments. However, Manchester shows the complexity of a bottom-up approach to regional governance, often termed 'localism'. In 1974, the government of the United Kingdom created a Greater Manchester County Council to oversee planning, transport planning and operations and the provision of leisure services in the Greater Manchester region. The metropolitan council was directly elected. The region contained 10 boroughs, which themselves were newly created in 1972 from 62 local governments as part of a sweeping local government in the United Kingdom. These top-down reforms, creating larger units of local government, forced local cooperation through amalgamation. In 1986, the UK government abolished metropolitan county councils, although the metropolitan county continued to exist as a geographical frame of reference. Many functions of the Greater Manchester County Council were devolved to the ten constituent boroughs, but emergency services and transport continued to be managed on a county-wide basis by a newly formed Association of Greater Manchester Authorities (AGMA). The AGMA was also responsible for shaping regional policies, lobbying central government and the European Union and undertaking research related to the region. The executive board consisted of representatives of the ten constituent boroughs, the Greater Manchester Integrated Transport Authority, the Greater Manchester Police Authority, the Greater Manchester Fire and Civil Defence Authority and the Greater Manchester Waste Disposal Authority. Four neighbouring local authorities became associate members. Their representatives could participate in debates and meetings, but were excluded from the voting process.

Following the start of the global economic crisis in 2007–2008, the ten local councils agreed to a 'multi-area agreement', which is a government initiative to encourage local governments to work together in exchange for greater local autonomy

(Hetherington 2008). In 2009, the Local Democracy, Economic Development and Construction Act of 2009 led to work to create a combined authority for the Greater Manchester, responsible for public transport, economic and skills development, housing regeneration, waste management and carbon neutral development and planning, subject to approval by proposed member councils. Each member borough was to be represented by one local council member, appointed by the local council. The councils approved the proposal in early 2010. After a 15-week public consultation, the Department for Communities and Local Government approved the creation of a Greater Manchester Combined Authority. The creation of the authority required the approval of the UK parliament, which was received on 22 March 2011. The new authority came into force on 1 April 2011, replacing the AGMA. At the same time, a new local transport authority was created, the Transport for Greater Manchester Committee. This is managed by 33 local councillors, one for each 75,000 local residents. The Combined Authority: "pay to subsidise local bus services, pay concessionary fares, maintain passenger services and facilities, and support finance costs."

Policymakers in the United Kingdom have held up Manchester as an example to other metropolitan areas, arguing that not only did the AGMA maintain a sustained vision for the region but also developed a mature partnership model which engaged with the private sector. However, they have also noted that 'city region status is . . . a negotiating platform for local government to bid for more powers from central government. Therefore, central government is deciding who 'deserves' power, and who doesn't' (Forjan and Shakespeare 2009, p. 8). In the UK context, this poses the question, should devolution be 'earned' or viewed as a right?

Supporters of the Greater Manchester Authority argue that localism enables better decision-making for economic development, transport and development of the local skills base. However, they have also argued that greater financial freedom is needed for local bottom-up cooperation to succeed. Interestingly, paralleling the arguments in Oates (1999, 2006), they have argued that localism 'could potentially increase innovation and the spread of innovative practice. Local people know best about their local area. The most efficient decentralized governance structure captures and applies local knowledge and expertise to improve processes, spread innovation and improve outcomes' (Forjan and Shakespeare 2009, pp. 9–10).

Perhaps most importantly, some supporters of localism view it as being in opposition to 'equalisation': the idea that funding levels and economic opportunity should be the same for all regions. For example, Forjan and Shakespeare (2009) argued that in prioritising 'spatial geography' over 'economic geography', it ignores local economic circumstances and the impact localities can have on surrounding areas.

As Manchester shows, bottom-up governance can be successful, but does not always develop quickly. Greater Manchester has a 40-year history. While the Greater Manchester Combined Authority is an outgrowth of a less formal alliance of local governments, they did not spring up spontaneously, but arose as an informal response to the abolition of a metropolitan government.

Reasons for and Against Decentralisation and Devolution

Proponents of decentralised governance argue that policies designed at the national level are too inflexible to meet local needs. Local governments need the tools to respond to local conditions in a timely and consensual manner (see, e.g. Oates 1999, 2006; Forjan and Shakespeare 2009). Local residents, the argument goes, are more aware of local problems and needs than an often distant central government and are better qualified to decide local development priorities. This view has become increasingly widespread after the apparent failure of traditional regional development policies (e.g. Blitz and Chung 2003; Chung and Roberts 2003). In the United Kingdom, assistance to lagging regions dates back to the 1930s (Scott 1997, 2000). Despite three-quarters of a century of regional aid, most lagging regions continue to lag. As well, when local governments depend mainly on the national government for their funding, local issues become national ones. London, discussed in Chap. 6, is a good example. Despite being one of the richest cities in Europe, London cannot fund transport projects such as Crossrail One and Crossrail Two on its own. It must compete with local agendas elsewhere in the United Kingdom. London's gain is viewed as a loss for the rest of the country. The promise to provide money to build London's Crossrail is then rephrased as, 'if we invest in Crossrail, we must invest in rail elsewhere in the United Kingdom' (see, e.g. Bounds 2013). In the same vein, Pickford (2013) wrote that per capita spending on transport in London is greater than all spending in the rest of the United Kingdom. Eighty nine percentage of planned capital investment went to London and the southeast of England. London received £2,600 per capita, as opposed to just £5 per capita in the north. However, London's mayor argued that much of the money spent in London went to firms based outside London, creating 40,000 jobs.

Oates (2006) listed four reasons for decentralised governance. Firstly, decentralised levels of government provide goods and services whose consumption is limited to their own jurisdictions. They can tailor outputs of goods and services to the preferences and circumstances of their constituents. Consequently, economic welfare is higher than if such services are provided according to a national 'one size fits all' policy.

Secondly, when households are mobile, they can move to jurisdictions that provide outputs which meet their tastes and their needs, increasing potential gains from the decentralised provision of public services. This is discussed in Tiebout (1956). He argued that in a political system characterised by multiple local jurisdictions, prospective residents can be viewed as consumers who select a place to live based on the goods and services it provides, on the basis of their ability to pay for those services.

Thirdly, Oates argued that decentralised local authorities compete with their neighbours for investment, businesses and residents. This constrains their ability to increase taxes. Doing so would make them less attractive to prospective residents and investors. Instead, they must become more efficient in providing public services, offering a wider range of services of better quality at lower cost.

Finally, Oates argued that decentralisation encourages experimentation and innovation. Local authorities can adopt new approaches and innovations in public policy, making it a 'laboratory' for fiscal experiments (Oates 2006, p. 2). Innovation spreads, improving local finances and making local government more efficient.

Decentralised governance can foster economic as well as political development. Local autonomy, where local governments and local taxpayers pay for their local services, is thought to encourage a push for economies and efficiencies (Tiebout 1956; Oates 2006). Cross-border competition can also force local governments to innovate in order not to fall behind. For example, there have recently been calls to allow Northern Ireland to reduce its tax levels below those of the rest of the United Kingdom so as to compete with tax rates in the neighbouring Irish Republic (Houlder 2011).

Events in Central and Eastern Europe have been influential. Several international agencies, particularly the European Bank for Reconstruction and Development, have graded governments in the region since the early 1990s (European Bank for Reconstruction and Development 2013). The World Bank publishes an annual ease of doing business assessment (mainly at the national scale, but in the case of a few countries, at the regional scale) which looks at barriers to investment and development, arguing that poor governance inhibits growth (World Bank 2013). Good governance – that is, the quality of governance – makes a difference. The extent to which governance embraces the rule of law, the existence of independent institutions, transparency and an independent civil service directly affect growth and are more important than location, regional policies, taxation or other local attributes.

However, there are strong arguments against decentralisation. Prud'homme (1994) argued that decentralisation, particularly fiscal decentralisation, can cause inequality. If local authorities must rely entirely on their own local resources, worse-off regions will not be able to provide the same range of services as better-off regions, causing permanent inequality. Regions which are attractive benefit from new investment; those which are not fall behind, resulting in a permanent downward spiral. In the United States, Detroit and Flint, Michigan; Newark, New Jersey; and New Orleans are examples of this kind of negative growth trap. In Europe, capital city regions, such as London and Bratislava, often grow faster than the rest of the country. Other regions find it difficult to catch up (see, e.g. Pickford 2014; Centre for Cities 2014). National governments redistribute wealth through transfer payments to ensure that the quality and range of services are similar across the country (equalisation policy), but higher costs of living in high-growth regions can shut out potential migrants. On the other hand, a policy of redistribution can generate resentment in high-income regions, such as London and Catalonia (see, e.g. Economist 2012f). Boris Johnson, Mayor of London, is just one of the many who have argued that London's economic success has been good for the rest of the United Kingdom (the debate is briefly summarised in Centre for Cities 2014, chapter 2). Johnson has been particularly keen to see London retain more tax revenue. He has argued that £2,500 for every Londoner goes to other parts of the United Kingdom saying that London is a 'cash cow'. He has described London as 'the powerhouse of the UK economy', stating that Londoners should 'get the benefit of their contribution' (Crerar 2012).

Strong independent local governments can be inefficient, particularly micro-jurisdictions such as those in Hungary and the United States. Smaller governments lack the scale economies needed to reduce costs and needlessly duplicate services. Strong local governments can be parochial, making decisions which maximise their benefits and ignore negative externalities passed on to their neighbours. Parochialism can cause harmful competition among local governments and regions. Local authorities underbid each other to attract investment and development. However, they often gain less in tax revenues and other incomes than predicted and face unanticipated higher costs from new development leading to what has been described as a zero-sum game. A recent survey of the Chicago metropolitan region argued that competition among local governments was hindering rather than promoting growth (Organisation for Economic Cooperation and Development 2012, pp. 20–21).

Many argue that excessive autonomy encourages excessive local government spending, leading to high levels of indebtedness and even bankruptcy. In 2012, in the United States, San Bernardino and Stockton, California, declared bankruptcy, as did Harrisburg, Pennsylvania (see, e.g. Economist 2012d; Walsh 2012). In 2013, Detroit followed suit. Similar to local governments in Hungary, discussed in Chap. 13, US local governments borrowed money against anticipated tax revenues which turned out to be extremely over-optimistic. In Spain, during the current financial crisis, controlling spending in Spanish regions has been far more difficult than controlling central government spending (Economist 2012a, e). It has been suggested that even in China, lack of control over local government borrowing, much of it outside the formal banking system, may destabilise the entire economy (Magnus 2012). In 2011, loan repayment deadlines were extended as a result (Rabinovitch 2013). However, it was estimated that as much as 25 % of local government projects had failed to generate any revenue (Financial Times 2012). More recently, Chinese central government authorities noted that 99 cities out of 391 examined, 195 county-level administrations and 3,645 townships had debts exceeding 100 % of their economic output (Economist 2014).

Governance as a Learning Process

As the chapters in this book show, governance is a learning experience shared among institutions and people. Governments work against constantly changing economic and political conditions. They must constantly innovate, changing what they do and how they do it. Local governments are often unprepared to take on new responsibilities. As Hungary and Slovakia show, discussed in Chaps. 13, 14 and 15, they can find themselves out of their depth. Local governments were often at the mercy of more sophisticated investors and developers who were able to call on a range of consultants and experts to negotiate the way through relatively young legal systems and relatively inexperienced local government staff. London offers a good example of how the learning process can be managed. After the Greater London Authority and the Mayor of London came into being in 2000, responsibility for the transportation system was only gradually transferred to them. They did not take control of the Underground until 2003 and later added other modes of transport, including the Docklands Light Railway, Tramlink and the London Overground. This allowed them to build up management capacity and institutional experience gradually.

Reorganisation of local governments leads to a loss of expertise and a loss of institutional memory. However, many argue that it can also improve the democratic process. If local governments do not respond to voters or are controlled by entrenched elites, a new administrative system can remove entrenched political elites, depoliticise decision-making and enable more responsive governance to develop, much as privatising state-owned firms was viewed as a way to foster a new management orientation (see, e.g. O'Leary 1987, pp. 381–3). However, as London shows, the abolition of a regional government, such as the destruction of the Greater London Council in 1986, can eliminate a regional champion, causing the loss of a regional vision.

Reorganisation of local governments is often undertaken with the aim of rationalising service delivery and approaches to regional problems. However, no one local government or collection of local governments can deal with all regional problems. One might argue that problems create regions rather than the other way around. Attempts to reorganise administrative units to meet a particular set of regional needs may work in the short term, but over time, populations disperse, metropolitan areas grow and new problems arise which fail to fit into existing regional boundaries. As Poland shows (discussed in Chaps. 5 and 4), it is difficult if not impossible to create purpose-built regions which can deal with all regional problems. The alternative is to foster intergovernmental relations among local and regional governments, among different agencies and between local and regional governments and national ministries, departments and agencies. The Organisation for Economic Co-operation and Development (OECD 2011) has made such a recommendation for Poland, arguing that Poland needs to strengthen the system of multilevel governance. They stressed the need to introduce specific incentives for intra-metropolitan cooperation and the need to shift strategic planning from the municipal or local government scale to the functional scale. They also argued that to enhance economic growth, more cooperation was needed across levels of government and with private actors and that local governments needed greater powers to design, implement and monitor development programmes (OECD 2008). However, in Australia (discussed in Chap. 3), Hungary, Poland and elsewhere, efforts have been made to develop such relations and linkages with limited success.

The United States is perhaps furthest along the path of building relations among governments. State boundaries, and for the most part, county and local government boundaries are considered inviolable. Administrators and legislators must work around boundary constraints, creating informal and formal coalitions to meet specific local needs – sewage management, water treatment and management, air quality management, transport management and education – but can be seen elsewhere as well. The result is often a kind of messy pluralism. Government responsibilities may overlap. For example, a local government may belong to several overlapping jurisdictions – one being responsible for delivering educational services; another, with different boundaries, responsible for environmental and waste management; and another, again with different boundaries, responsible for water management. Higher-level agencies may duplicate agencies at lower levels: in the United States, there is a federal department of transportation, but there are also state departments and, in many large cities, city departments of transportation. Frey (2005) and Eichenberger and Frey (2008) have described such complicated systems as 'Functional, Overlapping and Competing Jurisdictions' and view them as a common feature of federal systems. In the European Union, similar problems have arisen among local and regional authorities in different countries, discussed in Part II (Chaps. 8, 9, 10 and 11). There, cross-border cooperation bears some similarities to difficulties which arise between US states.

Thus, governance involves more than administrative and legal frameworks. It is about cooperation, collaboration, consensus building and institutions. It is about fostering dialogue, exchanging views and reaching agreement. Transferring powers to lower levels of the government may not produce the expected results if local governments are unable to exercise them properly.

Local and regional governments not only need to learn to cooperate among themselves but need to learn to work with the private sector. Often, when issuing bonds, taking financial advice, managing budgets and implementing spending plans, local governments need outside expertise. They must learn how to cope with sophisticated opponents who have extensive experience and considerable resources to deal with them. Raising money to support the government, through taxation, bonds, selling property and leasing property, is a new skill for most local governments. It is difficult to acquire expertise in these areas, particularly when local governments are small in size and have limited resources. The learning process often takes longer than expected, and often, mistakes occur.

Conclusion

Localism is important. Some have described this century as 'the century of the city', arguing that in years to come, most people will live in cities. They will become the locus of political action and the focus of political debate (Rossant 2013; Pierce et al. 2008).

Local and regional government reforms often take place in periods of economic and political change (OECD 2011). This was the case in Central and Eastern Europe and Spain. Local governments faced the simultaneous problems of economic restructuring and political transformation: privatisation, the closure of factories and businesses, rising unemployment and rising geographical inequality. In theory, greater autonomy and greater fiscal independence offered extensive benefits. However, in many cases, too much was expected of the government reform, too quickly. In a rapidly changing world, even experienced governments can find themselves overstretched. New laws and administrative systems require training and experimentation. Time is needed to determine the rights and obligations of local governments, which depend not just on the written law but on its interpretation and on what eventually becomes accepted practice. Legislators and government employees at national, regional and local levels need time to get to grips with new competences and learn how to successfully manage them. Elected officials and local government employees often lack the skills to manage their new responsibilities. Equally important, this learning process takes place against a moving background: legal decisions, changing politics and constantly changing fiscal conditions mean that learning becomes a continuous process. As London shows, reform need not take place all at once, but can be gradual, allowing time for local governments to acquire new skills, develop new relationships with existing governments and agencies and gain confidence and experience.

References

Blitz R, Chung J (2003) Back-slapping and handshakes turn to failure. Financial Times, Monday June 23

Bounds A (2013) Call to allocate more money to regional cities. Financial Times, 14 March 2013

Caulfield J (2000) Local government finance in OECD countries. Paper presented to 'Local Government at the Millennium' international seminar, University of New South Wales, 19 February 2000. Available at: http://www.cgc.gov.au/__data/assets/pdf_file/0015/4650/Dr_Janice_Caulfield-LG_Finance_in_OECD_Countries.pdf. Viewed 26 June 2012

Centre for Cities (2014) Cities Outlook 2014. Centre for Cities, London. Available at: http://www.centreforcities.org/research/outlook14.html. Viewed on: 18 January 2014

Chung J, Roberts D (2003) Regional grants not delivering jobs. Financial Times, Monday June 23

Crerar P (2012) 'Give London back its tax': Boris Johnson demands return on cash that capital makes for Britain *Evening Standard*, 4 April 2012. Available at: http://www.standard.co.uk/news/mayor/give-london-back-its-tax-boris-johnson-demands-return-on-cash-that-capital-makes-for-britain-7618555.html. Viewed on 10 April 2012

Economist (2012a) Spain's regions: the centre tries to hold, 4 February

Economist (2012b) Bagehot. Until London's mayor has to raise money as well as spend it, a great global city cannot grow up, 31 March 2012

Economist (2012c) Local pay in the public sector is a good idea, brutally hard to implement, 31 March 2012

Economist (2012d) Stockton's Bankruptcy. California's Greece, 30 June 2012

Economist (2012e) The Beleaguered Governments in Madrid and Rome are battling with their highly indebted regions, 28 July 2012

Economist (2012f) Catalonia Europe's Next Independent State? 22 September 2012

Economist (2012g) Innovation in government Britain's local labs. Something rather exciting is happening somewhere rather improbable: local councils, 22 September 2012

Economist (2013) English cities freedom at last: England's big cities are getting a surprising taste of autonomy, 2 February 2013

Economist (2014) Local government debt. Counting Ghosts, 4 January 2014

Eichenberger R, Frey B (2008) Chapter 6: functional, overlapping and competing jurisdictions (FCJ): a complement and alternative to today's federalism. Handbook of fiscal federalism. Edward Elgar, Gloucestershire, pp 154–181

European Commission (2012) EU enlargement: the accession process. Available at: http://ec.europa.eu/enlargement/enlargement_process/accession_process/criteria/index_en.htm. Viewed 26 June 2012
Eurostat (2014) Volume indices per capita 2010-2-12, EU-28. Available at: http://epp.eurostat.ec.europa.eu/statistics_explained/index.php/GDP_per_capita,_consumption_per_capita_and_price_level_indices. Viewed: on 31 January 2014
Financial Times (2012) Leader: China's Debt Bill: China must push ahead with financial reform, 14 February 2012
Forjan S, Shakespeare T (2009) Can Localism Deliver? Lessons from Manchester London: Localis/Policy Exchange. Available at: http://www.localis.org.uk/images/articles/Can%20Localism%20Deliver_Localis%20&%20Policy%20Exchange.pdf. Accessed 1 Mar 2013
Frey B (2005) Functional, overlapping competing jurisdictions: redrawing the geographic borders of administration. Eur J Law Reform 5(3–4):543–555
Hetherington P (2008) Cities cooperate to accumulate. The Guardian (London), Wednesday 21 May. Available at: http://www.guardian.co.uk/society/2008/may/21/localgovernment. Viewed on 12 June 2013
Houlder V (2011) Call to let Belfast set business tax rate. Financial Times, 25 May 2011
Magnus G (2012) China's national problems start in local government. Financial Times, 20 May 2013
O'Leary B (1987) British farce, French drama and tales of two cities: Reorganizations of Paris and London governments, 1957–1986. Public Adm 65:369–389
Oates WE (1999) An essay on fiscal federalism. J Econ Lit 37(3):1120–1149, September
Oates WE (2006) On the theory and practice of fiscal decentralization. Institute for Federalism and Intergovernmental Relations IFIR Working Paper 2006–2005. Available at: http://www.ifigr.org/publication/ifir_working_papers/IFIR-WP-2006-05.pdf. Viewed 26 June 2012
Organisation for Economic Cooperation and Development (2008) Territorial review Poland. OECD, Paris
Organisation for Economic Cooperation and Development (2011) Urban policy reviews, Poland. OECD, Paris. Available at: http://www.oecd-ilibrary.org/urban-rural-and-regional-development/oecd-urban-policy-reviews-poland-2011_9789264097834-en. Viewed on 15 September 2013
Organisation for Economic Cooperation and Development (2012) OECD territorial reviews: the Chicago tri-state metropolitan area. OECD, Paris. Viewed on 18 September 2013
Pickford J (2013) Transport spending skewed to London, says think tank. Financial Times, 24 June 2013
Pickford J (2014) Recovery likely to widen wealth gap between London and regions. Financial Times, 19 January 2014
Pierce N, Johnson C, Peters F (2008) The century of the city. Rockefeller Foundation, New York. Available at: http://www.rockefellerfoundation.org/blog/century-city-no-time-lose. Viewed on 20 September 2013
Prudh'homme R (1994) On the dangers of decentralization. World Bank, Policy Research Working Paper no. 1252. Available at: http://www-wds.worldbank.org/external/default/WDSContentServer/WDSP/IB/1994/02/01/000009265_3961005225011/Rendered/PDF/multi_page.pdf. Accessed 26 June
Rabinovitch S (2013) China's banks stave off local government loan default. Financial Times, 30 January 2013
Rossant J (2013) The century of the city will change the way we do politics. Financial Times, 25 June 2013
Ryder A (1990) Growth pole policy in Poland and the Lenin steel-works. Geoforum 21:2
Scott P (1997) British regional policy 1945–1951: a lost opportunity. Twent Century Br Hist 8(3):358–382
Scott P (2000) The audit of regional policy: 1934–1939. Regional Studies 34(1):55–65

Solof M (1998) History of metropolitan planning organizations. North Jersey Transportation Authority, Newark. Available at: http://www.njtpa.org/getmedia/b95661af-dfd4-4e3d-bb87-39e617619c7b/MPOhistory1998.pdf.aspx. Accessed on 3 Jan 2014

Statistical Office of the Slovak Republic (2014) Total import and total export by continents and economic groups in the year 2012. Available at: http://portal.statistics.sk/showdoc.do?docid=71912. Viewed 2 January 2014

Texas Department of Transportation (2013) On-line manuals. Available at: http://onlinemanuals.txdot.gov/txdotmanuals/ppl/metropolitan_planning_area_boundaries.htm. Viewed 10 June 2013

The European Bank for Reconstruction and Development (2013) Transition report. European Bank for Reconstruction and Development, London. Available at: http://www.ebrd.com/pages/research/publications/flagships/transition.shtml. Viewed on 2 December 2013

Tiebout C (1956) A pure theory of local expenditures. J Polit Econ 64:5

Unison (2007) Unison Bargaining support factsheet, Local Statistics, November 2007. Available at: http://www.unison.org.uk/acrobat/B693.pdf. Accessed on 26 June 2012

United States Department of Commerce, Bureau of the Census (2012a) Statistical abstract of the United States, Washington, DC. Available on line at: http://www.census.gov/compendia/statab/. Accessed 26 June 2012

United States Department of Commerce, Bureau of the Census (2012b) Census of Governments. Available on line at: http://www2.census.gov/govs/cog/2012/formatted_prelim_counts_23jul2012_2.pdf. Accessed 30 June 2013

United States Department of Transportation (2013) Metropolitan area data base (2013). Available at: http://www.planning.dot.gov/mpo.asp. Viewed on 20 August 2013

United States Department of Transportation, Federal Highway Administration, Metropolitan Planning Organization (MPO) (2015) Database: http://www.planning.dot.gov/mpo.asp viewed 9 January 2015. Information also available at Association of Metropolitan Planning Organizations (AMPO), Washington, DC. Viewed at ampo.org, 9 Feb 2015

Walsh MW (2012) Crushed by promises. The New York

World Bank. 2013. *Doing Business 2014: Understanding Regulations for Small and Medium-Size Enterprises*. World Bank Group, Washington, DC

Part I
Governance, Intergovernmental Relations, and Development

Chapter 2
Top-Down and Bottom-Up Metropolitan Integration in Poland

Tomasz Kaczmarek and Andrew Ryder

Abstract Since the start of the 1990s, the population of Polish cities has decentralised, forming constantly expanding metropolitan agglomerations. Core cities have lost population, retail activity, and employment to peripheral areas, but despite a declining asset base, they must still provide a range of services to meet the needs of their region's inhabitants. The comprehensive reorganisation of Poland's administrative structure in 1999 replaced 49 voivodes (regions), organised mainly around a leading town, with 16 larger regions, and created a new level of local government, the powiat, analogous to the county or kreis, leading to fragmented metropolitan development. The long period between the restoration of local government powers in 1990 and the creation of the new administrative system left a vacuum in metropolitan management just when suburbanisation was most rapid. The new, 1990 administrative system contained no measures for metropolitan management. As a result, no single government within metropolitan areas has been able to effectively or efficiently deal with region-wide problems. Smaller units of government fear domination by core cities, and core cities fear the further loss of jobs and residents to suburban municipalities. National efforts to create metropolitan units go back to 2003, when a law was passed mandating physical planning in metropolitan regions. This was followed by attempts to create units of metropolitan governance, mandatory for Warsaw, the Katowice conurbation, and the Gdansk-Gdynia region, and voluntary for Poland's other main metropolitan regions, particularly between 2007 and 2009. However, these top-down attempts failed. Instead, a growing number of metropolitan regions have turned to informal agreements among local government units. Many are informal forums which serve mainly to exchange information. However some, often created in response to more or less urgent needs, are single purpose, aimed at managing waste, water resources, or transport. A growing number have a broader, multifaceted character, including those in the Wrocław,

T. Kaczmarek (✉)
Institute of Socio-Economic Geography and Spatial Management, Adam Mickiewicz University in Poznań, ul. Dzięgielowa 27, 61-680 Poznań, Poland

Center for Metropolitan Research, Adam Mickiewicz University, Poznań, Poland
e-mail: tomkac@amu.edu.pl

A. Ryder
Formerly Department of Geography, University of Portsmouth, Portsmouth PO1 3HE, UK
e-mail: andrewrydernyc@gmail.com

© Springer Science+Business Media Dordrecht 2015
J. Buček, A. Ryder (eds.), *Governance in Transition*, Springer Geography,
DOI 10.1007/978-94-007-5503-1_2

Katowice, and the Gdansk-Gdynia metropolitan regions, which aim to coordinate regional economic development, transport and urban planning, and even marketing. Perhaps experience in Poland shows that although metropolitan cooperation might be desirable, it is often best organised from below, in response to local needs, rather than from above. Equally, it is possible that over time, as a culture of intra-metropolitan cooperation grows, acceptance of national legislation to create metropolitan government will grow.

Keywords Metropolitan governance • Top-down metropolitan governance • Bottom up metropolitan governance • Local government associations

Introduction

Urban agglomerations and metropolitan areas are motors of growth in relation to production and services, labour markets, innovation and technology, and social and cultural life (OECD 2011). Big cities and their surroundings generate high standards of living and can transform their entire hinterland. The development of spatial relations between a city and its surroundings fosters more intensive administrative cooperation among municipal units, including the development of institutional forms of governance. Metropolitan integration is increasingly vital for socioeconomic growth, particularly in the face of growing international competition. Local cooperation can generate economies of scale and agglomeration. More important, it can help to create a critical mass by linking together disparate units of government in a metropolitan region. For example, in the case of Upper Silesia, no one city has more than 310,000 people, but together the region has almost three million city dwellers. Similarly, the British Chancellor of the Exchequer has argued that cities in the north of England are 'individually strong but collectively not strong' and called for the creation of a 'supercity' in the north of England, made up of four midsized cities: Manchester, Sheffield, Bradford, and Leeds which together have a population of nine million, greater than London (Groom 2014). In the face of competition, local cooperation can foster more effective management, the coordination of spatial planning, and more effective provision of services, raising the quality of life throughout a metropolitan region. Increasingly, when looking at metropolitan areas, policymakers think in terms of critical mass. They argue that too many local governments can result in fragmented policy formation and obscure regional strengths (see, e.g. OECD 2012, pp. 20–21, 290–99).

The adaptability of metropolitan management to changing socioeconomic conditions is essential to ensure that cities and metropolitan areas remain competitive and foster innovation. The creation of new and more effective methods of metropolitan area management, either through new metropolitan governments or other metropolitan-wide institutions, is a common response to local changes caused by suburbanisation and to global changes caused by changing technology, manufacturing, and business practices.

Metropolitan Integration in a Polish Context

Until 1989, local governments in Poland were little more than state enterprises. They were responsible for providing and managing housing and public transport for their inhabitants, for preparing land for construction, and for managing local building programmes to meet the needs of industrial development. They had no taxation powers and relied on the central government for all their funding. They were under no compulsion to make a profit on the enterprises they managed and laboured under a perpetual shortage of resources (Ryder 1990). After the collapse of communism in 1989, local governments gained autonomy and broader powers over their own affairs. They gained limited financial autonomy and tools to manage their own budgets. Local administration reform reinstated local governments in cities and communes in 1990, giving local governments political and moral legitimacy. Consequently, they became more vocal in protecting their administrative and fiscal rights. Today, Poland has 2,479 local governments. 1,576 are rural communes, and 1,585 have populations less than 10,000 (Główny Urząd Statystyczny 2012).

In the early 1990s, Polish politicians, social scientists, economists, and political scientists began to discuss how to reorganise the administrative structure of the country, in existence since 1975. The result was the most comprehensive administrative reorganisation in central-eastern Europe, exceeding even that in the former German Democratic Republic. After several years, it was decided to replace 49 administrative regions – *województwo* (voivodeships) – with just 16 larger ones. These came into being at the start of 1999. The reform established two other levels of government: the *powiat* (secondary local government units, similar to German *kreis*) and the *gmina* (commune). The voivodeships are part of the central government administration. The two lower levels, the powiats and the *gmina*, are relatively autonomous. This 'new' administrative system was a return to a three-level system which existed until 1939.

The pre-1999 voivodeships were usually focused on one large city or town. Some had encompassed entire metropolitan areas. This was particularly true of voivodeships which surrounded the country's largest cities – themselves voivodeships – including Kraków, Poznań, Wrocław, Warsaw, Łódż, and the Gdańsk-Gdynia conurbation The post-1999 voivodeships embraced more than just a large city and its hinterland. In contrast to the previous administrative system, the largest city did not always dominate the economic and political life of the voivodeship. The decision on where to draw the boundaries of the new regions involved extensive consultation. Residents in areas which bordered two regions were consulted on which region they wished to adhere to.

The creation of larger voivodeships and powiats complicated the administrative system and the management of metropolitan regions. Increased local government powers added to this complexity. Cities were recognised as distinct units, but no effort was made to create administrative structures which embraced metropolitan regions. By the early 2000s, it was evident that metropolitan areas were rapidly growing in size and population even though many core cities were losing population (OECD 2011, p. 42; Mikuła 2013, slides 5–6). Thanks to a post-communist explosion of car ownership and growing suburbanisation, more and more local governments were drawn into metropolitan orbits. In 1990 there were only 5,261,000 registered passenger cars in Poland. By 1999 the number had reached 9,282,800, and in 2012 it reached over 18,744,000 (Eurostat 2012, p. 84). A first attempt at metropolitan management, the Spatial Planning and Spatial Management Act of 27 March 2003 (*Ustawa o planowaniu i zagospodarowaniu przestrzennym z dnia 27 marca 2003 roku*) (Dziennik Ustaw 2003), was taken in 2003, with the passage of an act requiring metropolitan planning. However, this was restricted to physical planning and did not include the delivery of services and other functions. Subsequently, the Polish government attempted to create metropolitan area councils to deal with region-wide issues, to improve service delivery, and to reduce costs. Unfortunately, efforts to claw back local government powers and reallocate them to metropolitan area governments failed due to strong resistance among local governments and their residents.

This paper describes some of the proposals first made by the Polish government to foster metropolitan governance. National government proposals were first made in 2008–2009, but, as of 2014, were still not implemented. Proposals for changes came from two sources: administrative circles at the national level (through proposed legislation) and local governments (through the establishment of partnerships, associations, and joint metropolitan strategies).

Concepts and Methods of Metropolitan Integration

'Metropolitan integration' can be defined as the creation or adaption of local management structures to deliver tasks and solve problems which embrace an entire metropolitan area. An analysis of systems of metropolitan management in Europe shows a range of degrees of integration (Kaczmarek and Mikuła 2007a). They can be condensed into two basic forms (Fig. 2.1). In the first, the implementation of institutionalised metropolitan area management is led by and stems from the goals of the state, i.e. 'top-down metropolitan integration'. In the second, integration stems from local government initiatives and often depends on informal cooperation and joint action, i.e. 'bottom-up metropolitan integration'.

In practice, this division is not so clear-cut. The European experience shows that metropolitan integration results from both top-down and bottom-up activities. It is the result of, and often a compromise between, different visions of metropolitan reform at the national level and among local communities and the political powers which represent them. In countries where these visions are relatively consistent, metropolitan reforms have been successful. In those where they are incompatible and where there is a shortage of political or social will, metropolitan management remains a matter of debate.

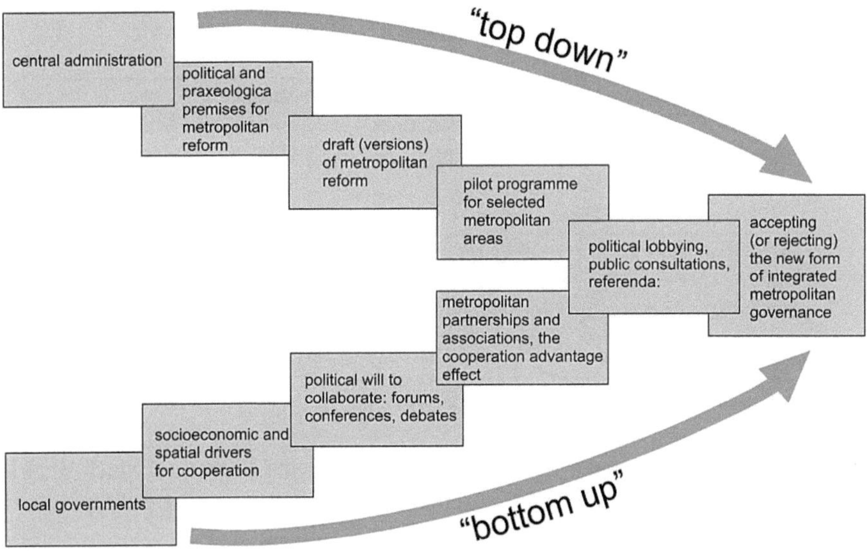

Fig. 2.1 Creating metropolitan integration: 'top-down' and 'bottom-up' method (Source: Author)

The 'top-down' method of achieving integrated metropolitan management is often a response to low levels of grass-roots cooperation. A demand for improved region-wide infrastructure and economic and social problems often lead central governments to introduce legally binding top-down solutions. Low levels of local government cooperation have been typical in Poland over the past two decades. This is partly because the reorganisation of local governments fragmented relations among governments at the same time as they gained administrative independence. In addition, the long time needed for the reorganisation of the administrative system inhibited the growth of personal and institutional relations among local governments, some of which were soon to disappear. After reorganisation occurred, it took time for new local governments and their employees to come to grips with their new powers and with problems in their newly formed regions. Many local governments were reluctant to cede their powers to a higher-level metropolitan organisation. Moreover, in suburban and rural areas, new administrative structures often resulted in new local councils with political compositions which differed from those of their central cities.

The work of the central government, referred to as 'metropolitan reform', is related to the concept of 'new regionalism' or 'new regionalisation' (Keating 1998). Outside of Poland, since the 1980s, this has been viewed as a process of building (or reinstating) metropolitan regions responsible for a vast range of tasks and with concomitant competences, with their own sources of income. Such regions often operate independently from the central government. They are accountable to the communities to whom they provide services and on whose behalf they implement their own development policies. As well, metropolitan management requires the cooperation and participation of entities from the private and public sector (Metropolitan governance, Fürst 2005). Thus, the construction of a new territorial

	Top down cooperation	Bottom Up cooperation
Strong cooperation / high institutionalisation	Integrated metropolitan area government	Voluntary partnerships among local governments, often in the form of legal agreements
Weak Cooperation / low institutionalisation	Obligatory cooperation among local governments to deliver set tasks	Informal forums, conferences, meetings, and advisory bodies

Fig. 2.2 A classification of metropolitan area cooperation

government in metropolitan areas often depends on the degree to which a civil society has developed and on the acceptance of market economy principles and democratic standards of political and social life.

One can categorise cooperation in European metropolitan areas into four basic forms, ranging from strong, formal, top-down cooperation to weak, informal, bottom-up integration (see Fig. 2.2). These forms are based on the legal form of cooperation, the degree of institutionalisation, the degree of autonomy of executive bodies, and the nature and method of electing (or selecting) people who manage metropolitan governments, their competences and responsibilities, their accountability, and their degree of financial autonomy (see Norris 2001; Fürst 2005).

Strong top-down cooperation involves the establishment of a new administrative level which gives metropolitan areas the status of units of local government. Weak top-down cooperation consists of a city and its surroundings joining up to deliver one or a few specific tasks. Strong bottom-up metropolitan cooperation is based on voluntary municipal partnerships and agreements among those administrative units which form the metropolitan area. It may be supported by a legally binding agreement, approved by the central government. Such cooperation embraces a wide range of activities. Weak bottom-up cooperation consists of the establishment of informal or weakly constituted structures such as regular forums, conferences, and consultative bodies, which play an advisory role (Heinz et al. 2004; Walter-Rogg and Sojer 2006). The division between strong and weak depends on the degree to which the agglomeration-wide or metropolitan institutions are developed, their legal status, their degree of political legitimacy, their competences, and the number and nature of tasks they deliver. In this context, ‘collaboration’, ‘cooperation’, and ‘integration’ are most frequently used to describe the co-working of the elements within a system.

Premises for Metropolitan Region Integration

As noted, in Poland, big cities and their hinterlands share a growing number of problems related to suburbanisation and to globalisation processes. Since the start of the 1990s, increased car ownership has fostered the spread of metropolitan areas.

New roads, new peripheral retail, warehousing, manufacturing, and new residential developments encourage further decentralisation. In Poland, the best known examples are the gmina of Tarnowo Podgórne and the motorway junction near Wrocław known as Węziel Bielany (Matykowski and Tobolska 2005; Czerwiński 2012; Smętkowski et al. 2009).

As in most of Western Europe, the situation is complicated by lack of a definition of what constitutes a metropolitan area. In the United States and Canada, for example, national census bureaux compile data on urban regions. These areas are based on a range of factors, including density, size of the central city, and the share of residents who commute to work in the central city (see, e.g. US Census Bureau 2012; Statistics Canada Statistiques 2013). Poland does not define functional metropolitan areas and does not maintain a national statistical programme to monitor population and economic growth in metropolitan areas.

In Poland, the situation is more confused. There is no national definition of a metropolitan area. Each voivode planning office delineates metropolitan regions using its own method, generally based on density, and commuting into the core city or cities. This has resulted in metropolitan areas in some voivodes being divided into three regions: a core city (or core cities in the case of the 'Tri-City' (Trojmiasto) and Katowice regions); an inner, or first-generation suburban zone consisting of municipalities alongside or close to core cities; and an outer metropolitan zone, with more recent settlement. Some voivodeships identify only two zones: a core city and a metropolitan region. Data from individual voivodeships is shown in Table 2.1. As can be seen in the table, in many metropolitan areas, suburbanisation has spread far beyond the inner ring. In Poznan, just over one-third of the metropolitan population is in the outer ring and just over 40 % in the core city.

In Poland, core city populations are declining, whilst suburban populations are growing (OECD 2011, pp. 42–46; Mikuła 2013). Growth in suburban areas is

Table 2.1 Basic data on metropolitan areas in Poland (2008)

		Population			
City region	Core city	Core city	Suburban inner ring	Core city + suburban inner ring	Metropolitan area
Warszawa	Warszawa	1,707,981	910,905	2,618,886	2,972,262
Śląsk	Katowice	311,179	305,871	2,279,771	2,829,593
	14 cities of GZM[a]	1,973,900			
Kraków	Kraków	765,441	284,014	1,040,455	1,461,981
Łódź	Łódź	750,125	376,475	1,126,600	1,126,600
Wrocław	Wrocław	635,803	177,279	810,082	1,171,424
Poznań	Poznań	559,458	307,000	866,458	1,351,180
Trójmiasto (Tri-City)	Gdańsk	455,717	236,942	981,391	1,192,108
	Gdynia	249,787			

Source: Own compilation based on voivodeship definitions of metropolitan areas. Populations compiled from GUS (Central Statistical Office) website: http://www.stat.gov.pl

[a]GZM – Górnośląski Związek Metropolitalny (Metropolitan Union of Upper Silesia)

relatively uncontrolled due to weak regional planning. This has led to the spontaneous growth of housing and settlements in suburban communes Wild (1983, pp. 169–180; 195–7) described in some detail as a similar process which took place in (western) Germany in the 1970s. He noted that expansion was due to the lower cost of land in fringe areas, the rise of car ownership, the dispersal of employment and retailing, and the consequent under-provision of infrastructure and services in suburban areas. Shaw (1983, pp. 110–121) described the process of retail decentralisation in German cities during the same period, noting the rise of new forms of retailing, particularly hypermarkets, referring to the lack of cooperation between authorities. He noted that retail activities 'zoned out' by one local authority could be accepted by a neighbouring one (Shaw 1983, p. 120). More recently, in eastern Germany, a similar process of retail and industrial decentralisation took place after the collapse of the German Democratic Republic. In Poland, the number of jobs in out-of-town locations has grown in the manufacturing and service sectors. This is due to lack of regional planning and due to competition among local governments for investment. Land is cheaper in the suburbs and gaining planning permission less difficult. The result is a growing amount of travel between one's place of work, place of residence, and commercial facilities, from the city to the outer districts and vice versa. Most of this is by car: public transport is poorly integrated, which inhibits non-car commuting (OECD 2011, pp. 14–15; 74–6). As a result, the deconcentration of core cities' functions, including commercial, recreational, and cultural functions, has taken place to the advantage of suburban areas. In tandem, the emigration of taxpayers to outer districts of metropolitan regions, especially people from the higher tax rate group, has reduced the purchasing power of residents in cities and increased it in the suburbs. Central cities have suffered a loss of tax and other income and reduced economic power as businesses have moved to suburban areas. However, central city responsibilities and the costs associated with them have not declined and have often grown. Infrastructure and development needs are often higher in cities, which are also affected by problems in the wider area such as water management or road construction. A common problem arises when the residents of suburban areas use services provided by the core city and pay taxes elsewhere, leaving core cities to finance services and infrastructure of supra-local importance. These problems are not unique to Poland. They have been experienced by cities in Germany (Wild 1983) and the United States for half a century or more (Reynolds 2003).

Boundaries constrain the operations of local governments. Poland has several management levels, each with different legally fixed tasks and competences. The result is a need for cooperation between different administrative levels, such as the commune and the powiat, the commune and the voivodeship, or the powiat and the voivodeship, as well as among governments at the same administrative level.

Thus, metropolitan areas in Poland are characterised by strong fragmentation and competing local interests. Poland's largest cities are usually separate from geographically larger territorial units, i.e. the commune, powiat, or region. Often, they are themselves administrative regions, either the equivalent of powiats in their own right or denominated 'cities separated from the powiat'. When the new three-level administrative system was formed at the end of the 1990s, the intention was to guarantee the autonomy of communes in fulfilling their municipal tasks. In practice, this has inhibited the creation of agreements between core cities and the communes and powiats in which they are located. Moreover, the administrative fragmentation of metropolitan regions sometimes results in numerous, overlapping intergovernmental compacts. Some are between just a few governments, and some are more wide ranging. Some embrace the provision of several services and some just one.

Intra-metropolitan cooperation is needed to handle problems and provide services which have an impact beyond the borders of individual local authorities. Some cooperation in areas such as public transport is due to their 'networked' character. Some cooperation arises to ensure better and more cost-effective service provision, such as education, healthcare, and cultural institutions. Cooperation can also relate to services and activities which are more effective over a larger area, such as the promotion of tourism and inward investment.

Current national legislation makes no provision for legally binding cooperation among local governments at the same administrative level, such as among communes, or between governments at different administrative levels, such as powiats and communes or powiats with voivodeships. Despite this, the cooperation of cities with their surrounding regions is not a new phenomenon in Poland. For many years, municipal agreements and partnerships have been used to support specific tasks related to infrastructure development and the supply of utilities, especially in areas like water management, waste management, and public transport. Other reasons for cooperation among governments in metropolitan areas arise when a problem, such as unemployment, exceeds the competences of individual local governments or exceeds the territorial limits of individual local governments. This includes the management of regional parks, river catchments, or tourist regions. The need for cooperation also arises in extraordinary situations such as floods and industrial disasters (Kaczmarek 2005) or to maintain disaster preparedness. Cooperation can also arise from technical advances, such as those in new-generation transport or telecommunications which require wider area coordination and cooperation.

There are many advantages in establishing metropolitan areas as management units. Operations to deal with problems which extend beyond the boundaries of individual local units can be more easily coordinated. Integrated management can secure a region-wide tax base needed to pay for metropolitan services. Providing services on a metropolitan scale can reduce costs, increase efficiency, and reduce tax levels. This can give a metropolitan area an advantage over competing metropolitan regions (see, e.g. Oates 2006; Tiebout 1956). Within the European Union, another important factor is an enhanced ability to lobby for and obtain external financial resources, including EU funds, to supplement local budgets and meet local goals. Intra-metropolitan area cooperation can also allow participating governments to develop and harmonise statistical databases on a metropolitan-wide scale. This can facilitate socioeconomic analyses, making it easier to establish development strategies.

However, integration can bring drawbacks. Member communes can lose some independence in managing their affairs. They need to compromise and reach a consensus regarding joint development policies and spatial planning and need to jointly contribute to costs related to the agglomeration area as a whole. They may also need to build a new type of regional identity and agglomerative thinking among residents ('think metropolitan, act local').

Metropolitan Reform: Government Proposals

In 2003, the Polish government passed an act mandating regional physical planning in urban areas or *obszar metropolitalny* (*Dziennik Ustaw* 2003). These bodies created under this act are in many ways analogous to so-called Metropolitan Planning Organisations in the United States, which have existed in various forms since the 1970s, described in Solof (1998). Such organisations are responsible for ensuring that local infrastructure plans among different municipalities match up properly. Although they were first organised in 1962, they did not gain real powers until 1973. From that time onwards, the US federal government made the organisation of such planning organisations a requirement for obtaining national funding for highways and public transport. In 1991, they took on new powers, becoming the equal of state transportation departments and assuming lead authority for transport planning projects in their urban regions. However, such organisations do not cross state lines and remain limited to transport planning.

In 2008–2009, the government of the Republic of Poland started work on a bill to introduce new forms of cooperation between local government units within metropolitan regions. The latest project (from 2009) was titled 'National Urban Policy and Cooperation of Local Governments Act'. Although the new act remains a work in progress, there is a general consensus on the need to foster metropolitan governance and the forms it should take.

The goal of the various draft acts, generally referred to as 'metropolitan acts', was the establishment of a legal basis to tighten the collaboration of administration units in Poland's leading metropolitan regions. These are often referred to as 'agglomerations' (*aglomeracja*) or, more widely, as 'metropolitan areas' (*obszar metropolitalny*). The draft acts have aimed to foster socioeconomic and spatial cohesion and effective metropolitan area management. In the proposals, it was agreed that integrated management should be mandatory in three metropolitan areas: Warsaw, the largest metropolitan area in Poland, and the conurbations of Upper Silesia and the Tri-Cities (Trojmiasto) – Gdańsk, Gdynia, and Sopot. In other cases, the establishment of metropolitan units (in Polish – *zespół metropolitalny*) was to be voluntary. To some extent, the proposed legislation would have complemented the legislation passed in 2003 mandating regional physical planning in urban areas (*obszar metropolitalny*) (Dziennik Ustaw 2003).

The proposed responsibilities of metropolitan units included obligatory and optional tasks. Mirroring the 2003 mandate on metropolitan physical planning, the legislation called for local government units to coordinate spatial planning; cultural heritage protection; environmental protection; power, heating, and gas supply; the management of the water sources and the water supply system; sewage disposal and treatment; the management of waste disposal sites and municipal waste; traffic and public transport management; and labour market oversight. Other requirements included the joint development of infrastructure-related projects. Metropolitan governments were expected to initiate and develop intergovernmental cooperation to deliver metropolitan-wide services. They were also expected to market the entire metropolitan area and were to be responsible for obtaining and managing the public and private funds required to deliver their tasks.

In general, the structure of a metropolitan unit was to resemble an inter-municipal partnership. The unit was to consist of an assembly, which would be a decision-making and controlling body, and a board, which would be the executive body. The work of a metropolitan unit would be administered by a separate administration, consisting of a unit bureau headed by a unit executive manager.

A metropolitan unit assembly would have consisted of delegates from all the local governments which comprised the metropolitan unit. Each local government would have been represented by one delegate who, by force of law, would have been the commune/town/city/district mayor. However, under the proposals, the metropolitan government would have the exclusive right to adopt statutes which applied to the entire metropolitan area, and to amend them, to adopt a metropolitan area spatial development and management strategy, to decide on directions for the metropolitan unit board's operations, to accept the board's operation reports, to adopt the metropolitan unit's budget, and to decide on the standards and costs of those metropolitan-wide public services with which it was entrusted.

The metropolitan unit assembly would have been able to pass legislation only on a double absolute majority basis. A resolution would need to be approved by the majority of the local government units and also by governments which represented the majority of the metropolitan area's residents. The executive body of a metropolitan unit was to be a collective board chaired by the mayor of the city with the highest population.

To ensure the cohesion and coordination of the work of voivodeship governments and metropolitan units, the bill proposed the formation of a metropolitan commission with oversight powers, consisting of five representatives from the voivodeship government and five from the metropolitan unit.

As of 2013–2014, the legislation remained dormant. Proposals for some kind of metropolitan governance remain on the political agenda, and as noted below, recent changes in how European Union structural funds are allocated have led to greater metropolitan cooperation. However, metropolitan areas continue to decentralise. Any formal structure created today will become obsolete in a few years. Any attempt to foresee future expansion would mean incorporating mainly rural communes into larger urban areas, subjecting them to land use and other controls, and effectively freezing their land uses and local status.

Bottom-Up Initiatives of Metropolitan Integration

In the absence of national legislation, local governments in metropolitan areas have turned to less formal bottom-up cooperation. This has generally taken the form of single-purpose associations which are among the most popular forms of cooperation between local governments in Europe (Kaczmarek 2005).

Currently, metropolitan cooperation in Poland generally takes one of several forms. Three forms have legal standing. The strongest are unions (in Polish, *związki*). They are followed by 'associations' (*stowarzyszenia*) and 'understandings' or 'agreements' (*porozumienia*). In addition, there are informal and often transient agreements among local governments which have no legal standing. Finally, as in the case of Wrocław, the Wrocław Development Agency has the status of a limited company under corporate law. The national government collects information only on the number of registered unions in Poland. Unions need the approval of the Ministry of Digitisation and Administration (Ministerstwo Administracji i Cyfryzacji), part of the central government, and must also be approved by statute passed by the voivodeship government (see, e.g. Województwo Śląsk 2007).

In 2009, there were 233 municipal associations in Poland, of which 44 operated in metropolitan areas. In 2014, there were 234 (Ministerstwo Administracji i Cyfrizacji 2014). Nearly 35 % of the associations cover neither core cities nor neighbouring 'first-ring' municipalities. Only ten included metropolitan core cities. In 2009, the most popular form of cooperation was environmental protection (66 %). This often ties in with water and sewage management (55 %) and waste management (45 %), in addition to tourism and recreation. To a smaller extent cooperation concerns the development of other types of infrastructure, for example, gas supply and heating (14 %) and public transport (16 %).

Among the 44 associations within major metropolitan areas, only a few were metropolitan in character, including core cities and suburban municipalities. A map of Poland's leading metropolitan agglomerations and the nature of the informal integration in those areas, whether relatively strong or weak, are shown in Fig. 2.3. They include the Metropolitan Transport Association of the Bay of Gdansk, the Municipal Transport Union of the Upper Silesian Industrial District, and the Metropolitan Association of Upper Silesia, later renamed Metropolia Silesia (Silesian Metropolis). These two metropolitan regions are unique among Polish metropolitan areas in that they are not dominated by a single large city. The Gdansk region consists of two large cities and several smaller ones. Upper Silesia consists of a cluster of medium-sized cities. The lead city, Katowice, contains less than 16 % of the population living in the region's core cities and only 11 % of the population of the broader metropolitan area.

In 2003, following the initiative of the marshal of the Pomerania Region (voivodeship) and the mayor of Gdansk, the Metropolitan Council of the Bay of Gdansk was established by ten local governments. The main goal of the council was to resolve common development problems in the Tri-City Metropolitan Area. In 2006, the marshal of Pomerania Region; the mayors of Gdansk, Gdynia, and Sopot; and nine

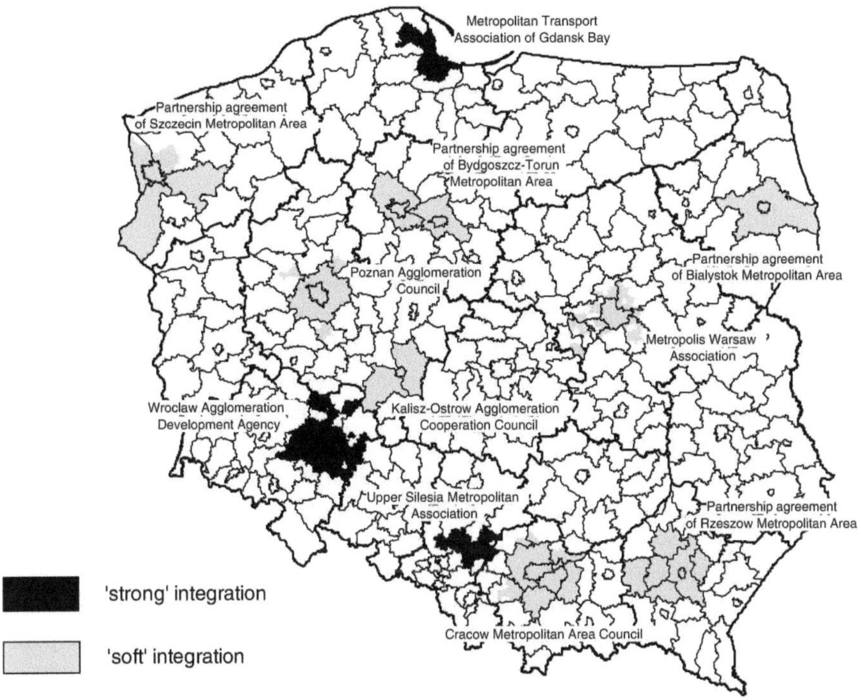

Fig. 2.3 Metropolitan areas in Poland – 'bottom-up' integration

other local governments signed a Memorandum of the Gdańsk Metropolitan Area (Majewski et al. 2006). This was followed a year later by the Tri-City Charter, a declaration of intent to prevent destructive competition whilst maintaining respect for the traditions and identities of the three cities. The charter included a catalogue of major tasks to be implemented jointly by the local Tri-City governments. The declaration was followed by the creation of the Metropolitan Transport Union of the Bay of Gdansk (*Związek Komunikacyjny Zatoki Gdańskiej*), a regional transportation authority that was registered with the Ministry for Administration and Digitisation (*Ministerstwo Administracji I Cyfrazacji*) on 5 June 2007, embracing 14 municipalities of the Tri-City Metropolitan Area. Its goal was to jointly develop local transport policy and manage local public transport, including integrating the fare system. In 2008, the association introduced a common metropolitan travel ticket to be used across the whole area. Initially, the agreement had more ambitious goals, but disputes between the two main cities, Gdańsk and Gdynia, resulted in the formation of two smaller associations, formed around the two cities. One source of contention was the name of the association: both Gdynia and Gdańsk wanted their city to be part of the association name (Lackowska and Zimmerman 2010). In the end, the two cities agreed to describe the region as the 'Bay of Gdańsk' (*Zatoka Gdańsku*). However, in 2011, Gdynia withdrew from the agreement, and the association was renamed the Gdański Obszar Metropolitalny (Gdańsk Metropolitan

Region) which included 17 cities, 8 powiats, and 24 gmina. Its aims included joint purchasing, coordinating local plans, and sharing best practice. Despite the exclusion of Gdynia, there is a strong regional identity. Another informal initiative dates back to 1998: an Internet site with the name Trojmiasto.pl. This commercial site started by advertising cultural events in the region. In 2008, it became a limited company. Today, it offers general information about the area to tourists and residents and maintains a catalogue of local firms and employment opportunities (http://www.trojmiasto.pl/historia.php).

Upper Silesia is an example of a stronger association. In June 2007, 14 local governments in the Upper Silesia (Górno-Sląsk) Voivodeship registered the Górnośląski Związek Metropolitalny (the Upper Silesian Metropolitan Union) with the Ministry of Administration and Digitisation. This union consists of 14 city governments, all of which had county status. One of the justifications for the union is that the region needs to work together to generate a critical mass of talent and capital and strengthen the region's competitive advantage. The aim of the union is to establish a common growth strategy for the region, manage roads and public transport, support the development of the local labour market, support and enhance local education, conduct economic research, and sell the region to prospective investors. The view of the member governments was that although each city individually was relatively small, the agglomeration is the largest in the country in terms of population, with 200,000 firms, a special economic zone, 10 % of Poland's research capacity, and generating 8 % of GDP.

As far as Poland is concerned, a unique concept of metropolitan integration has been implemented in the Wrocław Metropolitan Area. The focus here is on cooperation in promoting economic development. This has been institutionalised by the Wrocław Agglomeration Development Agency (Agencja Rozwoju Aglomeracji Wrocławskiej). The agency was established through the initiative of the mayor (Prezydent) of Wrocław City government in 2005, initially by eight local governments as the Wrocław Regional Development Agency (Wrocławska Agencja Rozwoju Regionalnego), a non-profit joint-stock company, with shares owned by member governments. However, the origins go back to 1999, shortly after the creation of Poland's new administrative system. The agency was meant to operate as a supra-local organisation, above municipalities and powiats. Its goal was to support economic development and entrepreneurship within the agglomeration; promote education and the dissemination of information about the region, to encourage foreign direct investment; and provide training. It served, and serves, as a 'one-stop shop' to provide investors with a full range of services. Currently, it has 30 governments including the city of Wrocław and the Dolnosląsk (Lower Silesian) Voivodeship government and 28 local communes. The organisation also has several private-sector partners, including a human resources firm and two international property firms. This spontaneous creation resembles similar groupings of local governments in the United Kingdom, particularly that in Greater Manchester (Economist 2013). As in Poland, none of Britain's local governments embrace an entire metropolitan area. In the absence of national legislation, they have had to cooperate informally to promote broader regional goals. In the Greater Manchester area, ten local governments formed a combined entity as long ago as 1986, which controls the entire transport budget and is run by a 'cabinet' consisting of one official appointed by each member municipality. As well, Greater Manchester's governments have acted in concert to promote the economic development of the entire region, in a way not dissimilar to that seen in Wrocław (see the discussion in Sect. 1.2 of the Introduction).

The metropolitan 'governance' concept puts a strong emphasis on informal metropolitan integration. Coordination mechanisms, although not being legally binding, often constitute the start of the road towards more formalised cooperation methods (Kaczmarek and Mikuła 2007a, b).

The expansion of local government cooperation can also be observed in the Kraków Metropolitan Area. Since 2004 it has regularly held forums of the municipalities of the Kraków Metropolitan Area. These have focused on cooperating to create a spatial development plan for the Kraków Metropolitan Area. A subsequent step towards greater integration was the signing of a declaration of cooperation and the establishment of the Kraków Metropolitan Area Council in November 2007. The goals of the council included the construction of ring roads and motorway connections, upgrading roads which connect areas of economic importance, investments in the field of environment protection, waste management, sporting and recreation facilities, as well as projects concerning education, culture, territorial marketing, and public policy. Apart from Kraków, the council consisted of 52 suburban municipalities.

Similar to the Kraków Metropolitan Area, in May 2007, the city of Poznań, 17 suburban municipalities, and the powiat of Poznań signed an Agreement of Cooperation among the Local Governments of the Poznań Agglomeration (*Porozumienie o współpracy pomiędzy samorządami aglomeracji poznańskiej*). By 2010, the number of suburban municipalities had increased to 20. The stakeholders of the agreement created an Agglomeration (Metropolitan Area) Council. Initially, this acted as a forum for exchanging information among the local governments and for determining potential areas of cooperation. The intent was to cooperate in supporting business initiatives, marketing activities, public transport, education, healthcare, and spatial policy. The members of the Agglomeration Council unanimously supported the founding of a Metropolitan Research Centre. Each municipality provided resources for this purpose, in proportion to its respective population. In June, 2011, the council published a development strategy for the Poznan metropolitan region, titled Poznan Metropolis 2020 (Metropolia Poznań 2020) (Centrum Badań Metropolitalnych 2011). At that time, the organisation took on a new form, renaming itself the Poznan Metropolis Association (Stowarzyszenie Metropolia Poznań).

Although relatively recent, bottom-up agreements should lead to the development of more specific administrative models for metropolitan areas. After 20 years of municipal autonomy, Polish metropolitan areas are starting an integration process in the field of metropolitan governance which has become necessary in response to dynamic processes of suburbanisation. In 2011, the OECD found that around 60 % of all municipalities engaged in some kind of inter-municipal cooperation, although the scope of cooperation was often relatively narrow (OECD 2011, p. 18).

They also noted that cooperation was hampered by a lack of legislation regarding agreements and financial commitments. Currently (2013–2014), there are five metropolitan associations in Poland: Gdańsk-Gdynia, Kraków, Poznan, Upper Silesia, and Wrocław. Similar initiatives have also been undertaken by smaller cities and areas, for example, Bydgoszcz-Toruń, Kalisz-Ostrów, Białystok, Rzeszów, and Szczecin.

Problems of Metropolitan Integration

Although discussions are continuing, no decision has been taken regarding the creation of formal metropolitan units. In Poland, integration remains a largely bottom-up phenomenon, with collaboration often organised around single issues. Individual voivodeships remain responsible for defining metropolitan regions on their territory.

In Poland, the level and rate of metropolitan integration are affected by a range of administrative, political, demographic, financial, and social barriers. However, the basic barrier to integrated metropolitan area management is the multiplicity of stakeholders involved. There is a lack of tight hierarchical connections among them. Potential partners include units with different populations and local wealth, as well as different demographic and socioeconomic characteristics. They include administrative units at the commune level or the intermediate (powiat) level. As Soja (2000) noted, the problem of administrative fragmentation in metropolitan areas becomes especially severe when the core city cannot extend its limits. In the administrative reform at the end of the 1990s, city boundaries remained unchanged, even as they were folded into new administrative regions. Today, it is almost impossible for cities to expand their boundaries.

Administrative fragmentation has led to a multiplicity of decision-making entities within increasingly functionally connected areas. This has resulted in conflicting and competing development priorities among local governments and different policies related to the expansion of housing areas, land use, or priorities for extending the transport infrastructure. Decisions taken by individual governments can affect development throughout an entire metropolitan area. Different development potentials among individual governments affect the amount of interest in cooperation and consequently cooperation itself. This is not unique to Poland. A recent OECD study of the metropolitan Chicago region (OECD 2012) found that fragmented government 'limited capacity to think and act regionally' and that local government officials, public stakeholders, and even private-sector actors often failed to recognise their interdependence. Instead, local decisions were based on 'a narrow sense of self-interest' (OECD 2012, p. 294). The report noted that trying to increase the number of firms in a jurisdiction by luring them from neighbouring jurisdictions with tax incentives was self-defeating, adding that 'It is preferable by far to attract firms by showing that a pool of talent and organisations exists in the region that can help a newcomer exploit it' (OECD 2012, pp. 20–21).

In Poland today, there are growing political disagreements between those who live in core cities and those who live in the suburbs. Suburban residents are often characterised by an elevated social status and have political preferences that are typical of the middle classes (McKenzie 1933). These residents are often strongly linked with the core city through their workplace or the places where they consume leisure and other services. Even when the political profiles of the core city and suburban communities are similar, perspectives on the development of the metropolitan area as a whole can be quite different. These are often expressed in local elections which favour candidates who emphasise the strong independence and competitive position of a suburban location and favour highly autonomous and individualistic policies. An increase in conflicts has also been observed between existing local residents and newcomers. Increased population creates a need for new housing, higher expenditure on infrastructure, and traffic problems and can overload public service facilities.

Intra-regional inequality is intensified by what Keating (1995) termed 'spill-overs'. This occurs when residents pay taxes in one area, where they live, but use services provided elsewhere, leading to what Keating called 'freeriding' – the financing of services and infrastructure of metropolitan significance by the core city. This phenomenon, which is difficult to measure, complicates the establishment of cooperation in maintaining the infrastructure and institutions of metropolitan reach.

Recent Developments

As noted, in 2014, 234 unions among local governments, and four among counties, were registered with the Ministry of Administration and Digitisation. Some dated back to 1991. Most were extremely specialised, with a focus on sewers and water or waste management. Another 85 had been disbanded.

However, in connection with the European Union 2014–2020 funding round, the Polish government mandated that local governments work together on applications for structural funds. A requirement is that they develop metropolitan area strategic (but not spatial) plans. This planning must include a social dimension. To apply for funds, they must create an 'integrated territorial investment' (ITI) organisation (*Zintegrowane Inwestycje Terytorialne* organisation) or designate an existing organisation to serve as one. The organisation must be approved by the voivodeship government and registered with the Ministry of Infrastructure and Development (Ministerstwo Infrastruktury i Rozwoju).

In response, on 14 February 2014, the Gdańsk region created an ITI Union, consisting of 36 local government members, including 12 cities with county status, 18 gmina, and 6 powiats, and 25 governments with observer status. The new organisation took the name Tri-City metropolitan region (Obszar Metropolitalny Trojmiasta). In March 2014, the city of Kraków and 13 neighbouring local governments announced the creation of a Kraków Metropolitan Association of Local Governments which would serve as the local ITI organisation.

Thus, although top-down efforts to foster metropolitan integration have failed in the past, the use of financial incentives, in the form of EU structural funds, has led to the spread of intra-metropolitan cooperation and metropolitan organisations. Recent Polish experience resembles what took place in America in 1991, when the US federal government passed the Intermodal Surface Transportation Efficiency Act. This act, and its successor acts, required all federally designated metropolitan areas (so-called metropolitan statistical areas) to establish a Metropolitan Planning Organisation to coordinate transport planning. They had to include not just governments of already urbanised areas within a metropolitan region but those which were likely to become swallowed up by the area over a 20-year period. For the most part, these organisations consist of officials appointed by local governmental jurisdictions; representatives of different types of transportation; including rail, public transport, and road transport; and representatives of state agencies and federal agencies. ISTEA directed additional federal funding to Metropolitan Planning Organisations, expanded their authority to select projects, and strengthened the requirement for a region-wide planning. To ensure compliance, the federal government made the creation of a Metropolitan Planning Organisation a requirement for receiving federal transportation funding.

However, in Poland, although the Ministry of Administration and Digitisation is responsible for approving and overseeing some forms of inter-municipal cooperation, such as unions (Związki), the Ministry of Infrastructure and Development oversees physical planning and is therefore responsible for monitoring newly created Integrated Territorial Investment Organisations.

Conclusion

European experience suggests that strong local leadership advances metropolitan integration (Jouve and Lefevre 2000). Such leadership often comes from core city authorities or from persons at the next highest administrative level. They often initiate cooperation and negotiate its expansion. This has been the case in Manchester and appears to be the case in Wrocław. As Diller (2002) noted, lasting collaboration requires political leaders who are focused on delivering current goals and think of the longer term and within a wider metropolitan framework. They include the political leaders of the core city, such as mayor, but also political leaders in neighbouring local jurisdictions which are functionally integrated with the core city. They may want to tighten their connections with the core city or with other local governments in the metropolitan area. Regional political leaders from a higher level of government (the voivodeship in Poland) may also foster collaboration.

However, in metropolitan regions focussed on a single large city, the fear of the core city's supremacy and the fear of negative financial and investment-related 'backwash effects' can create more barriers to cooperation than in polycentric metropolitan systems (Walter-Rogg and Sojer 2006). Handing some competences to a metropolitan government erodes the powers of larger cities, subjecting them to a hierarchically higher unit: they enter into what Benz (2003) referred to as the 'shadow of hierarchy'. It is therefore not surprising that the bottom-up metropolitan integration process is most advanced in the polycentric Upper Silesian region (the Silesia Metropolis) and least advanced in the strongly monocentric Warsaw Metropolitan Area.

Regional integration is often more successful if integration efforts are supported by a group of experts who organise and manage the cooperation programme (the German *Fachpromotoren* and *Prozesspromotoren*). They include researchers and practitioners, as well as managers who are viewed as technocrats and therefore above local politics. This can be clearly seen in the Silesia Metropolis and the Poznań Metropolis, where research units are designing metropolitan strategy alongside local governments. In the latter case, a special research facility, the Metropolitan Research Centre, has been created to assist in policy formation.

Often, new administrative structures are needed to initiate the process towards metropolitan management. Initially, these may be informal, including forums, round tables, conferences, councils, and working groups (see Brake 1997). Collaboration of local government units is usually 'horizontal', grouping governments at the same administrative level. However, this often requires 'vertical' support from a higher level of government which encompasses the local units. This kind of integration can form the basis for obtaining external financial assistance. In Poland, the cooperation of local units with voivodeship governments is increasingly apparent in several metropolitan areas, including Poznań, Wrocław, and Upper Silesia. Successful metropolitan governance also requires grass-roots public support. Metropolitan integration, regardless of its form or sophistication, must be accepted by those who reside in local jurisdictions. The initiators of collaboration need to clearly explain why it is necessary and what benefits it can bring. Public support can be fostered by a marketing campaign to make residents aware of the advantages of the proposed changes. Without such a campaign, the integration process may fail. Finally, as the experience of Wrocław and Upper Silesia shows, building cooperation takes time and often occurs in small, incremental steps. These are the challenges that face metropolitan reform in Poland, regardless of whether it is a 'bottom-up' or 'top-down' process.

The OECD (OECD 2011) has developed a synthetic measure, based on urban labour market areas (ULMA). In Poland, powiats were used as the basic building block. In a manner similar to that used in defining census metropolitan areas in the United States, the delineation of a metropolitan area was based on a core area, determined by urban status and population density, and adjacent areas with a significant level of commuting into the core area. Urban municipalities were defined as those with population densities greater than 300 km^{-2}, rural ones below that level. Powiats in which 85 % or more of the population lived in 'urban' gminas were defined as urban; those in which under 50 % lived in 'urban' gminas were defined as rural; and those in between were defined as 'mixed' (OECD 2011, p. 32). Travel to work information was obtained from the Main Statistical Office. Using this method, it was estimated that about 70 % of Poland's population was urbanised, as opposed to official estimates of about 61 % (OECD 2011, p. 42). However, this process is complicated, and data on commuting is not collected regularly or comprehensively.

References

Benz A (2003) Regional governance mit organisatorischem Kern. Das Beispiel der Region Stuttgart. Informationen zur Raumentzwicklung 8(9):505–512

Brake K (1997) Städtenetze – ein neuer Ansatz interkommunaler Kooperation. Archiv für Kommunalwissenschaften I/97. Stuttgart, 98–115

Centrum Badań Metropolitalnych (2011) Metropolia Poznań 2020: Strategia Rozwoju Aglomeracji Poznańskiej. Available at: http://metropolia2020.poznan.pl/tresc-dokumentu. Viewed 12 June 2014

Czerwiński S (2012) Kształtowanie się struktury przestrzennej "Węzla Bielańskiego" w rejonie autostrady A4. In: Cioka S, Krzysztofa J (eds) Z Problematyki Reginoalnej Dolnego Śląska. Rozprawy Naukowe Instytut Geografii i Rozwoju Regionalnego Uniwersytetu Wrocławskiego 23.

Diller C (2002) Zwischen Netzwerk und Organisation. Raumforsch Raumordn 2:146–154

Dziennik Ustaw (2003) Ustawa z dnia 27 marca 2003 r. o planowaniu i zagospodarowaniu przestrzennym (Dz. U. z 2003 r. Nr 80, poz. 717). Available at: http://www.administrator.elk.pl/akty-prawne/zagospodarowaniu_przestrzennym.pdf. Viewed 12 June 2014.

Eurostat (2012) EU transport in figures statistical pocketbook 2012. Publications Office of the European Union, Luxembourg

Fürst D (2005) Metropolitan governance in Germany. In: Heinelt H, Kübler D (eds) Metropolitan governance: capacity, democracy and the dynamics of place. Routledge, London, pp 151–168

Główny Urząd Statystyczny (2012) Mały Rocznik Statystyczny 2011. GUS, Warsaw

Groom B (2014) Chancellor to propose east–west, HS3. Financial Times, 23 June 2014, p 3

Heinz W et al (2004) Interkommunale Kooperation in Baden-Württembergischen Stadtregionen. Stuttgart, Karlsruhe und Freiburg. Deutsches Institut für Urbanistik, Berlin

Jouve B, Lefevre C (eds) (2000) Local power, territory and institutions in European metropolitan regions. Frank Cass, London

Kaczmarek T (2005) Struktury terytorialno-administracyjne i ich reformy w krajach europejskich (The territorial-administrative structures and their reforms in European states). Wydawnictwo Naukowe UAM, Poznan

Kaczmarek T, Mikuła Ł (2007a) Ustroje terytorialno-administracyjne obszarów metropolitalnych w Europie (Territorial and administrative organisation of metropolitan areas in Europe). Bogucki Wydawnictwo Naukowe, Poznań

Kaczmarek T, Mikuła Ł (2007b) Metropolitan areas in Poland – towards a new scale of urban governance? Quaestiones Geographicae 26B:97–105

Keating M (1995) Size, efficiency, and democracy: consolidation, fragmentation, and public choice. In: Judge D, Stoker G, Wolman H (eds) Theories of urban politics. Sage, London

Keating M (1998) The new regionalism in Western Europe. Edward Elgar, Cheltenham

Lackowska M, Zimmerman K (2010) New forms of territorial governance in metropolitan regions? A Polish–German comparison. Eur Urban Reg Stud XX(X):1–14

Majewski W, Pawłowicz L, Pisarewicz P (eds) (2006) Memorandum of the Gdańsk Metropolitan Area. The Gdańsk Academy of Banking at the Gdańsk Institute for Market Economics, Gdańsk. Available at: http://www.pomorskie.eu/res/wizytowka/wersje_jezykowe/memorandum_en_2006.pdf. Viewed 12 June 2014

Matykowski R, Tobolska A (2005) Nowe inwestycje przemyłowe i mieszkaniowe w gminie Tarnowo Podgórne (w strefie podmiejskiej Poznania). In: Marszał T (ed) Gospodarka w przestrzeni polskich miast – Economic Activity in the Space of Polish Cities. Biuletyn/Komitet Przestrzennego Zagospodarowania Kraju Polskiej Akademii Nauk; 216 Warszawa: Państwowe Wydawn. Naukowe 2005

McKenzie RD (ed) (1933) The metropolitan community. McGraw-Hill, New York

Mikuła Ł (2013) Planning for urban regions in Poland: top-down failures, bottom-up struggles. Available at: http://www.unil.ch/files/live//sites/igu-urban/files/shared/Mikula_planning_urban_regions_Poland.pdf. Viewed 12 June 2014

Ministerstwo Administracji I Cyfrizacji (2014) Wykaz Związków Międzygminnych. Available at: http://mac.bip.gov.pl/rejestry/zwiazki-powiatow-i-zwiazki-miedzygminne_12_12.html. Viewed 12 June 2014

Norris D (2001) Whither metropolitan governance? Urban Aff Rev 36:532–550

Oates WE (2006) On the theory and practice of fiscal decentralization. Working papers 2006–5, University of Kentucky, Institute for Federalism and Intergovernmental Relations. Available at: http://www.ifigr.org/publication/ifir_working_papers/IFIR-WP-2006-05.pdf. Viewed 1 June 2013

Organisation for Economic Cooperation and Development (OECD) (2011) Urban policy reviews: Poland. OECD, Paris

Organisation for Economic Cooperation and Development (OECD) (2012) Territorial policy reviews: the Chicago metropolitan tri-state area. OECD, Paris. Available at: http://www.oecd.org/washington/49912798.pdf. Viewed 12 June 2014

Reynolds L (2003) Intergovernmental cooperation metropolitan equity, and the new regionalism. Wash Law Rev 78:93–160

Ryder A (1990) Growth pole policy in Poland and the Lenin steel-works. Geoforum 21(2): 229–244

Shaw G (1983) Trends in consumer behaviour and retailing. In: Wild T (ed) Urban and rural change in West Germany. Croom Helm, Kent

Smętkowski M, Jałowiecki B, Gorzelak G (2009) Obszary metropolitalne w Polsce – diagnoza i rekomendacje (Metropolitan areas in Poland – diagnosis and recommendations). Studia Regionalne i Lokalne 35(1):52–73

Soja E (2000) Postmetropolis. Blackwell, Oxford

Solof M (1998) History of Metropolitan Planning Organizations. North Jersey Transportation Planning Authority, Inc., Newark. Available at: http://www.njtpa.org/getmedia/b95661af-dfd4-4e3d-bb87-39e617619c7b/MPOhistory1998.pdf.aspx. Viewed 13 Mar 2014

Statistics Canada Statistiques (2013) Census metropolitan area and census agglomeration definitions. Available at: http://www.statcan.gc.ca/pub/93-600-x/2010000/definitions-eng.htm. Viewed 1 June 2013

The Economist (2013) English cities: free at last, 2 February 2013

Tiebout C (1956) A pure theory of local expenditures. J Polit Econ 64(5):416–424

United States Census Bureau (2012) Statistical abstract of the United State 2012s. US Government Printing Office, Washington, DC

Walter-Rogg M, Sojer M (2006) Metropolitan governance reform in Germany. Paper for the conference "Governance and spatial discontinuities: reterritorialisation or a new polarisation of metropolitan spaces?". Montreal, INRS-Urbanisation, 24–25 April 2006

Wild MT (ed) (1983) Urban and rural change in West Germany. Croom Helm, Kent

Województwo Śląsk (2007) Obwieszczenie Wojewody Śląskiego z dnia 27 czerwca 2007 r. w sprawie ogłoszenia statutu Górnośląskiego Związku Metropolitalnego (GZM) z siedzibą w Katowicach DZIENNIK URZĘDOWY WOJEWÓDZTWA SLĄSKIEGO, 28 June, pp 10598–10604. Available at: http://www.katowice.uw.gov.pl/_pliki/urzad/dzienniki_urzedowe/2007/dz108p.pdf. Viewed 12 June 2014

Chapter 3
The Sydney Metropolitan Strategy: Implementation Challenges

Patrick Fensham

Abstract The New South Wales (NSW) Labour Government released the Metropolitan Strategy for metropolitan Sydney in December 2005 (entitled City of Cities: A Plan for Sydney's Future). This was the first comprehensive strategy for accommodating Sydney's growth and development since 1994. The strategy was reviewed in 2010 and a new version (entitled Metropolitan Plan for Sydney 2036) released. With a new (conservative) coalition government elected in 2011 a further strategy (entitled A Plan for Growing Sydney) was released in 2014. This paper provides an overview of the successive strategies, which have had consistent elements. There has been a particular focus on creating stronger employment centres in suburban Sydney to 'balance out' the strong growth of central Sydney, thereby reducing commuting time and congestion while providing employment and other activities closer to where people live. The paper also examines the implementation arrangements and the extent to which they have succeeded. The implementation of the plan has been hampered by several factors. The State Government is responsible for metropolitan planning and for setting regional development strategies, but lacks the fiscal resources to implement the plan. The Sydney metropolitan region stretches 65 km east to west and 40 km north to south and includes over 40 local governments. There is no metropolitan level of governance, which means there is no 'independent' champion of metropolitan planning aims. Since local governments also face tax constraints, they tend to focus on meeting local needs rather than metropolitan ones.

A proposed Greater Sydney Commission holds the promise of more effective implementation of metropolitan planning aims, though weaknesses in infrastructure funding arrangements and ensuring 'whole-of-government' support, are enduring.

Keywords Regional planning • Regional planning processes • Intergovernmental relations • Sydney, Australia

P. Fensham (✉)
SGS Economics and Planning, 209/50 Holt Street, Surry Hills, Sydney, NSW 2010, Australia
e-mail: PFensham@sgsep.com.au

© Springer Science+Business Media Dordrecht 2015

J. Buček, A. Ryder (eds.), *Governance in Transition*, Springer Geography,
DOI 10.1007/978-94-007-5503-1_3

Introduction

The last 10 years has seen a revivial of metropolitan planning for Sydney with three plans or strategies released by the New South Wales (NSW) Government in that time. The first of these was released by the Labour State Government in December 2005 entitled *City of Cities: A Plan for Sydney's Future*. This was the first comprehensive strategy for accommodating Sydney's growth and development since 1994. A key aim of the 25-year strategy was to encourage growth in nominated suburban centres, to create residential and employment concentrations in locations other than central Sydney and thereby reduce commuting time to the city while providing employment and other activities closer to where people live. The strategy was reviewed in 2010 and a new version (entitled *Metropolitan Plan for Sydney 2036*) released. A Liberal-National coalition (conservative) government replaced a long-standing Labor administration in NSW in March 2011. In December 2014 this new government released its own document entitled 'A Plan for Growing Sydney' with a 15 to 20 year time horizon.

This paper provides an overview of the key elements of these strategies (mostly consistent between the 2005, 2010 and 2014 versions) and in particular examines the implementation arrangements and the extent to which these are succeeding or failing.

This paper contains eight sections. The first is an overview of the roles and responsibilities of the levels of government in New South Wales in relation to metropolitan strategic planning. The second is a discussion of the problems and challenges facing the Sydney metropolitan region. The third is a discussion of the key elements of Sydney's recent metropolitan plans or strategies. This is followed by a discussion of the implied and important distinction between the state and local levels of strategic interest. It then summarises the most recently proposed implementation framework and discusses how this 'rates' against suggested 'ideal' implementation elements. It concludes with an evaluation of the problems facing metropolitan governance in Sydney, and in Australia in general, and looks at how these have influenced the fulfilment of recent Sydney metropolitan planning strategies.

Australian Metropolitan Strategic Planning at a Glance

Before addressing some of the details in Sydney's Metropolitan Strategy, released in December 2005, an overview of some of the basic and key elements about metropolitan planning responsibilities in Australia is necessary.

Australia has a federal system of government with six states (in order of population size from biggest to smallest, New South Wales, Victoria, Queensland,

Western Australia, South Australia and Tasmania), and two territories (the Australian Capital Territory which contains the capital Canberra and the Northern Territory), which have a limited form of self-government. Relevant points regarding metropolitan planning include the fact that the states prepare the strategic planning frameworks for their major metropolitan areas and regions and approve local planning and development control ordinances which theoretically should be consistent with the state frameworks. The states have the key infrastructure responsibilities, for example, main roads, public transport, education, health and policing. By contrast, local councils are comparatively weak compared to many countries, but have control over most local-level developments (though the state retains control over some designated projects). Moreover, local government is generally 'fragmented', though this is more so in some city areas than others. For example, in the Sydney metropolitan area (well over four million people), there are over 40 local councils, while in the metropolitan area of Brisbane (around 1.5 million people) there are only three large councils. Despite fragmentation in New South Wales and other states, there is no metropolitan level of government. State governments have the key metropolitan governance and strategic planning responsibilities in cities within their jurisdiction but also have rural and regional responsibilities. By contrast, the federal (national) government has the principal taxing power and control over defence, immigration and macroeconomic policy (with an independent reserve bank setting interest rates) but has few direct powers related to urban planning and development.

Within Australia, about three-quarters of tax revenue goes to the federal government. About one-fifth goes to the states, and the rest goes to local governments. However, own-source revenues as a percentage of own-purpose outlays are 142 % for the Commonwealth, 50 % for the states and 80 % for local governments. This is often referred to as 'vertical fiscal imbalance'. Local governments rely on the state and federal government for funding to carry out many of their responsibilities, and state governments rely on federal government support. In New South Wales, local governments are also financially constrained because they have limited powers of taxation (Worthington and Dollery 2000). The main sources of locally based revenue are taxes on property (or municipal rates); fees and fines (referring to user charges imposed for services rendered and fines associated with regulatory functions); net operating surplus of public trading enterprises (normally utilities); grants from the Commonwealth or respective state government; and interest received from council investments. Under state law, property rates in NSW are 'pegged'. This means that they generally cannot be raised beyond an amount set by the state government, typically related to the rate of inflation (Dollery and Wijeweera 2010). As a consequence, it has been argued that councils are constrained in providing local services, they cannot address infrastructure backlogs, and they have to impose higher service fees and charges to make up for budget shortfalls (e.g. Jopson 2010).

Sydney's Growth and Development Challenges

Sydney is in the state of New South Wales and is the largest city in Australia – 4.1 million people at the time of the 2005 Strategy (and 4.3 million in 2010), anticipated to rise to 5.9 million by 2031. The city has grown strongly in the recent era of globalisation – it is Australia's global city and home to international financial headquarters and most of the advanced business services jobs in Australia (though in recent years both Brisbane and Melbourne have been growing at comparable rates in net population terms). The historic heart and settlement of Sydney (1788) was on its eastern edge. The Central Business District (CBD) is on Port Jackson at Circular Quay where there was freshwater. The higher-density areas of the city are around this historic core. Although lower-density growth has proceeded to the north and south, growth has been mainly to the west, typified by detached housing in post-war car-based suburbs. As a result, the metropolitan area is now 65 km from east to west and around 40 km from north to south (see the development of Sydney, outlined in Fig. 3.1).

Sydney's regional problems are characteristic of many larger cities where urban growth has spread beyond local boundaries. Although the city and its neighbouring jurisdictions are subordinate to the State Government, the State Government has limited ability to compel local governments to cooperate and address a regional vision.

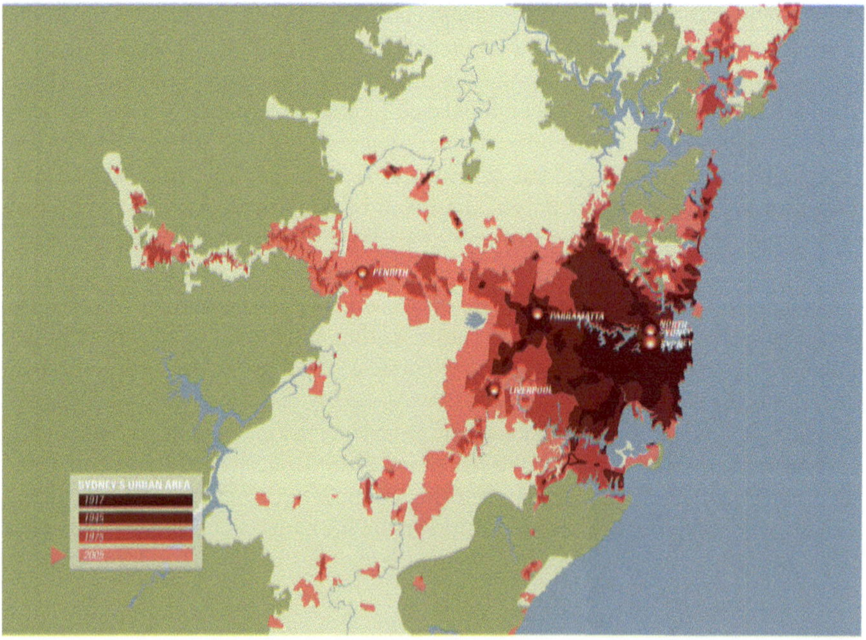

Fig. 3.1 Sydney's urban development over time (Source: NSW Government – Department of Planning 2005)

The Sydney Morning Herald

The Eagles and their never-say-die comeback

Three men and their babes: the new blokey drama

Sydney is a sprawling, gridlocked, polluted mess. It's time to fix it

CAMPAIGN FOR SYDNEY A HERALD SERIES

CITY IN CRISIS

Car mania puts quality of life up in the air

Fig. 3.2 Press view on Sydney's development (Source: Goodsir and Dick 2005)

It lacks the spending power to implement regional plans on its own, although it can stimulate development through investment in infrastructure and through the work of state agencies, particularly UrbanGrowth NSW, formerly Landcom, which is the state corporation tasked with developing land for housing. Founded in the 1970s, Landcom was originally meant to help provide affordable housing for the Sydney region, developing property on the urban fringe. However, from 2002 it operated as a for-profit development corporation, managing the redevelopment of state-owned land, both brownfield and greenfield. As UrbanGrowth NSW it is now mainly focused on leading renewal projects in infill areas, including preparing and wholesaling public land for development.

The 2005 Strategy had to consider four key development challenges (Fig. 3.2 indicates attention in the printed media to the crucial issues). The first was building on Sydney's role as a global city and building on the strengths that have attracted high-value jobs, but spreading the benefits to and creating more jobs in the western part of the metropolitan area while sustaining the high-growth east. The second was to provide housing for 40,000–50,000 people per year and meet a diverse range of needs as households get smaller and older on average (providing for around 600,000–700,000 homes in the 25-year planning period). The third was to contain or control the urban footprint to protect valuable rural and agricultural lands (constrained by geography to a certain extent anyway), and the fourth was to promote the sustainable development of the metropolitan region by reducing car dependence and congestion, reducing per capita water and energy use, reducing waste and planning for climate change.

The 2010 Strategy highlighted the same or similar challenges. By 2014, with the change to a more conservative government, the development challenges that are highlighted in the metropolitan plan are similar, with even more people per year to accommodate, but there is much less explicit reference to social and environmental challenges, though there are limited actions included which address these issues.

Metropolitan Strategy Themes at a Glance

The 2005 Metropolitan Strategy contained eight key elements or themes. The first was creating stronger cities within the metropolitan area. Sydney and North Sydney, branded 'Global Sydney', were to remain the core of the region, but were to be complemented by the so-called regional cities, creating a multicentred city (Parramatta, Liverpool, Penrith and Gosford).

The second was the creation of a strong 'Global Economic Corridor' between North Sydney, the Sydney Central Business District, and the Airport. It was hoped that this would lead to the creation of 150,000 more jobs in Global Sydney, major centres, specialised centres and employment precincts, including Macquarie Park, Chatswood, St Leonards, Green Square, Randwick education and health precincts, Sydney Airport and Port Botany.

The third element was to create more jobs in Western Sydney. The aim was to accommodate almost 240,000 more jobs, including 'higher-order' jobs in regional cities and specialised centres and in employment precincts with access to the Orbital Motorway Network.

The fourth element, related to the previous three, was containing Sydney's urban footprint and limiting sprawl. This was to be done by ensuring that any new release of land met strict sustainability criteria and that green and rural areas were protected.

The last four elements are also related to sustainability. The fifth was encouraging transit-oriented development by creating major centres for jobs, service and residential locations which would be related to transit and have higher population densities. Element six, providing fair access to housing, jobs, services and open space, aimed at giving suburban towns, villages and neighbourhoods a public transport focus and developing new and diverse housing in existing areas. The seventh element was the creation of what is called connected centres through improvements in the public transport network, providing improved access to the Global Economic

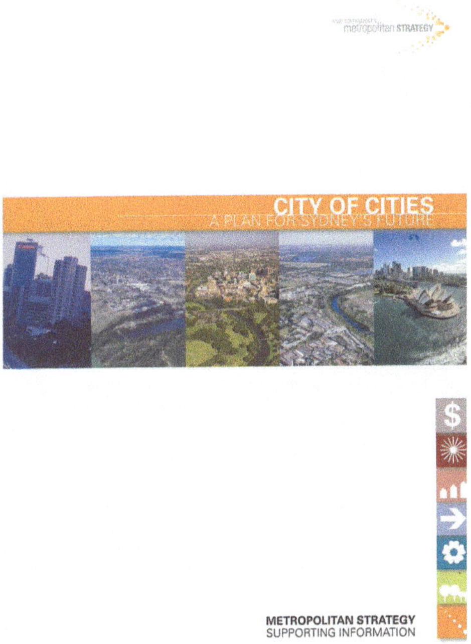

- 1,100,000 more people by 2031
- 640,000 new homes
- 500,000 new jobs
- 5 cities, 3 corridors, 1 global city

Fig. 3.3 City of Cities: A Plan for Sydney's Future – front cover and main parameters (Source: NSW Government – Department of Planning 2005)

Corridor and investing in the rail network and strategic bus corridors to provide faster and direct public transport linking suburban centres and nodes.

The eighth element was to create stronger regions. Jobs were to be linked with housing growth and improved communications and transport in the Greater Metropolitan Region including the Central Coast, Illawarra and Lower Hunter.

The *2010 Metropolitan Plan* contains these or similar elements, though with a more resolved transport plan and less emphasis on connections to the regions. The basic parameters also shifted over the five years between documents. Figures 3.3, 3.4 and 3.5 show the covers and key parameters for the 2005, 2010 and 2014 documents.

The 2014 plan is not as clear in describing its key elements. It contains 22 directions across four high-level goals. Summarising these directions suggests that the main elements are:

- Growing a more internationally competitive Sydney CBD as part of an 'expanded' Global Economic Corridor extending to Parramatta and along the North West Rail Link corridor to Norwest and Rouse Hill
- Growing Greater Parramatta as Sydney's second CBD
- Growing other strategic centres (including Penrith, Liverpool and Campbelltown-Macarthur as regional city centres)

- Transforming the productivity of Western Sydney through growth and investment
- Delivering well-planned greenfield precincts and housing
- Accelerating urban renewal across Sydney at train stations, providing homes closer to jobs
- Enhancing capacity at Sydney's gateways – Port Botany and Sydney Airport and at the proposed new Sydney second airport at Badgerys Creek
- Promoting Sydney's arts and culture, tourism and entertainment industries
- Creating a network of linked green spaces
- Protecting the natural environment and biodiversity

State and Local Levels of Interest

The key map from the 2005 Metropolitan Strategy in Fig. 3.5 made a statement about the strategic elements from the State Government's perspective. This can be compared to the 2014 map (see Fig. 3.5b) which has fewer elements, potentially indicating less state level ambition, or perhaps greater realism about the extent of state level influence. Other places and elements not on the maps 'make up'

a

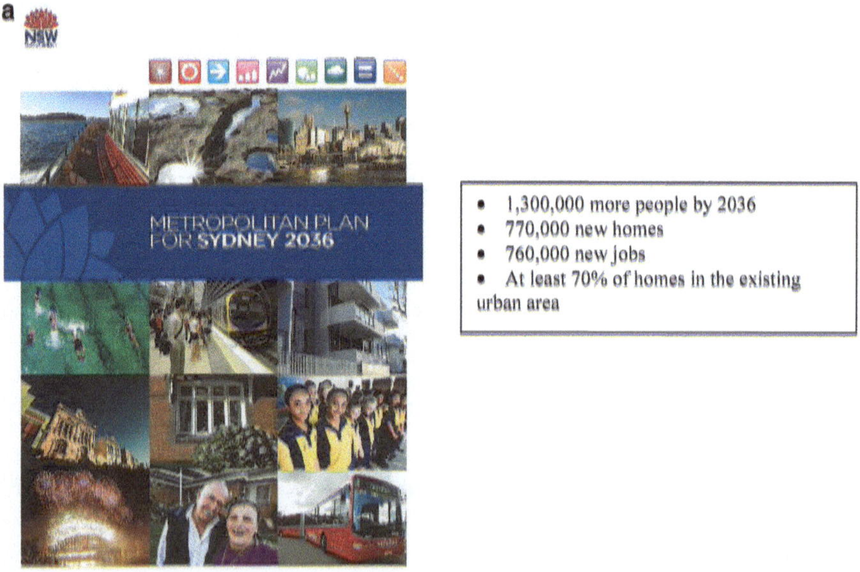

Fig. 3.4 (**a**) Metropolitan Plan for Sydney 2036 – front cover and main parameters (Source: NSW Government – Department of Planning 2005). (**b**) A plan for growing Sydney – front cover and main parameters (Source: NSW Government – Department of Planning and Environment 2014)

b

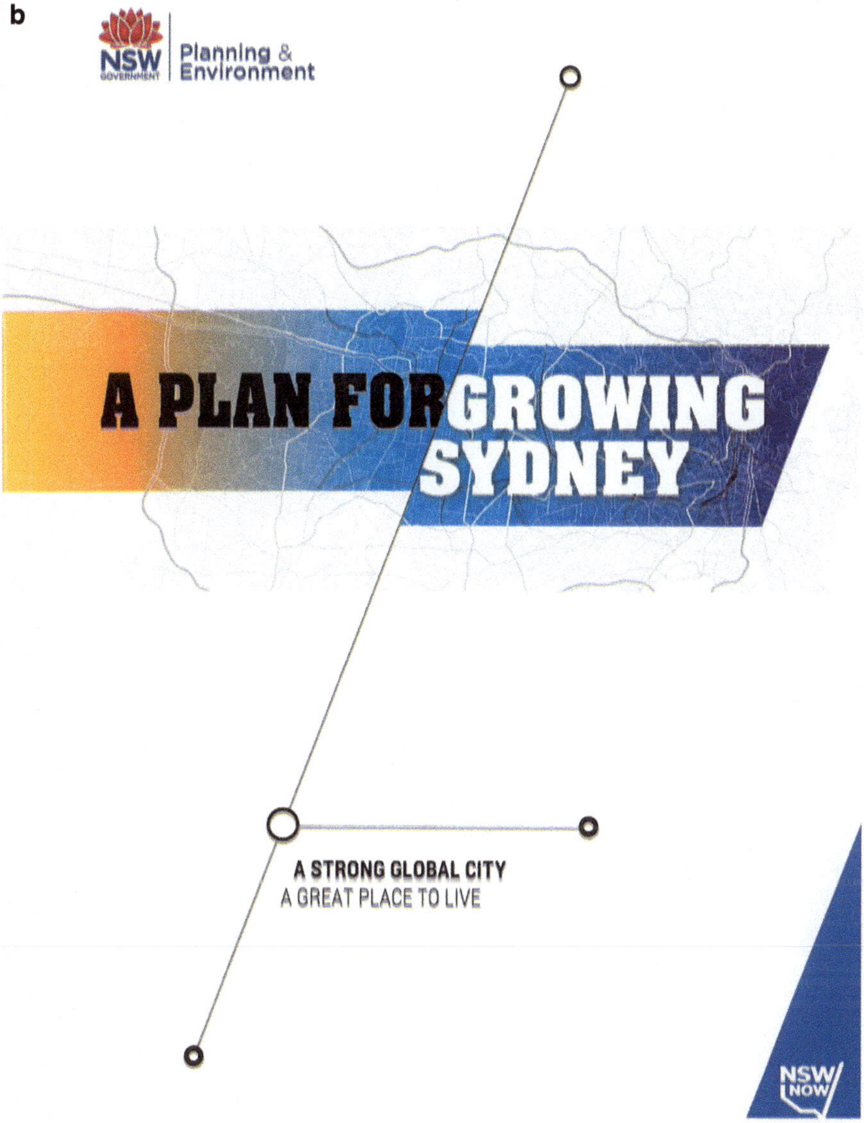

1.6 million more people by 2031
664,000 new homes
689,000 new jobs
2 CBDs (Sydney/North Sydney and Greater Parramatta), 3 Regional City Centres and 22 other strategic centres

Fig. 3.4 (continued)

a
GROWTH CENTRE
North West and South West land release areas.
REGIONAL LINK
Better connected and stronger regions linked to Sydney.
CORRIDOR
Areas around transport routes that connect centres, containing related activities. Corridor types: Economic, Renewal, Enterprise
RAIL NETWORK AND STATION
ORBITAL MOTORWAY NETWORK
The motorway and freeway system including the Eastern Distributor, M7, M5, M4 and M2.
NEW NW–CBD–SW RAIL LINK (POSSIBLE EXTENSION)
Links the major new growth and employment areas of the city.
BUS TRANSITWAY
Bus based rapid transit system providing fast, efficient bus services.
STRATEGIC BUS CORRIDOR
New direct and frequent bus services linking larger centres across Sydney.
EMPLOYMENT LANDS (PLANNED)
Contain various employment activities such as factories, warehouses, high tech manufacturing, transport logistics or major storage operations with some associated offices. These places are vital to our economy and ability to service the city.
POTENTIAL EMPLOYMENT LAND FOR INVESTIGATION
REGIONAL/STATE PARK
NATIONAL PARK
RURAL AND RESOURCE LAND
EXISTING URBAN AREA
Nepean River
Newcastle
Gosford
Warnervale
F3
North West Growth Centre
Rouse Hill
Hornsby
Castle Hill
Norwest
M2
Penrith
Mt Druitt
Blacktown
M4
Westmead
Macquarie Park
Chatswood
Brookvale / Dee Why
North Sydney to Macquarie Park
St Leonards
Parramatta River
Parramatta
Olympic Park / Rhodes
NORTH SYDNEY
SYDNEY
M7
Western Sydney Parklands
Western Sydney Employment Hub
Prairiewood
Fairfield
Parramatta to City
Burwood
Bondi Junction
Green Square
Randwick Education and Health
Badgerys Creek
Cabramatta
Bankstown
City to Airport
South West Growth Centre
Liverpool
Bankstown Airport / Milperra
M5
Sydney Airport
Port Botany
Leppington
Hurstville
Kogarah
Georges River
Sutherland
Campbelltown
Glenlee
F5
Canberra
F6
Wollongong
THE METROPOLITAN STRATEGY MAP
1,100,000 more people
640,000 new homes
500,000 new jobs
25 year plan
5 cities
3 corridors
1 global city
GLOBAL SYDNEY
Main focus for national and international business, professional services, specialised shops and tourism. It is also a recreation and entertainment destination for the Sydney region with national significance.
GLOBAL ECONOMIC CORRIDOR
The corridor of concentrated jobs and activities in centres from North Sydney to Macquarie Park and from the City to Airport and Port Botany, will remain the powerhouse of Australia's economy.
REGIONAL CITY
Parramatta, Liverpool and Penrith will provide for more lifestyle and work opportunities close to the growing parts of Sydney.
SPECIALISED CENTRE
Places such as hospitals, universities and major research and business centres that perform vital economic and employment roles across Sydney.
MAJOR CENTRE
The major shopping and business centre for the district, usually with council offices, taller office and residential buildings, a large shopping mall and central community facilities.
PLANNED MAJOR CENTRE
These places are currently being planned to become major centres.
POTENTIAL MAJOR CENTRE
These places may grow to take on the role of major centres in future.

b

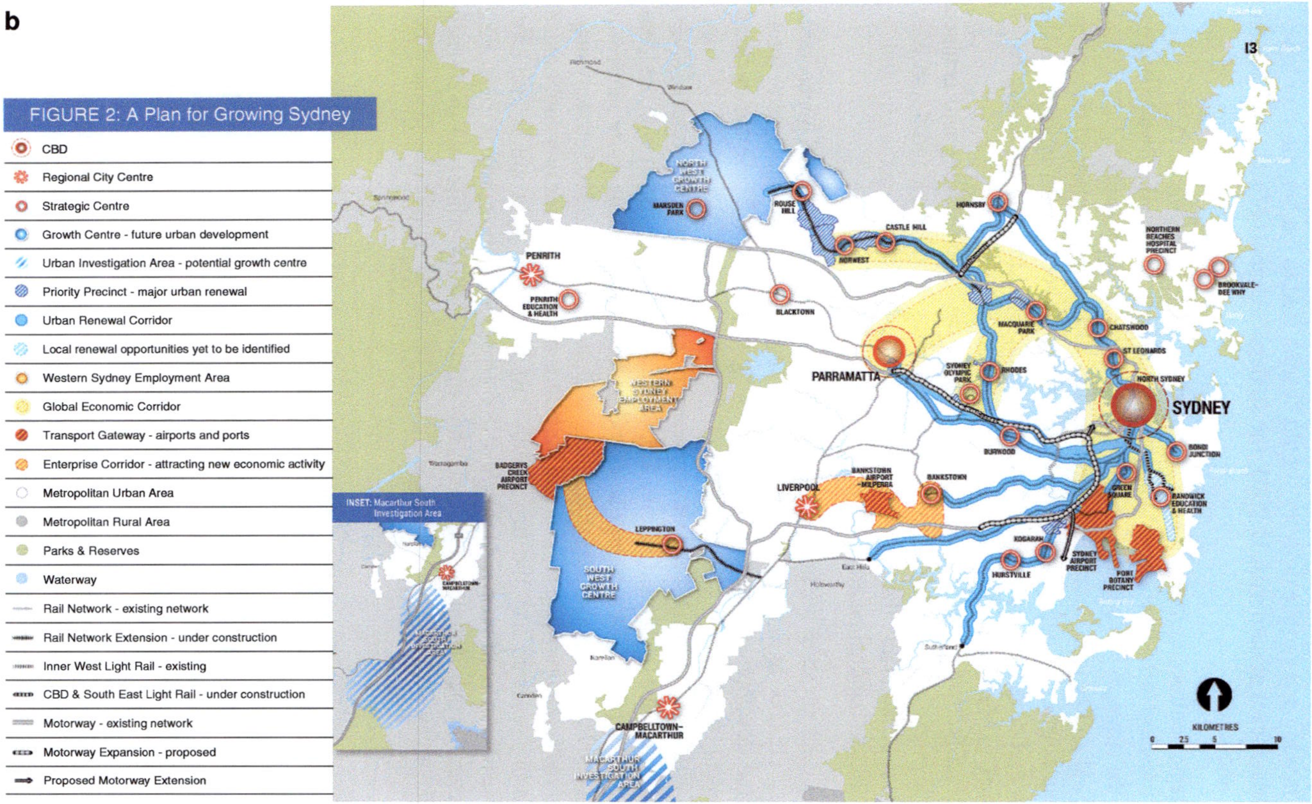

Fig. 3.5 (**a**) 2005 The strategy map (Source: NSW Government – Department of Planning 2005). (**b**) The 2014 plan map (Source: NSW Government – Department of Planning and Environment 2014)

the city, but, consistent with the 'subsidiarity principle', these may be viewed as being principally of local interest. This principle is based on the idea that policy-making and service delivery should be allocated to the 'smallest' or lowest appropriate level of authority. In 2005, the strategic elements of state government interest included:

- strategic centres (regional cities, major centres, specialised centres)
- key transport infrastructure including the existing and proposed railway network, and the strategic bus corridors, and the orbital motorway network
- renewal and economic development corridors including the planned major economic corridors along the M7 and M5 motorways, between Parramatta and the City, and development of the 'global economic corridor' encompassing North Sydney to Macquarie Park and the City to the Airport
- areas of focus for land use planning including the major existing and planned employment land areas, planned new release areas (North West and South West Growth Centres) and the major rural and resource lands framing the city and
- the links to regional centres

The map does not show the hundreds of smaller centres – the so-called towns, large villages, villages and neighbourhoods that were also to accommodate future residential development and local-level retailing in particular. These are of 'local government interest'.

The 2014 map highlights the strategic centres and gateways, new release areas, rail extensions and (expanded) Global Economic Corridor, consistent with the ideas in the 2005 strategy. Two enterprise corridors have replaced the more numerous renewal and economic corridors that followed motorways in the 2005 strategy map, while some small priority precincts and inner area rail corridors are now highlighted for housing renewal, and the new South-East inner city light rail is included. The Macarthur South area is highlighted for investigation as a new release area. The Western Sydney Employment Area is highlighted in 2014, while back in 2005 all industrial areas were highlighted indicating their perceived significance at that time.

Ideal Elements for a Successful Metropolitan Strategy

Having a plan or a strategy for metropolitan development is only part of the story. Many other drivers of success have to be managed. From the perspective of Mark Spiller (2006), these drivers, or key implementation elements, include the following (see also Fig. 3.6).

Appropriate Land Use Regulation Based on the Metropolitan Strategic Framework
Strategic settings need to be resolved and translated into land use regulations, and there needs to be a clear process to achieve this. This will be particularly important in relation to the objectives for housing mix and affordability. There may be a proactive role for planning instruments to address this issue.

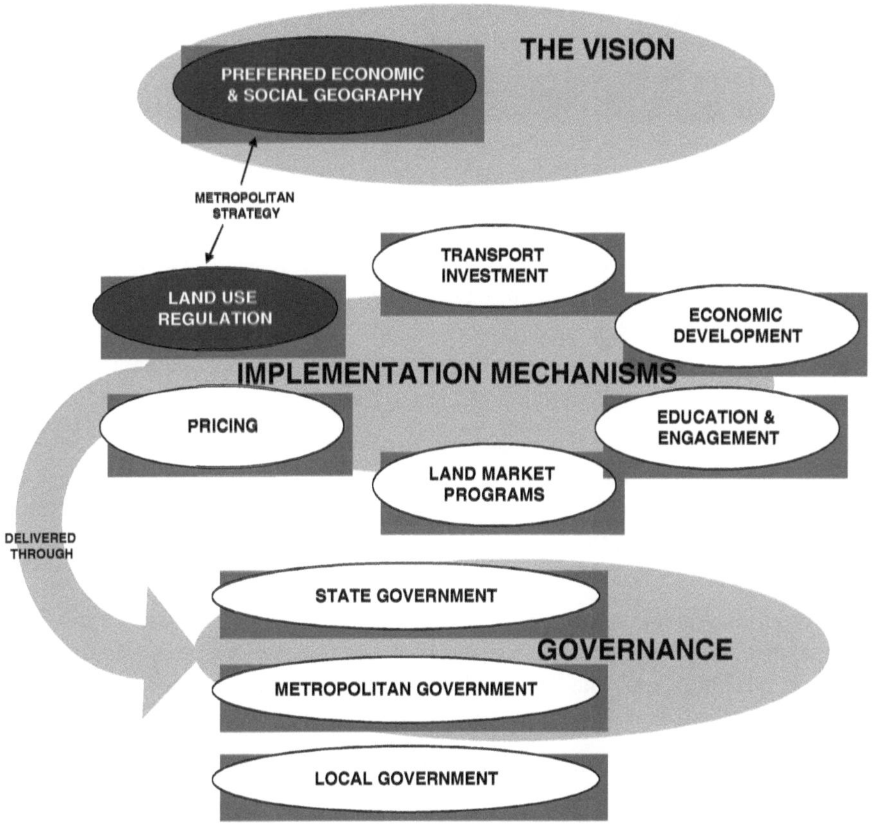

Fig. 3.6 Required elements for metropolitan strategy implementation (Source: Spiller 2006)

Pricing New pricing policies are needed for infrastructure, mechanised travel and energy and natural resource use. Australia's metropolitan areas have largely been built on the back of cheap land, cheap water, cheap petrol, cheap energy, unpriced road access and subsidised infrastructure. It is of little wonder that these cities perform poorly in world terms against environmental indicators (while performing strongly against quality of life indicators).

The types of pricing policies required include the introduction of electronic road tolling, not just for a few projects but for the whole network of arterial roads, geared to recovering the cost of congestion and other externalities. The revenues generated should be ploughed back into public transport. In addition, the pricing should include full up-front cost recovery for roads, water supply, sewerage, storm water management, rail extensions and other private benefit infrastructure in greenfield areas and infill areas alike, where these costs are locationally differentiated.

Investment in Public Transport Adequate investment in *public transport systems* is vital to realising the visions in Australia's metropolitan plans. Sydney in particular has recently suffered from underinvestment in public transport. While its extensive established network enables it to outperform other Australian cities in terms of public transport mode share, congestion pressures and growth mean there are likely to be diminishing returns without new investment.

While the current NSW Government has accelerated some public transport improvements and investment, building on the previous government's plans, it is unlikely that significant shifts in modes share toward public transport will be achieved in the short to medium term. To achieve the significant shift required a significantly expanded programme of investment will be necessary. This will also require innovation in investment procurement and infrastructure funding. Like many cities around the world, there is a need to look at land value capture strategies to help pay for improved public transport and other social infrastructure. At present, private landowners often enjoy significant windfalls as a result of public investment in rail lines and the like.

Measures to Address Land Market Failures To achieve major transformation, particularly in the 'strategic centres', the government needs to be prepared to intervene to overcome land market failures and constraints like fragmented property ownership, contamination or lack of innovation in the private developer sector.

Education and Community Engagement Community resistance to new or more intense development is strong. This affects local politics. There is an inbuilt bias against change and against future generations of residents. There is an important role for education and marketing programmes and much more sophisticated community engagement to help further shift public behaviour and preferences.

Economic Development Strategies Ambitious metropolitan strategies usually aim for shifts in the economic geography of the city, that is to say, the distribution of economic activities within the city region. Supported by major infrastructure investments, this can only be achieved if there are sound economic development strategies and policies in place to complement or underpin the preferred settlement pattern.

Metropolitan Governance As well as having these various implementation mechanisms tuned to the vision, good policy requires that *governance arrangements* support the implementation process, by mediating the inevitable political tensions between local community interests and regional community interests. Spiller has suggested that in Australia these tensions will only be resolved by establishing Metropolitan Commissions. A 'strong arm' state government level Minister is not sufficient because long-term strategic prescriptions need to be embraced by both sides of politics and by local communities, which is impossible if the Minister is setting himself/herself up as the political champion.

A Metropolitan Commission, advocated by Spiller (2006), would be a sphere of governance covering the metropolitan community of interest, separately mandated by the Parliament and independently funded – possibly through some form of property rate surcharge. There would be an element of democratic representation in this Commission. It could involve one delegate from each constituent local council (or group of councils) plus a list of appointed members responsible to the Parliament of the day. What the Commission could do is put some space between the Minister at State level and the formulation and implementation of metropolitan strategy. When politically pressed on the strategy, the Government could reasonably claim the protection of non-interference in the due process of a legitimately constituted quasi-autonomous planning body. For their part, councils would be relieved of the task of reconciling what are sometimes irreconcilable differences between local and metropolitan interests.

Spiller (2006) also argues that the Commonwealth or federal government needs to be involved in supporting state metropolitan strategies and the work of the proposed commissions, by providing incentive payments to the states for improvements in metropolitan performance congestion, sustainability and so on. This makes economic sense. Spiller argues that for every $3 in GDP growth, $1 goes in taxes, and two thirds of these taxes flow to the Commonwealth. So the biggest fiscal winners from implementing these plans would be the Commonwealth. Successive federal governments have missed the opportunity to simultaneously take Australia to a new level of competitiveness, make a positive contribution to the environment and fill up the coffers for social investment.

Sydney's Metropolitan Strategy Rated Against the Implementation Elements

Both the 2005 and 2010 documents outlined the way in which the strategy was intended to 'influence' key government functions in three areas:

1. Spatial planning – by informing planning and the preparation of plans and instruments at the subregional and local levels
2. State involvement in strategic places – using special purpose or land development agencies in areas and centres of state or strategic interest
3. State investment priorities – in the state infrastructure strategy, in asset management, in funding and pricing and in project delivery

The use of the term 'influence' may be significant in these documents. It might be expected that a whole-of-government adopted strategy would be expected to 'direct' or at least 'guide', implying that the strategy would be a key reference across government. Despite or perhaps because of their comprehensive scope, these metropolitan plans appear to be still viewed with suspicion outside of the planning agency and to be utilised with discretion.

The 2014 plan highlights similar delivery and implementation elements with the key and important addition of a proposed Greater Sydney Commission 'tasked with the responsibility to drive the implementation of the Plan' (NSW Government 2014, p.18). As mentioned, UrbanGrowth NSW is included as a government agency to deliver major urban renewal projects. Though the graphic summarising the delivery framework includes 'infrastructure investment', there is no framework nor any

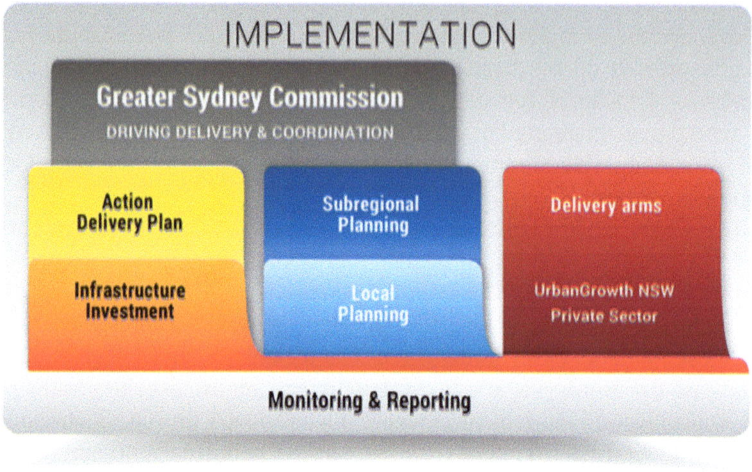

Fig. 3.7 A plan for growing Sydney delivery framework

further discussion on a related programme of investment or funding (see Fig. 3.7). Other government policy documents, notably the NSW Long-Term Transport Master Plan (NSW Government 2012) and State Infrastructure Strategy (Infrastructure NSW 2014), provide more coverage of investment and funding initiatives.

Table 3.1 below provides an assessment of how Sydney's recent Metropolitan Plans line up against the 'ideal elements'. There has been sufficient time to judge the performance of the 2005 Strategy but also to consider whether the 2014 Plan addresses any observed gaps or weaknesses.

In the 'pass' category is land use regulation, where processes of subregional planning and assistance with LEP (Local Environmental Plan) reviews have been established, though caveats are noted in the table. Draft subregional plans were prepared to reflect the 2005 Metropolitan Strategy, but were never finalised. The current government has committed to subregional planning again and this has revised hope in more effective strategic land use planning. A key issue is whether the housing and employment development capacity or potential provided via revised LEPs is actually able to be developed feasibly now or in the life of the plan.

Some progress has been made in implementing some key plan elements, but more needs to be done in several areas. The first is *State investment to support the strategy*. Failures to invest in public transport were a principal reason for the Labour government's downfall, though the 2010 Plan was much stronger than the 2005 Strategy in this area. Nevertheless, underinvestment in public transport has been a factor in poor rates of new housing supply and a lack of basic underpinning for core strategy objectives aiming to address agglomeration, sustainability and reduced congestion. While there are signs of improvement in transport investment, and land

Table 3.1 Assessment of Sydney's metropolitan strategy implementation elements

Implementation mechanisms	Comment	Pass ✓, some progress ~ or fail x
Land use regulation	LEPs to conform with subregional plans geared to Metro objectives – funding for LEPs[a] with the prospect of the feasibility of development potential provided in local plans as a threshold test. Councils at least conscious of the need for local plans to accommodate target growth – though state tests and sanctions for non-performance are still weak.	✓
Pricing	No reference in the transport section of the 2005 Plan; passing reference in an implementation objective (G4) but no further detail. Nothing explicit in the 2010 or 2014 Plans. There have been tentative moves to price access to the CBD by time of day with higher peak hour charges on the Sydney Harbour Bridge and toll roads are an ad hoc user pays system.	x
Investment in public transport	Principally backlog investment in rail, e.g., the North West Rail Link serves many already developed catchments; 'sibling' document NSW Long-Term Transport Master Plan sets out a comprehensive transport agenda, though this emerged prior to the 2014 metropolitan Plan for Growing Sydney; few of the priority transport projects are truly transformative, though the Parramatta light rail has been elevated above others by the government; no explicit value capture or other funding arrangements. Increasing though incomplete recognition of state-level elements in state infrastructure strategy (e.g. future investment in cultural facilities in Parramatta) but need commitment to a greater share of ongoing state investment in strategic centres.	~
Land market failures	2005 Strategy gives Landcom (the government land developer) an 'urban renewal' role, but no clear mandate to assemble or package sites. Landcom now effectively dismantled. State-level Sydney Metropolitan Development Authority (SMDA) established with the 2010 Plan as renewal authority but with very limited geographic coverage (Redfern-Waterloo and Granville) and impact. UrbanGrowth NSW formed (merger of Landcom and SMDA) to redevelop government-owned land and undertake planning in some limited development corridors. A difficulty will be influencing infrastructure particularly transport agencies. Challenge of land assembly and effective facilitation in centres without government land holdings are still to be addressed.	~
Education and engagement	No systematic communications or engagement on the elements and merits of the plan.	x
Economic development strategies	Strong feature of the 2005 Strategy documentation in particular, including integrated thinking re-housing and employment. Modest follow-up work in practice and few significant commitments in the 2010 and 2014 Plans particularly at the place level.	x
Metropolitan governance	Ultiamtely ineffective Metropolitan Strategy CEO group established via the 2005 Strategy, not retained in 2010. Greater Sydney Commission proposed in 2014 plan. Needs to be a genuine granting of independent implementation power with state and local interests represented to be a success.	x

Source: Author
[a]Local Environmental Plan

use integration, there is scope for a much stronger alignment of state investment with the metropolitan strategy – consistent with the idea of 'infrastructure investment' identified as a key implementation component.

Secondly, *land market failures* are becoming more common, as the readily developed older industrial sites in the established areas of the city become contested between employment or residential uses, and in shorter supply, and greenfield land becomes more expensive to develop. The need for a more coordinated and interventionist approach in infill areas has finally been acknowledged with the establishment of UrbanGrowth NSW which is focussed on key renewal sites. It has initial though limited funding to seed development through site preparation, planning and wholesaling. The previous Labour government established a Growth Centres Commission to address the aspects of market failure hampering the delivery of greenfield development at the time of the 2005 Plan, but it was abolished and absorbed within the state planning department prior to the release of the 2010 Plan. The withdrawal of Landcom from greenfield land development is another potential problem, though a positive property cycle can underpin a more healthy housing supply performance.

'Failures', on this assessment, are recorded for the following:

Policies of road and transportation pricing are missing from the successive strategies. Through iterations of reform to the mechanisms for the funding of fringe urban development, an unfortunate and ad hoc system has emerged. In relation to road and transport pricing, modest changes such as peak hour premium tolling on the Sydney Harbour Bridge are no substitute for a more considered and comprehensive approach. Nevertheless, the politics of road user charges are a minefield for state governments. Interestingly, and at a distance, federal-level advice has been stronger in advocating for road user charges.

Education and engagement on the merits of intensification and redevelopment is required, as development gets increasingly stymied and bogged down by what can sometimes amount to 'knee jerk' local community resistance.

Economic development is supported in both the 2005 and 2010 documents, but there has been modest follow-up activity – particularly in relation to the polycentric city agenda. The 2014 Plan does highlight continued work on Industry Action Plans, but there is less commitment to place specific economic development and coordinated investment activity. More needs to be done to support the development of the strategic centres identified in the strategy, particularly the regional cities of Parramatta, Liverpool and Penrith (and, in the 2014 Plan, Campbelltown-Macarthur in the outer south west). For Parramatta, a positive property market has assisted the achievement of some strategic development objectives, with stronger office and residential development. As mentioned, the new government has committed to investigating a light rail network for Parramatta, including funding early works.

Governance remains the ultimate 'Achilles heel' for the metropolitan strategy. Notwithstanding the clarity of the implementation rhetoric outlined in the strategy (spatial planning, state involvement in strategic places and state investment priorities), the tensions between state and local interests, between current and future communities and between innovation and inertia are not able to be resolved without governance reform that establishes a robust representative agent for a metropolitan-wide constituency.

The new strategy and new government has committed to the establishment of a Greater Sydney Commission (GSC). Its form, composition and ultimate powers have yet to be defined but it holds the promise of a much more coherent implementation platform for the metropolitan strategy

Additional Reforms

The recent metropolitan strategies provide a clear framework for the growth and development of metropolitan Sydney. In general they propose a shift to a more polycentric urban form and contained outward expansion though the 2014 Plan is less explicit on this aim and includes a number of initiative for strengthening the central city. While the plans are strong, the implementation progress has been weak – with key elements not addressed. There is not the space here for a detailed discussion on the reforms required to address the implementation gaps. But it is worth highlighting projects and reforms that would address two areas of particular weakness. These are a public transport investment programme in support of the urban structure implied by the plan and also a coherent programme of governance reform that elevates the role of metropolitan strategy in decision making.

Public Transport Investment: Linking Regional Cities to and Networking the Global Centre

In the area of public transport, the government has largely been scrambling to address backlog. Even the extensions to the rail network to service the north-west sector (construction well underway) and to the south-west sector (recently completed) are really only bringing the system and service up to the standard expected of a reliable, functioning metropolitan network.

The investment in strategic bus corridors including the 'metro' buses responds to changed travel behaviour and needs and has provided some much needed cross-regional transport options linking centres. Implementation on this is now lagging and in any case will not provide the high-capacity and high-frequency corridors that were originally implied. A proposed Parramatta focused light rail system is an exciting initiative, and a further harbor crossing as part of a plan to convert part of the network to a European metro-style system is bold (and expensive). Nevertheless, neither will radically transform intra-metropolitan connectivity or linkages.

What the strategy lacks are proactive public transport initiatives that support the key land use planning and employment objectives. In particular, more attention needs to be given to the services in, and relationships between, the regional cities, the specialised centres and Global Sydney, with its concentration of skilled workers and advanced business services jobs. Better access to these jobs and services is critical for the accelerated development of these suburban business centres.

In England, the centres within the orbit of London – such as Reading, Stevenage and Watford – connected by fast rail to the city, have enjoyed significant growth over the last 20 years. These centres have grown as residential hubs but also as significant employment centres in their own right – trading on good access to central London's advanced jobs pool. These growing centres are all no more than 30 min away from central London and are playing the 'regional city' role expected of Parramatta, Penrith, Liverpool, Campbelltown, Gosford, Wollongong and Newcastle in Sydney's greater metropolitan region.

Within London (and other global cities), the underground rail system 'networks' a multiplicity of centres and nodes and sustains a dense core of businesses and activities. While further debate is required as to the desirability and feasibility of improved fast rail connections between strategic centres in metropolitan Sydney, the following connections should be considered for investigation:

- A Liverpool 'Y' link connecting Liverpool to the East Hills line and the airport and CBD
- Links between Bankstown, Bankstown Airport/Milperra and Liverpool
- A fast rail service between Penrith and Parramatta and onto the CBD (the 'western express' upgrades proposed in the 2010 Plan were consistent with this but are no longer supported in the 2014 plan)

In the longer term, an outer urban rail loop connecting the proposed second Sydney airport with the North West and South West Rail Lines.

In terms of better networking centres in an expanded global core, consideration should be given to:

- The development of an 'inner' loop service connecting Greater Parramatta and Sydney Olympic Park (where there is significant renewal potential) with the Sydney CBD and then using the proposed new harbour tunnel and existing link on the north shore between Chatswood and Epping, connecting back to Parramatta with a new link from Epping. This would connect many thousands of jobs at key universities and health precincts.
- Additional rapid bus services connecting outer areas such as the Northern Beaches with the Global Economic Corridor.

Coherent Governance Reforms

Coordination and enforcement, both formal and informal, are needed for metropolitan plans to succeed. This should embrace not just local governments within the region but state agencies within the region as well as quasi-governmental organisations responsible for delivering services in the area. Even a Metropolitan Commission as recommended by Spiller and now proposed by the NSW Government is probably not enough to meet the planning and development needs of the Sydney metropolitan region. The evidence from the recent metropolitan strategies and plans in Sydney is that they are not key reference documents across government. There is a need to breakdown the 'silo' mentality and approach that hampers the whole-of-government action. At the moment, agencies forward their bids (Total Asset Management Plans) to the Treasury without particular reference to the Government's strategic agenda. The central agencies 'play king' without being accountable to relevant strategies. However, several reforms, among others related to place-based renewal (not discussed here), are required. For the strategy to work, the Metropolitan Strategy needs to be viewed as a key reference at the central (i.e. state) agency (Treasury and Premier's Department) level to complement the work of a Metropolitan Planning Commission which should be established to advise on implementation, make detailed policy in relation to topics and places of state significance and assess developments of state significance.

While local-level forward infrastructure planning exists, for a plan to succeed, there is a need for a long-term, infrastructure programme consistent with the metropolitan strategy (with at least 10 years of funding). The current government is now proposing to fund major infrastructure via a sale of state electricity assets rather than sysematically locking in a funding stream. Planning has been separate from infrastructure development. Agency and cross-agency planning and infrastructure and asset bids should be consistent with the Metropolitan Strategy. There is a role for the Metropolitan Planning Commission in assessing and setting priorities.

The success of the plan is also contingent on the creation of meaningful targets and deadlines for the implementation of policies. Among these is the creation of comprehensive, consistent, place-based housing and job targets/projections for use across the government and within local government. This should be based on subregional planning undertaken with local councils on the future role and development of all the centres which might accommodate future population and jobs. This includes large villages, villages and neighbourhoods, as well as the 'strategic' centres. The new subregional planning arrangements, to be led by the Greater Sydney Commission, provide some promise of meaningful reform in this regard.

Another need, related to the previous paragraph, is to develop key performance indicators for the whole of government that reflect the Metropolitan Strategy aspirations (e.g. an increased share of expenditure to strategic centres nominated in the Metropolitan Strategy). This would allow not just policymakers but the general public to measure investments and developments against the plan targets.

Finally, local council LEPs (local government development control frameworks) should be quantitatively assessed against the Metropolitan Strategy targets, including an assessment of the capacity and potential provided in the LEPs.

Again with the experience of recent failure, there is some prospect of a better alignment between the metropolitan plan and local plans with regard to housing aims, given the development of better capacity measurement and feasibility tools.

Conclusion

Australia is a federal system with three basic levels of government: the federal or national level, the state level and the local level. Local governments are 'creatures of the state' which can reorganise them or restructure them at will. However, the State Government tends to respect local government boundaries. People identify with local authorities, and this makes administrative reorganisation politically difficult though the current government has established a local government amalgamation and reform agenda. Most councils, unsurprisingly, are resistant to any change.

From 1945 to 1964, the Cumberland County Council operated as a form of metropolitan government for the Sydney region. The county council was abolished by the State Government because it opposed the state development plans for the Sydney region. It was replaced by a succession of state bodies, in a manner similar to the replacement of the Greater London Council by Mrs Thatcher in the 1980s. New South Wales also has regional organisations which bring together local governments, called Regional Organisations of Councils. There are four ROCS covering metropolitan Sydney. These manage a limited range of regional issues which embrace several local authorities and are based on inter-municipal agreements. However, the ROC are not regional governments. They have a relatively weak advisory and advocacy role, although they do manage some brokering of shared service agreements.

As a result of vertical fiscal imbalance, the Sydney metropolitan plan (indeed, all state governments given their significant service responsibilites) is underfunded. The state lacks the resources to undertake large projects which require substantial capital investment and lacks the ability to foster plan implementation by withholding money from councils for non-compliance. The winding down of Landcom, originally established to provide for more coordinated development and innovation in new release areas, leaves a gap in the implementation architecture for the metropolitan plan. The establishment of an apparently well-resourced UrbanGrowth NSW to focus on urban renewal at least provides the promise of a stronger performance in housing supply in the established areas of the metropolitan area.

In Australia and in New South Wales in particular, the state governments have most of the relevant cities shaping powers so local government shouldn't be 'blamed' for failing to implement regional or metropolitan strategies. Until now there has been no effective metropolitan-wide organisation in Sydney with even limited democratic and financial legitimacy (or at least an effective 'compact'), and there has not been a region-wide forum in which governments can exchange views and debate policy. Moreover, intergovernmental relations make long-term planning complicated.

The current NSW Government appears to have learned some of the lessons of recent history when it comes to metropolitan planning and implementation. It has attempted to better integrate transport and infrastructure investment with metropolitan strategic land use planning, though there is still a long way to go with the transport agencies in particular still typically leading the agenda. Notwithstanding new investment in rail extensions to growth areas and an exciting initiative for a Parramatta light rail system – some 15 km from the Sydney city centre – there is still a lack of transformative infrastructure investment which would achieve the long-held metropolitan planning aims to 'rebalance' the metropolitan area to provide more opportunities for the growing Western Sydney population (now more than two million in a city of well over five million).

Recent governance reform has been both promising and disappointing. The urban renewal agency UrbanGrowth NSW is bringing a more coordinated and proactive focus to redeveloping precincts in the established areas of Sydney, but it is focussing on areas of surplus government-owned land rather than necessarily where intervention would support planning aims. At the same time, Landcom is being wound down, and its important role in innovative new release housing initiatives will be lost, with no replacement agency suggested.

The proposal for a Greater Sydney Commission represents a positive breakthrough. For too long metropolitan planning has been bogged down in politics and short-sighted actions, by both state and local governments. The new metropolitan wide commission, though yet to be established and have its membership, powers and resources confirmed, will hopefully represent a turning point for Sydney.

The great roadblock though remains the financial capacity of state governments to invest in forward infrastructure programmes to deliver against long-term planning aims. On the one hand, they (NSW included) have been too timid in innovations where they do have powers such as value capture (or betterment) taxation, notwithstanding demonstrable windfall gains made by landowners and developers following rezoning and infrastructure investment decisions. On the other hand, federal-state financial relations are antiquated and work against state autonomy in addressing the problems and achieving the aims included in recent metropolitan strategies and plans.

Governance reforms which integrate planning and infrastructure investment decisions at the heart of NSW Government decision making, led by an independent metropolitan commission, innovation in infrastructure funding and an ambitious agenda of federal-state financial relations are required to improve prospects for delivering what has otherwise been a robust and mostly consistent set of directions for Sydney in recent metropolitan plans and strategies.

References

Butson T (2012) End of line for North West Rail Link says Treasurer Wayne Swan. Express Advocate, Gosford, New South Wales, 8 May. Available at: http://express-advocate-gosford.whereilive.com.au/news/story/end-of-line-for-north-west-rail-link-says-treasurer-wayne-swan/. Accessed 9 May 2012

Dollery B, Wijeweera A (2010) An assessment of rate pegging in New South Wales local government. Commonw J Local Gov, Issue 6, July, pp 56–76

Goodsir D, Dick T (2005) Sydney is sprawling, gridlocked, polluted mess, it's time to fix it. Sydney Morning Herald, 30 May 2005, p 1

Infrastructure NSW (2014) State Infrastructure Strategy Update 2014, Recommendations to the NSW Government (November 2014). http://www.infrastructure.nsw.gov.au/media/43622/inf_j14_871_sis_report_book_web_new.pdf

Jopson D (2010) Gone to pot councils on brink of slashing services. Sydney Morning Herald, 5 April 2010

Landcom (2011) Annual report 2011, Sydney, NSW. Landcom. Available at: http://www.landcom.com.au/AnnualReport2011/default.aspx. Viewed on 26 June 2012.

Landcom (2012) How we operate. Available at: http://www.landcom.com.au/AnnualReport2011/introducing-landcom/how-we-operate.aspx. Viewed on 26 June 2012

New South Wales Government (2010) Metropolitan plan for Sydney 2036

New South Wales Government – Department of Planning (2005) City of cities – a plan for Sydney's future. New South Wales Government, Sydney

New South Wales Government (2014) A Plan for Growing Sydney, December 2014

NSW Government (2012) NSW long term transport master plan

Spiller M (2006) Metropolitan planning policy: the implementation trilogy. In: Planning for Australia's major cities: Sydney, Melbourne and South East Queensland. Seminar jointly convened by SGS Economics and Planning and UNSW City Futures, Museum of Sydney, 30 May 2006

Worthington A, Dollery B (2000) The debate on Australian federalism: local government financial interrelationships with state and commonwealth governments Australian. J Public Administration 59(4):12–22

Chapter 4
Governmental Devolution as a Motor of Local Development

Andrew Ryder

Abstract Since the start of the 1990s, governments worldwide have experimented with decentralisation or devolution of powers to lower levels of government. Within the European Union, this was related to the concept of subsidiarity, but elsewhere it has been linked to the enhancement of democracy and to the promotion of economic development (European Union 2015). The United Kingdom and Korea were and are two of the most centralised countries in the Organisation for Economic Cooperation and Development, but even they have experimented with decentralisation and devolution of powers. In the United Kingdom, London was given a regional government at the start of this century. During the same period, the Korean national government promised to give greater powers to Seoul, the capital of the Republic of Korea. In the case of London, devolution has been extremely successful, allowing London to formulate its own development agendas and, through its elected mayor, giving it a voice on the national and international stage. Through several initiatives including a congestion charge for cars entering the city centre and a successful campaign to host the Olympics, London's government has been able to alter the political discourse in the United Kingdom. However, in the case of Seoul, devolution was less successful, and the mayor of Seoul used the commitment to decentralise as a springboard to national politics and was subsequently elected President. Despite this, Seoul also sponsored several major policy initiatives including greening of the city centre through the reconstruction of the Cheongyecheon river. The success of London is also due to the fact that devolution of authority and the handing over of powers to London's government took place gradually, allowing the Greater London Authority to build up its management skills. However, Greater London's powers remain extremely limited. The regional government has no independent tax-raising authority and has control only over planning and related policies. In addition, London's boroughs, the local governments, still retain considerable administrative powers. Despite this, the Greater London Authority and London's mayor have become new voices in the policymaking process, sometimes complementing and sometimes working in opposition to national government. Perhaps, one lesson is that decentralisation and devolution are not just about financial empowerment but also about leveraging limited resources to foster growth. A second lesson is that gradualism is

A. Ryder (✉)
Formerly Department of Geography, University of Portsmouth, Portsmouth PO1 3HE, UK
e-mail: andrewrydernyc@gmail.com

© Springer Science+Business Media Dordrecht 2015
J. Buček, A. Ryder (eds.), *Governance in Transition*, Springer Geography,
DOI 10.1007/978-94-007-5503-1_4

often better than a big bang reform: in the case of London, it has allowed the Greater London Authority to build up the skills and competences needed to manage a large international city.

Keywords Devolution • Local government finance • Local government expertise • Metropolitan governance

Introduction. How Can We Activate Local Economies?

Since the 1920s, numerous countries have offered assistance to lagging or distressed regions. These range from aid to distressed areas in the United Kingdom from the 1920s onwards (described in detail in Scott 1997a, b, 2000; Scott and Rooth 1999) through to concentrated regional aid programmes like the Tennessee Valley in the United States or the Centralny Okręg Przemyslowy in Poland in the 1930s. More recently, the European Union created the European Regional Development Fund, providing monetary assistance to so-called lagging regions, defined either by lower than average per capita gross domestic product or higher than average unemployment. However, many theorists have argued that monetary aid alone is not enough. Perhaps, the best context for this view is Africa, where thousands of millions of dollars in aid given since the 1960s appear to have produced few lasting results (Moyo and Ferguson 2009).

Growth Pole Theory was widely popular in the 1960s and 1970s and argued that if heavy industry were placed in a region, up- and downstream activities would follow. However, in the case of Southern Italy, where the *Cassa per il Mezzogiorno* managed a development programme (described in Carey et al. 1955), the programme failed in its aims and was eventually abandoned.

It is also argued that development depended on good transport and infrastructure (e.g. Taaffe et al. 1963). However, in Eastern Germany, despite the construction of new motorways, railways, and fibre-optic links since 1990, unemployment has remained stubbornly high, and outmigration persisted (Bundesministerium 2006). More recently, some theorists such as Porter (1990) have suggested that activation stems from "clusters" which foster growth often led by local "champions". Porter has therefore argued that growth comes from an increase in human capital support for networks. Florida (2002) has suggested that diversity fosters growth. Others (Carlino and Saiz 2008) have suggested that amenities and even branding lead to economic development. Pointing to "cities of culture", many argue that arts and culture have been motors of development. Liverpool is said to have gained £800 mn in economic benefits after a year as a European Capital of Culture (Groom 2010b). Still others such as Thrift (1986) and Johnston (1986) have suggested that regions lag because they are helpless before global forces. However, in general, regional policy appears to consist of monetary aid to so-called lagging regions, however they are defined.

Does Money Work?

There is a long tradition of supplying money for economic development through grants or low-interest loans. The European Union has used such a system since the 1970s, supplying matching grants for a variety of aims in lagging regions. However, a growing body of evidence suggests that this does not work. In an analysis of economic development grants awarded to prospective UK investors at the turn of the century, Blitz and Chung (2003) and Chung and Roberts (2003) found that much of the money was spent but wasted: firms to which money was given closed and in some cases even became bankrupt. Few of the promised social benefits materialised. Moreover, in many national development assistance systems, lagging or poor regions seem to occupy more than half the country. As a result, there is often a lack of focus on development, and few projects seem to achieve a critical mass. In their review of French regional planning policies, Tuppen and Thompson (1994) noted that this was the case in France. More recently, John D Rockefeller IV, US Senator for the state of West Virginia, said that the late Senator Byrd of West Virginia knew that "before you can make life better, you have to have a road to get in there, and you have to have a sewerage system, and all those things" (quoted in Clymer 2010).

As noted, spending on transport and communications does not always work, at least on their own. Some countries, particularly Germany, have invested in infrastructure, particularly motorways and high-speed railways but also in new power lines, fibre-optic networks, and new environmental protection equipment. In the early 2000s, despite over a decade and a half of investment, the former German Democratic Republic still lagged behind Germany's western regions (Bundesministerium 2006). Annual evaluative and progress reports on regional assistance programmes for the former German Democratic Republic noted that despite extensive reconstruction of infrastructure, the region still suffered from a skill deficit and recommended that spending on training and education be increased to improve human rather than just physical capital.

How Else Can We Activate Local Economies?

It is often argued that one of the best ways to foster local development is to devolve power to local governments. In the European Union, the principle of subsidiarity has embraced this view, arguing that decision-making should be devolved to the lowest practicable administrative level. Subsidiarity is the idea that decision-making ought to be handled by the smallest, lowest, or least-centralised competent authority. Central government should perform only those tasks which cannot be performed effectively at a more immediate or local level (European Union 2015).

It is a fundamental principle of EU law. According to this principle, the European Union may only act (i.e. make laws) where action of individual countries is insufficient. The principle was established in the Maastricht Treaty of 1992 and was already a key element of the European Charter of Local Self-Government, an instrument of the Council of Europe promulgated in 1985 (see Article 4, Paragraph 3 of the Charter).

The United States and Switzerland are federal systems in which local (as opposed to regional or national) governments play the main role in shaping local policy and supporting local growth. However, by the standards of OECD countries and by global standards, they are exceptional. Local governments in the United States and Switzerland have elected councils and mayors. They raise a substantial amount of their income from local sources and have a great deal of autonomy in imposing or waiving taxes, determining spending programmes, and shaping policy. Moreover, they can raise money through property taxation, income taxes, sales taxes, excise taxes, and fees. As a consequence, local governments can provide incentives to prospective investors, can waive taxes, can reduce taxes, can alter or simplify planning regulations, and can bargain.

Oates (2006) has argued that devolved governments bring a range of benefits to national politics and to national economies. He gives four main ways in which this is so. First of all, regional and local governments can adapt public services to the preferences and circumstances of their constituencies rather than a central solution which presumes that one size fits all. Second, when governmental powers are decentralised, this encourages competition among jurisdictions to attract investment and residents, which means that local jurisdictions try to restrain tax growth while also trying to improve the extent and quality of services they offer to local residents. Third, decentralisation encourages experimentation and innovation among different jurisdictions, which can provide a valuable laboratory for experimentation and ways of reducing costs and improving services. Finally, when households are mobile and can select the jurisdiction in which they want to live, it is easier for them to find a match between the services they want at a price they can afford. This last is often referred to as the "Tiebout Hypothesis", which argues that "the consumer voter may be viewed as picking that community which best satisfies his preference pattern for public goods. [T]he consumer-voter moves to that community whose local government best satisfies his set of preferences. The greater the number of communities, and the greater the variation among them, the closer the consumer will come to fully realizing his preference position" (Tiebout 1956, 418).

The United States and Switzerland are not alone in harnessing the power of local government to activate economic development. Barcelona and Catalonia are a case in point. In 1976, upon the death of the Spanish *Caudillo*, Franco, Barcelona was a sprawling, neglected city with an economy based on outdated industry. Local devolution fostered local economic restructuring. Since 1979, the city has been transformed. Local and regional autonomy has allowed the city to restructure and grow. This was done in several ways. Thanks to powers gained since 1980, the city has become a leading service centre and is rapidly becoming a centre of biotechnology. To begin with, the city was rebranded by local and regional authorities. Barcelona became a flagship of Catalan versus Castillian culture. The city and region improved the environment, redeveloped the waterfront, and promoted new economies. Beaches replaced old wharves. A new biomedical centre was built on the waterfront – the

Institute for Research in Biomedicine – and a science park created by the Generalitat de Catalunya, Department d'Innovació, Universitats i Empresa and the Departament de Salut (Departments of innovation, universities and Enterprise and Department of Health). The metro system was upgraded and expanded, and new tram lines were built. Since Spain joined the European Union, Catalonia has consistently had higher-than-average levels of income and per capita gross domestic product and lower levels of unemployment. Similarly, the Basque Country gained responsibility for collecting and setting nearly all the taxes it pays, paying central government for those services it receives (The Economist 2010). This has enhanced local accountability but also local satisfaction with government and governance.

However, most countries are centralised, and some are extremely centralised. In such cases, local governments have few fiscal or administrative powers and are often merely agents of the state administration. Even in countries with strong local governments, fiscal resources are often limited because local governments have few taxation powers and only limited local revenue sources. To some extent, this is a result of a mistrust of local government. Local governments are often considered too parochial, too politicised, irresponsible, and unable to manage finances. Therefore, they need strict supervision and control. This view is particularly common in the United Kingdom but is also found in countries like Canada, where provinces can create or abolish local governments at will.

Seoul and London: A Tale of Two Cities

Korea and the United Kingdom are among the most centralised countries in the OECD. In each country, the central government controls most functions and provides most local government finance. Until recently, neither London nor Seoul had true local government. London had a regional council, the Greater London Council, until 1986, when it was abolished, but there was never a London mayor in the traditional sense of the word. The City of London, one of the 33 administrative regions within Greater London, has a lord mayor, as do individual boroughs, but this is a largely ceremonial role. London's 33 districts (32 boroughs plus the city of London) have elected local councils, but their powers were and are extremely limited.

Seoul had no elected mayor or council before 1995. Councillors and mayors were appointed by central government. However, Seoul is exceptional among Korean cities. It is several times larger than the next largest city and occupies a place of overwhelming importance in the national economy. It contains 25 districts or "gu", more than any other Korean city, which range in population from 139,000 (Jung-gu in 2003) to 634,000 (Nowon-gu in 2003). Since the restoration of Korean independence, Seoul's role in the country and its role in the national economy have steadily grown. In 1950, it contained just 8 % of the national population, but by 1990, it had reached 24.4 %. The total share of the population in Seoul itself had fallen to just 21.0 % by 2003, but when one takes into account the Seoul metropolitan region including Incheon City and Gyeonggi Province, the region's share of the national population is approximately 40 %. During the post–World War II era, Seoul's role in the country and its role in the national economy has steadily grown. The city's share in national GDP peaked in the period between 1985 to 1994, when it averaged about 24–25 %. In 2002, the city's share in national GDP was 21.8 %, slightly higher than its share of population, but the share of the Seoul metropolitan region in national GDP was approximately 47.7 %, up from 42 % in 1985.

The city has a highly educated labour force and plays an overwhelmingly important role in national research and development. 29.7 % of the population has at least a high school education (versus a national average of 29.0 % in 2003), but in 2003, an additional 23.4 % had a university degree as opposed to a national average of just 17.3 %. The city contains 43 of the Republic's 199 4 year colleges and universities, 24.1 % of the country's university students, and 26.3 % of the total faculty – some 12,800 in all. As well as that, in 2002 the city had 2,729 research and development institutions, 36.1 % of the national total. When the surrounding metropolitan region is included, Greater Seoul contained 4,974 establishments or 65.8 % of the national total. Most of these are in the private sector: 2,619 in Seoul proper and 4,758 in the Greater Seoul region including Seoul. The private establishments have risen sharply since 1995, thanks to changes in central government regulations, from 670 in Seoul and 1,548 in Seoul and its suburbs. Within Seoul proper, the workforce engaged in research and development was 35,309 in 2002, a doubling of the figure since 1995, and in the metropolitan area (again including Seoul), it was 114,461 out of a national total of 189,888. Since 1995, half of the national growth in research workers has taken place in the Seoul region.

Between 1998 and 2003, the city absorbed about one-third of all direct foreign investment coming into Korea. Most of this investment went into manufacturing and services, although in the city as a whole, the workforce is mainly engaged in services. In 2002, 87.34 % of businesses were engaged in the service sector, employing 79.16 % of the labour force. 10.07 % of businesses were in manufacturing, accounting for 15.02 % of employees. In the metropolitan area including Seoul city and the suburbs, the share of employees in manufacturing has dropped from 27.2 % in 1995 to 11.98 % in 2002 and the share in services risen from 63.9 % to 85.5 %. This parallels a national trend which saw service sector employment climb from 64.6 % to 71.2 %. In 2002, Seoul contained 23.5 % of the country's firms (735,258) and 26 % of the national employed labour force. The metropolitan region contained 45.6 % of the country's businesses (1,429,546) and 49.6 % of the national employed labour force. Despite declining population in Seoul city itself, the number of businesses has risen steadily since 1995, when it was 708,025, although the number of employees fell from 3,874,597 to 3,805,462. During this time, the number of businesses in the Greater Seoul region has risen by 15 % from 1,234,966 to 1,429,546 versus a national average increase of 13 % during the same period. The regional employed labour force grew by some 8 % during the same period, reflecting a 23 % growth outside Seoul as opposed to a national growth rate of just 7.1 % (OECD 2005).

Therefore, London and Seoul are wealthy cities which are relatively badly off. Neither city can freely spend money on basic infrastructure. Both pay more in tax to central government than they receive. In what seems a dysfunctional approach, both have been discriminated against by central government on the grounds that they are too wealthy, too large, and too important to the national economy. This treatment is not unique to these two countries: similar biases have existed in Poland regarding Warsaw, in Hungary regarding Budapest, in the Czech Republic regarding Prague, not to mention French efforts to decentralise Paris in the 1960s and 1970s, described in (Tuppen and Thompson 1994). Until the end of the 1970s, it was official policy within the United Kingdom to discourage new industry or office employment from locating in London and even in the London region. All proposed developments had to go before a national commission to justify any planned expansions or new developments. As a result, ageing industrial and service sectors were not replaced by new ones. Office tasks were often unnecessarily split among a number of different locations. Although London was not closed to new residents, the lack of employment in new sectors meant that critical mass was lost in some sectors. As well, reluctance to leave London meant that even as jobs were decentralised, the premium on those jobs which remained in London increased: employees preferred London and the London region to other UK regions.

In addition to restricting the growth of cities, post 1945 UK policy also aimed to reduce the powers of local government. By 1939, UK local governments were quite powerful. Many ran their own transport systems, their own water companies, their own gas and electric companies, and their own telephone companies. After the war, a series of nationalisations of industry meant that locally owned firms in most sectors were taken over by national corporations. Subsequently, in the late 1970s, the system of local government throughout the United Kingdom was simplified. Hundreds of local governments were merged into just over 500 administrative districts, many of which lacked a regional identity. England accounts for most of them: 39 counties, 296 districts, and 36 metropolitan areas. At the start of 2007, the United States, by comparison, has almost 90,000 governments (United States Bureau of the Census 2010, Table 416). Until 1947, British local governments managed the local planning system with very little supervision and set rates (or taxation) on property with little central government interference. Changes to the planning system after 1947 meant that national government played a central role in local planning policy formulation and implementation, having the right to over-ride local planning decisions and to impose national policy at the local level. Finally, at the start of the 1990s, the right of local governments to value property and set rates was removed and replaced initially by a so-called poll tax but later by a council tax for residential property based on 1990 valuations. Business property was taxed at a business rate set by national government. Local governments depended on central government for 80 % or more of local spending.

Since the late 1990s, first Labour and then Conservative governments have promised to devolve more powers to local governments (The Economist 2006). Moreover, a growing number of UK local governments have called for greater autonomy and powers (for example, Centre for Cities 2006). In the United Kingdom, despite lip service to the idea of strengthening local government, central government manages everything! Local governments need parliamentary approval for new tram, underground, and rail lines. They need central government money for capital projects since they have none of their own. They are unable to raise money through bonds because they have no revenue streams from which to repay them. Budgets are an annual process, and long-term planning is difficult if not impossible since there is no carry-through or long-term spending commitment in many areas. In addition, local government is forced to keep track of spending, to justify spending to central government overseers, and to report and be evaluated on over 1,000 performance indicators. This has been described by Eric Pickles, then Tory Shadow Minister for Local Government, as "Stalinism without the violence" (The Economist 2006, 34). In many ways, local government simply administers public services on behalf of the state: these include education, streets, rubbish collection, and council housing, although this last has often been privatised and sold to tenants or to private housing associations.

Ultimately, in the United Kingdom, the main role of local government is to manage the planning process but under central government supervision. The central government sets planning policies and guidelines but allows local government some latitude in implementing those policies. Planning policies are published in the form of Planning Policy Statements. Planning Policy Statements (PPSs) (and their predecessors Planning Policy Guidance Notes – PPGs) are prepared by the government "after public consultation" to explain statutory provisions and provide guidance to local authorities and others on planning policy and the operation of the planning system. They also explain the relationship between planning policies and other policies which have an important bearing on issues of development and land use. Local authorities must take their contents into account in preparing plans. The guidance may also be relevant to decisions on individual planning applications and appeals.

However, most local government planning administration consists not of making town plans and development guidelines but of micromanagement: permission for a house extension, for a house alteration, for a tree to be cut down, or for changes to road layouts. Central government sets housing allocation targets for local governments, assigning them a share of projected growth in any region, but leaves it to them to decide where to put housing and what kind of housing should be built.

Korean Local Government

Since 1945, the Republic of Korea has been, on paper, a democracy. In 1949, after the election of the first Korean President, a local autonomy law was passed which provided for local governments consisting of directly elected councils and mayors appointed by the central government. Between 1956 and 1960, the laws were changed to allow direct elections of local mayors. After a military coup in 1961, local autonomy was eliminated, and democratic rule was suspended. In 1963, with the proclamation of a new constitution, democracy was theoretically reinstated, but local autonomy was not restored. A new constitution in 1972 suggested that local autonomy should remain suspended until the country was reunified, but in 1980, the constitution was amended to provide for the gradual restoration of autonomous local government which was to be based on the degree of financial self-sufficiency. In 1987, democracy was finally restored, and a new president was elected in 1988. The new government of Rho Tae Woo made a public commitment to restore local democracy. Direct elections to local councils were restored in 1991. However, it was not until 1995 that mayors and local governors were directly elected. Many Korean observers argue that this is the true beginning of local democracy (Kim 2004a, 15). However, despite the direct election of local councils and local chief executives, local government has limited policymaking and decision-making powers and, for the most part, extremely limited financial resources. The smaller and less urban the level of government, the more dependent it is on transfers from central government. In 2002, for example, rural districts (or counties) received 71.4 % of their revenue from central government in the form of transfers, revenue sharing, and specific grants. Provinces received 58.9 % of their total revenue from central government, but cities received just 43.6 %, and Seoul received just 5.9 % (Kim 2004a, 21). Even these figures are misleading. The limited extent of local government competences means that budget figures for local governments are misleading. Although approximately half of the local budgets of cities and provinces come from local revenue sources, the overall budget is relatively small because the number of competences and the cost of managing those competences is relatively low. At the same time, one should note that since the 1990s, local government and regional governments have developed good working relations with central government authorities.

Limited fiscal autonomy has been used as an excuse to defer the devolution of greater authority to local governments, and at the same time, limited local authority and competences have served to justify limited fiscal powers.

According to Kim (2004a), central government has been extremely slow to devolve authority. In 2002, out of some 41,603 competences exercised by government, 73 % were performed directly by central government, and 27 % were performed by provincial and local governments. Of these, only 6,306 (15 %) were treated as purely local competences. Another 3 % were delegated by central government to local governments, and the final 9 % were shared among all three levels of government. Moreover, although a special law was enacted in 1999 to accelerate the devolution of central government powers, 3 years later, only 100 functions had been decentralised (Kim 2004b). The slowness of devolution has been partly due to a mistrust of local government, partly due to a reluctance of central government authorities to cede powers to lower levels of government, and partly due to a perhaps misplaced desire to determine the "best" or most "efficient" distribution of powers.

However, central control remains strong. In South Korea, the central government has even specified what cars local governments can use and what should be served in staff canteens! Planning is also constrained in other ways. To give an example, cities manage roads but do not manage traffic signalling. In the case of Seoul, the city could not implement bus-friendly traffic signals because traffic lights were controlled by national, not local government! Can national governments ever effectively micro-manage in such detail, or do they become overwhelmed by minor details? In the face of local diversity, do national governments develop one-size-fits-all policies which result in a kind of least harm situation in which a single policy actually meets the needs of no local governments? And when national governments take over local decision-making, issues of mainly local importance become politicised at the national level, and local areas and regions fight with one another for funding in the national arena. Minor issues become national ones.

However, for many local governments, the number and extent of restrictions is a kind of straitjacket hampering innovation and the development of localised solutions to local problems.

London Since 2000

London consists of 33 local administrative districts: 32 boroughs plus the City of London. As noted, the London region had never had a mayor with executive powers before 2000. Since 2000, London has had a mayor who acts as a local executive who is responsible for transport management and transport planning, strategic planning, and economic development. However, the mayor has no taxation powers. His economic development budget comes from national government and is administered through a development agency which is independent of the mayor's administration, although he appoints the board members.

The first elected mayor, Ken Livingstone, was formerly head of the Greater London Council, a management body for all of London, which acted like a multi-borough local authority but was abolished in 1986. As well as creating a Mayor of London, the UK central government created a London Assembly with 25 elected members which would act as a local council or legislature to oversee budgets and policy. In its words, "The London Assembly holds the Mayor of London to account and investigates issues that matter to Londoners. Its 25 members are elected at the same time as the Mayor. The Assembly works by directly questioning the Mayor about his activities, strategies and decisions. It also publishes the findings and recommendations from its investigations and makes proposals to the Mayor" (London Assembly 2010).

As part of the reform process, a London Development Agency was created to manage development funding provided by national government for London. Income is mainly from the regional development budget for the London region formerly managed by national government. Over time, this body has come increasingly under the control of the mayor's office, but as Browne (2010) has noted, it answers both to the Mayor of London and the Department for Business, Innovation and Skills, a national government agency which has its own agenda and often tries to dictate what services the LDA should offer. In mid-2010, its role was to foster economic growth in London by offering investment aid and assistance including financial support. Board members were appointed by the mayor, and they were supposed to implement and support the mayor's plan for London. However, in June 2010, it was announced that the national government "office for London" would be abolished, and there were calls to abolish the authority altogether and fold most of its functions into the Greater London Authority (Sherwood 2010), and in July 2010, it was announced that all the regional development agencies in England would be abolished (Groom and Tighe 2010). Currently, the mayor chairs the London Investment Board of the Homes and Communities Agency, a national government agency which spends over one billion pounds annually in London, but the mayor has to seek approval from national authorities to implement local housing policies. Proposals for further devolution have called for giving the authority greater control over functions now managed by the national government's Home and Communities Agency, giving the GLA more control over housing and regeneration and making the mayor and the GLA responsible for the Port of London Authority and the management of the Royal Parks. Eric Pickles, now Communities and Local Government Secretary, suggested that "In London, this means transferring power and responsibility down from Whitehall and its quangos (Quasi-Non-Governmental Organisations) progressively downwards to City Hall, to London boroughs, and to neighbourhoods" (Sherwood 2010). Over time, the GLA's transport powers have also grown. The mayor initially had limited powers in this area, mainly in the areas of road management and transport and planning. However, over time, transport powers have also grown. Transport for London, answerable to the Mayor of London, now runs several railway lines, has taken over management of the underground (tube) network, and is likely to gain more influence on the award of franchises to private rail operators running services into London. Boroughs are also to be given greater powers over traffic control.

Despite relatively limited fiscal and administrative powers, since 2000, the mayor, first Mr Livingstone and later Boris Johnson, has instituted a number of distinctive and eye-catching policies.

Local Initiatives, London

One of the first aims of the Mayor of London was to improve London's environment. In 2002, a programme was established to create new parks and open space and refurbish 100 existing open spaces (New London Architecture 2005). As part of this process, the mayor appointed an adviser on architecture, Richard Rogers. This process of refurbishment and enhancement is described in *New London Architecture* (2007). A second programme aimed to reduce car use and road congestion by implementing a so-called congestion charge for vehicles entering a central zone. The aim of this was twofold: to reduce congestion and improve surface transport and to use the money gained from the charge to improve public transport provision. A third goal was to increase the amount of affordable housing in the capital through the use of existing planning instruments, mainly Section 106, a section of the planning laws which encourages local authorities to demand "planning gain" – that is, the provision of amenities specified by local government in exchange for planning permission – particularly for permission for developments which are denser or larger than average. A final aspect of the work of the mayor's office was to enhance London's world-class characteristics and build on them. This last aim was implemented in several ways. One was to lobby for funding to improve local public transport provision and to develop more transport-friendly policies. A second was to relax local planning controls so as to permit the construction of new office space in buildings which were both larger, in terms of floor plates or floor area and in terms of heights, than anything which had been permitted before. A third was to encourage private investment to develop relatively central but derelict areas such as those around King's Cross station or to redevelop central areas such as London Bridge station, providing modern infrastructure and enhanced accessibility. A final way was to better coordinate the marketing and foreign promotion of the city region to foreign investors.

Since 2000, London has allowed the construction of many new buildings and has encouraged the redevelopment of several derelict areas including the area around King's Cross station, London Bridge, Elephant and Castle, and, in connection with the Olympics, East London in the area around Stratford. The mayor has also had considerable success in his efforts to improve local transport. Bus services have been improved and boarding speeded up in the central area by requiring passengers to buy tickets before they board. New bus lanes have been created and new routes created. In addition, the GLA and numerous city stakeholders have successfully lobbied parliament to encourage the development of new rail links including Thameslink, a north–south main line railway which includes some existing rail lines, Crossrail, an east–west main line railway, and the East London Line, a project which has involved the reconstruction of an ageing underground line and its extension both north and south to create an eventual orbital route around the central London area. In the United Kingdom, in 2006–2007, central government met 76 % of the revenue needs of local governments (House of Commons 2007, 5). The Crossrail proposal required not only investment from central government but also changes to local taxation powers so that businesses located in areas along the new rail route would pay a tax supplement in exchange for better transport services. This required the creation of a new form of tax – at least a new form of tax for Britain known as the supplementary business rate. The power to set business rates, that is, a tax on business properties, was taken away from local governments at the start of the 1990s, and in real terms, the amount of tax collected on businesses has fallen by half (House of Commons 2007). Today, the Uniform Business Rate, also known as the National Non-Domestic Rate, accounts for just 20 % of local government revenue, down from 29 % in 1990–1991 (House of Commons 2007, 11), and while the average rate grew by 41 % during that period, the average council tax for domestic property increased by 132 %. However, in 2008, a new form of tax was created to enable local authorities like Greater London to tax businesses which are likely to benefit from infrastructure improvements paid for by local government. London was a driver in pushing for this reform. In some systems such as Canada and the United States, the increase in property values due to the improvement would result in higher tax revenue for the local authority, but in the United Kingdom, because tax rates are effectively fixed, it has been necessary to resort to a supplementary levy instead. In the case of this levy, the revenue is dedicated to paying for the improvement and cannot go into local government general revenue. Thus, Crossrail will cost £15.8 billion and will rely on a grant from central government of £5 billion, money from fares, £200 million from the Corporation of London (this is the "City" of London – the wealthiest of the area's 33 boroughs), £230 million from British Airports Authority, which runs Heathrow airport, £150 million from Canary Wharf, which is privately owned and has promised to help pay for the Canary Wharf tube station, and a payment in kind from Berkeley Homes, a major home builder and developer which has agreed to build a station box at Woolwich (Crossrail 2010b). Estimates are that the new line will allow 750,000 people to travel to central London and Canary Wharf, increasing the accessibility, attractiveness, and value of those areas (Murphy 2010). Benefits are expected to average approximately £70 million per year, and the overall cost–benefit ratio is estimated at 2 to 1 (Crossrail 2010a). The Mayor of London will levy a new business tax called the supplementary business rate on all companies within the Greater London boundary from 2010 until Crossrail is completed. The exact level of the levy is at the discretion of the mayor, but it will be capped at two pence in the pound – equivalent to a rate of 4.5 %. To protect smaller businesses from a disproportionate burden, properties with a rateable value of £50,000 or less will be exempt (UK Government 2010).

The mayor's office also successfully "sold" London to a variety of international audiences, most particularly the International Olympic Committee. The Greater London Authority sponsored and helped coordinate the project and helped develop the plan for the area in which the Olympic facilities will be located. As well, the GLA mobilised national government support and local popular support for the project and used the project proposal as a way to showcase London's development policies in regeneration, the promotion of green spaces, environmental protection, public transport, and housing.

Seoul After 1995 and the First Election of a Mayor

To some extent, the creation of a London government was portrayed as a return to the status quo before 1986 and was viewed by many as the recreation of the Greater London Council. It was not: the new government had fewer powers, but the new mayor was stronger than the head of the GLC. However, in the case of London, the principle of a regional government already existed. Planning, particularly transport planning, had become much more complicated after 1986, and at the very time when the city began to gain population and undergo a shift from a declining industrial centre to a leading financial centre, the 32 boroughs lacked a regional perspective and lacked the capacity to deal with the changes taking place. Seoul, by contrast, never had an elected government or mayor before 1995.

Seoul's local government manages transport, council or public housing, air and water quality, and planning and is also tasked with fostering economic development. However, work in all these areas is subject to central oversight. The city government has no control over taxation and no right to create its own local quality standards. At the same time, although national authorities have repeatedly argued that they want to devolve powers to local governments, the central government has also further constrained the city's power of action by trying to create lower levels of government resembling boroughs within the city.

Although Seoul elected its first mayor and council in 1995, many argue that the reform process started earlier, seeing the 1988 Olympics as a catalyst for change. Although the Olympics were promoted and managed by the central government, during the preparation for them, the Seoul city government and civic leaders measured Seoul against other cities in Korea and abroad. The Olympics fostered a development initiative leading to better transport and more open space. However, at the time, the city was still completely under the control of central government. The mayor and council were appointed, not elected, and the spending for the Olympics was a gift from central government spent as part of a national effort to promote Korea's image abroad rather than to meet city needs. Thus, a feature of the initiative was slum clearance as well as modernisation of the city's infrastructure. The use of Olympic investment to promote development is not unique to Seoul. In London, the Olympics have captured additional revenue from the UK national government to be spent in London on sports facilities, public transport improvements, and future housing which will first be used to accommodate Olympic contestants. Because national government rather than city government is the main source of funding (or, in the case of London, a mix of public and private sector money), the negative costs to the city are limited. Seoul was again able to mobilise investment for the 2002 World Cup. Again, most of the investment came from national government, but this time, the city had a greater say in how the money was spent and what it was spent on.

Despite various development initiatives, Seoul's share of GDP, population, and employment in the national economy have declined since 1995 as people and services have moved out to the surrounding suburbs. This gradual decline has been encouraged by the national government. Strong central controls have left the city relatively powerless to reverse these trends: it cannot offer tax breaks or investment incentives nor can it waive minor regulations in order to retain or attract new investment. As well, central government has encouraged the relative decline, arguing that Seoul itself is too big and too concentrated. The national government wanted to move the capital away from Seoul, arguing that the city was too crowded, too expensive, too strong economically, and too important to the Korean economy. Seoul authorities opposed the plan but found themselves powerless. In the event, although the then government approved the plan to move the capital to central South Korea, the country's Constitutional Court overturned the plan in 2004.

Since 1993, the population of Seoul city has declined from a peak of 10,925,464 to 10,276,968 by the end of 2003, a drop of almost 6 %. However, in common with other cities in advanced industrial economies, the number of households in the city has actually increased from 3,430,528 in 1993 to 3,714,697 at the end of 2003, a rise of over 8 %. Thus, even as population has been falling, demand for housing has grown, and this is perhaps a key factor in the medium term since demand for housing is likely to continue to grow regardless of population trends. This is no different from trends in London, New York, and other world-class cities, reflecting changing social mores as people marry later, the elderly maintain separate housing, and disposable incomes grow. However, changes in the city's overall population mask considerable underlying population shifts. For example, in 2003, some 633,000 people moved into Seoul city, while some 701,000 moved out. Within Seoul itself, some 1,670,000 people moved house in 2003, indicating an exceptional degree of housing mobility: total moves within the city amounted to some 15 % of the city's population, and moves in and out amounted to over 20 %.

In the case of Seoul, international sports events have acted as a motor of regional economic development. Seoul authorities and national government have worked together to promote venues for the 1988 Olympics and also for the 2002 World Cup. Although the Seoul metropolitan government claimed to have played a significant role, they relied on funding from central government and were chiefly responsible for coordinating planning including compulsory purchase and land clearances and ensuring that infrastructure was in place for the games. Part of this involved the expansion of the local subway underground system, or metro. This was started in the early 1970s and by 2010 had reached a length of 314 km (or almost 755 km if the lines run by the national railway and neighbouring subway and fixed rail systems are included).

Seoul is also exceptional because of the proactive stance of the city government. Under the leadership of Mayor Lee Myung-bak, the city government has led the country in terms of setting a development agenda for the city and region and has tried to "catalyse" the city's growth potential. A key feature of Mayor Lee's policies is the aim of moving Seoul into the post-industrial era, creating a better physical infrastructure through redevelopment and environmental management and so making the city more attractive to domestic and foreign investors. In creating a city development policy, the mayor and the city government have set up a counterpoint to central government, forcing a debate about national development policies and the future orientation of national development. Although he has worked closely with central government in setting policy to move Seoul into the post-industrial era, it appears that Mayor Lee's charismatic leadership has been fundamental to Seoul's drive to transform itself into a so-called world-class city. This involves changes at several levels aimed at several different groups of current and potential stakeholders. Broadly speaking, these groups consist of current Korean national residents (the electorate), current businesses and firms from within Korea, overseas residents, foreign businesses and firms, and future residents and businesses from overseas. For each of these, the Seoul Metropolitan Government delineated a series of development policies, although in several cases such as the environment and environmental management, the policies are aimed at meeting the needs of several different constituencies. The city government also aimed to create a new Digital Media City which would build on Seoul's existing strengths in media, film, publishing, and computer software.

The plans of city government were extremely ambitious, perhaps reflecting a lack of substantive powers: instead of being a blueprint for development, they were a list of development aims. They included the redevelopment of outdated housing stock, the demolition of obsolete office and industrial space, the renovation of congested central parts of the city, the removal of markets to less central but more accessible areas, and the encouragement of new office space in the city centre, particularly in the older central business district located north of the Hangang (river) in the vicinity of the City Hall. Other plans aimed at both foreign and domestic investors included improving the local transport system by expanding the subways, through better traffic management, by the creation of new busways and bus lanes, and by implementing various forms of congestion charging, reducing air pollution, creating more park space, and sponsoring the building of new housing, shopping, and office centres ("new towns" to serve as growth points in each of the city's 25 districts (gu)). The city government aimed to create major parks and open spaces in every one of the city's 25 districts. Plans called for an almost doubling of per capita space in the northern half of the city between 2002 and 2021 from 4.8 to 9.0 m^2 per capita and from 5.8 to 9 m^2 per capita in the southern half. Underlying much of this development is the aim of reducing inequalities among people and between districts. As the city government has put it,

> One of the most urgent tasks that Seoul faces is to narrow the gap between the different areas in Seoul. The gap between rich and poor has affected public amenities and other facilities, denying certain segments of Seoul citizenry equal benefit (Seoul Metropolitan Government 2002, *Vision Seoul*, 9).

Transport and the Environment: Key Priorities

While the government was able to use the planning process to encourage new office and factory construction, these relied on private sector investment, and the city was able to do little more than change land use and zoning codes and help with the assembly of land. However, it had more direct control over buses and subways – urban transport – and the environment.

The Seoul Subway started operation in the 1970s but underwent massive expansion in the 1990s. By the end of 2009, it had reached a length of 314 km (188.4 mi), but this only includes sectors operated by the firms operating the Seoul Metro. When the lines operated by the national rail company, Korail, the Incheon Subway, and other fixed rail operations in the Seoul region are included, the total length is 755 km, one of the world's longest. The system continues to grow, and as well, the city government reorganised the bus system, simplifying routes and enhancing services. However, the initial route revision of the bus services was badly managed. The resulting confusion led to public protest. Despite this, the aim of metro construction and the reorganisation of the bus system have the common aim of reducing car use, decreasing road congestion, and decreasing car-generated pollution.

In this regard, the city decided to demolish an ageing urban motorway and restore a long-covered-up river which ran underneath it. The Cheonggyecheon River Restoration project in the northern part of central Seoul was central to the city government's plans for the environmental transformation of the city. The Cheonggyecheon river flows in an east–west direction through northern Seoul, stretching from the heart of the old downtown to the Han River. Prone to flooding, the stream was widened and straightened in 1760, but by 1910, when the city was occupied by the Japanese, the river was used as an open sewer. In order to reduce flooding, the river was repeatedly dredged, and between 1937 and 1942, parts of the river were covered over. After liberation, the banks of the river became the location of squatter settlements. Plans to cover the remainder of the river were abandoned, although the process of covering the river resumed again in 1958, and within 20 years, it was entirely covered over and transformed from a river into an underground sewer. An arterial road was built along the course of the river, and subsequently, between 1967 and 1976, an elevated express highway was built on top of the new road. The transformation of the river into a roadway transformed the area through which it passed. It changed from being a district of poor housing to being a concentrated industrial area consisting of small factories and workshops selling goods in a variety of small shops concentrated in markets of regional, national, and even international importance.

By the late 1990s, the area was characterised by ageing, poorly built housing and infrastructure. It was suffering from population loss and in danger of becoming a physical slum. However, the area was important to the local and national economy and was dominated by a series of "markets", although many of these are located in multi-storey buildings. More than 200,000 merchants and 60,000 shops (many of which are little more than large stalls) were located in the vicinity of the "new river", and city authorities consulted widely before implementing the project. Grants of up to 80 % were offered to reconstruct and modernise existing markets. The city also created a special district in Songpa gu for businesses which wanted to leave the area. This is in the south-eastern part of Seoul and will accommodate shops, wholesalers, logistic facilities, a large discount store, and a cinema complex. Street vendors were encouraged to move to the nearby Dongdaemun Stadium area, creating a new street market.

The elevated highway was 5.8 km long and carried some 103,000 vehicles daily, while the road below carried an additional 66,000. Demolition of the highway and opening of the river resulted in a loss of road capacity, forcing traffic to be diverted to other areas. The city created new bus lanes and new bus shuttles to meet local transport needs and provided special freight routes and loading areas for goods transfers. Twenty-two bridges (four for pedestrians only) were created to cross the new stream to avoid dividing the city into two halves. However, long-term plans anticipate a reduction in traffic through the area.

The river itself is 5.84 km long and is supplied with fresh water from the Han River and a water treatment plant. Water pumped from subway stations is also emptied into the stream. While it is recognised that the river is not a self-supporting ecology, sewers and drains divert pollution away from the stream, and in conjunction with an aggressive environmental management programme, it is hoped that wildlife will gradually colonise the stream boundaries and that fish will recolonise the river (Seoul Metropolitan Government 2003, *Cheonggyecheon Restoration*). As well as being an ecological showcase, the river is expected to become an architectural showpiece through its bridges and its lighting.

As well as recreating the stream and encouraging some businesses to move to a new location outside the urban core, the city authorities hope to the gradual redevelopment of the district starting with the Seun Sangga Commercial District. This area was first designated a redevelopment district in the 1980s, but the large number of small landowners and leaseholders and associated land ownership rights issues made redevelopment impossible. In what is a good example of the benefits of decentralisation, the now autonomous Seoul Metropolitan Government was able to develop a city-run trust company based on the present rents for lease ownership in the area, effectively freezing existing lease rates. After the trust period is over, landowners will be given shares in the profits of the trust, and existing leaseholders will be given the first choice in purchasing new stores on an instalment plan while they are still trading from temporary stores. Although development will be private led, it will be supervised by Seoul Metropolitan Government and rely on extensive consultation with government-sponsored tenant consultation committees both in forming and implementing the plan. The city government will also provide administrative support for the project. It is hoped that development will spread from the restored stream and the original commercial district in a phased process which will gradually result in the extensive reconstruction of the city centre in the northern part of Seoul. The hope is that this will rely mainly on private investment, but it is expected that the planning regime will be stricter than that which prevailed in the past.

Review of the Effects of Devolution

In the case of London, the Mayor of London has gradually gained more powers. The Mayor worked with national government to encourage changes in local government finance and administration. At the same time, the Mayor of London was constrained in his freedom of action by the London boroughs, which strongly resisted any erosion of their already limited local powers. Thus, the development of the Mayor's powers and his exercise of them was a political process and involved substantial compromise and persuasion. In the case of Seoul, however, an outspoken and dynamic elected mayor used his time in office as a springboard to the presidency of the country. The post of mayor gave him a platform on which to voice his opinions and stand in opposition to the party then in power. While mayor, he used his access to the national stage to protect his city, but once he moved up to national office, decentralisation of governmental power was no longer a priority for his administration. Whereas the Mayor of London has had to rely on consensus building to implement his programmes, in Seoul the policymaking process was still led in a top-down way relying on a vision supported by one individual. The Mayor of Seoul came across as a wise leader implementing new policies. This resonated with the electorate, particularly since the Mayor of Seoul made the environment and transport key policy issues. These had appeared to be neglected by the national government, which had placed more stress on industrialisation and economic growth than environmental protection. This was not the same as in London. There, a long tradition of local government meant that pressure comes from below and the mayor's policies are more subject to checks and balances. The first London Mayor, Ken Livingstone, was voted out of office because Londoners in the outer boroughs felt that he was ignoring their needs.

In addition, the two cities, Seoul and London, are coping with different sets of problems. Seoul is coping with industrial restructuring and even industrial and population decline. The chief concerns are to reinvent the city, protect the city, improve the city, and foster growth. London, by contrast, is trying to cope with unexpected growth and is trying to channel growth and control growth.

In the case of Seoul, the process of devolving competences highlights a basic problem in Korean government: micromanagement, excessive regulation, and reliance on a set of specific rules for each contingency implemented by a formal agency rather than the use of a set of broad criteria which are implemented by local governments but monitored by national agencies.

London differs from Seoul in the richness of its institutions of local governance. These are informal and formal pressure groups which include local business associations such as London First, a business-lobbying organisation which describes itself as working "on behalf of the capital's leading employers to make London the best city in the world in which to do business"; the now defunct London Research Centre, which was set up after the abolition of the Greater London Council to undertake research on London-wide issues; and the London Planning Advisory Committee, which was made up of representatives from London's boroughs and London-wide agencies such as London Transport. This was later folded into the London Assembly, which now has a transport committee as well as a planning and housing committee. In addition, the national government had (and has) a Government Office for London and various department offices in transport, environment, and housing. Many of these groups were informal. They had no official standing and often represented particular sectoral interests. They were not elected and were responsible to no electorate. They might be described as "expertarian" rather than authoritarian or democratic, but they kept alive debates about policy within the London region and ensured that central government and borough governments could not completely ignore London-wide issues. It was partly due to their insistence on the need for a voice for all of London that the office of the Mayor of London and Transport for London were eventually created.

Seoul and London offer several lessons. Autonomy – even limited autonomy – can make a profound difference to local policies, the local quality of life, and local economic development. Autonomy allows local electorates to voice local concerns. And even weak decentralisation can affect economic growth. As the policy director for the Mayor of London recently noted, despite a small amount of autonomy, London remains relatively powerless, particularly when compared to other global cities. London's regional government raises just 7 % of its budget from its own sources of income as opposed to 50 % for New York City and 80 % for Tokyo (Browne 2010). London must still defer to central government in many areas including control of regional development, housing, and the development of employment and skills. The Mayor of London has only a limited voice to influence rail transport into London and in the Greater London region and has no control over the use of the Thames.

Despite this, giving local government a voice allows the creation and implementation of new policies. This can clearly be seen in the case of the congestion charge in London and the river restoration in Seoul. Devolution gives not just local government but local electorates a voice in policymaking: the city government becomes more accountable to local electorates and at the same time more responsive to them. Equally important, it acts as a counterweight to national government. In the case of Seoul, it acted as a political springboard for politicians to reach the national stage, and even in London, it has done so. Devolution has given a voice to local electorates and given them greater political choice. In the process, it has made local citizens more engaged with local management.

This is particularly true in the case of Seoul, which until recently was viewed as the capital of a so-called developing country. It is often said that people in developing countries are not concerned about pollution. The experience of Seoul suggests that this is untrue: they were concerned but had no way to voice their concern until a local mayor could serve as a spokesman. Limited local democracy means local citizens cannot articulate their demands, protest against adverse conditions, or fight for better ones. They have no forum in which to raise their concerns and are unable to hold policies up to public scrutiny.

Conclusion

Seoul and London are perpetually short of cash. They must constantly go cap in hand to national government to ask for funds to meet basic development needs. As a result, each request for funding is politicised and debated at national level, and each expenditure is weighed up against other local economic needs. To many, it seems unfair for London to be asking for money to pay for new infrastructure while former industrial cities like Liverpool or Birmingham seem to have equally just demands. If decision-making is pushed to the highest level, then all local decisions become politicised and national ones.

In addition, national governments often argue that strong cities unbalance growth. They claim that strong cities are too wealthy and too strong. They claim that strong cities disadvantage other regions, taking investment away from regions that need it more. As a consequence, they try to disable or hobble strong cities.

However, one might argue that idiot policies hobble growth. The idea of a one-size-fits-all regional policy makes little sense. To a considerable extent, such policies represent compensation for a lack of local government resources and a lack of local autonomy. Such policies try to meet the needs of all regions but also try not to leave any region feeling disadvantaged. Ultimately, regional policies decided by national parliaments aim for the lowest common denominator. One size fits all means that policies fit everyone badly and local needs are shoehorned into national agendas. In addition, much of the money spent on regional aid is wasted. As Scott

(1997a, b, 2000) noted, writing about the United Kingdom, transfers to lagging regions are often used to protect dying industries and failing sectors rather than promote new growth. Writing about Russia, McKinsey Global Institute (1999) argued that subsidising declining industries and regions actually cripples potentially successful ones, forcing them to pay a tax from their profits to subsidise loss-making regions and firms. Growing regions are stifled in the name of equalisation. Successful regions are deprived of the means to help them retain their edge.

Ultimately, congestion and crowding are signs of success. Growing cities are not a cause of poverty or a cause for alarm. Rather, successful cities are popular cities, and popular cities are successful ones. Migrants want to live there, offices want to locate there, driving up prices and the cost of living. Rather than trying to strangle growth, government should support it. Moreover, as Groom (2010a) has reported, it may be the case that money spent on growing regions generates a higher rate of return than money spent on lagging ones. Again and again, throughout Europe, the capital city is the national motor of economic growth. This is not just true of London and the United Kingdom but also Paris and France, Budapest and Hungary, Bratislava and Slovakia, Copenhagen and Denmark, and Dublin and Ireland, just to name a few places. In this case, the key issue is how to spread the benefits of development.

London has greater autonomy and more powers than Seoul. It has been more proactive in implementing new policies such as a congestion charge and has been more responsive to local electorates. Thus, at the end of 2010, the congestion area in central London is set to be reduced after the electorate voted against the original London mayor Ken Livingstone partly because of unhappiness over the extent of the congestion area. In the case of Seoul, devolution was extremely limited but did succeed in giving the city a voice and allowed the city to experiment with new policies and initiatives which were particularly strong in the environmental sphere. However, in contrast to the United Kingdom, the Mayor of Seoul Lee Myong-bak used his post as a springboard to national politics, becoming national president. By contrast, while the Mayor of London has a national presence, he has operated alongside the national government structure as a power centre in his own right.

London also provides a lesson in gradualism. It is often argued that devolution of power to local governments is risky because they lack the personnel, experience, and expertise to manage local issues, to develop consensus, and to manage local budgets. In the case of London, although the Mayor's office was created in 2000, the transfer of powers to his authority was a gradual one, Transport for London only slowly taking over the underground, Docklands Light Railway, Thames road tunnels, and in 2009 parts of the suburban railway system rebranded the "overground". Gradualism has allowed the Mayor's office and the London assembly to develop management skills and expertise and allowed the gradual growth of culture of political responsibility. Thus, devolution has been a learning process involving the gradual accretion of new powers. Consequently, as noted in Browne (2010) a more confident Mayor's office is now demanding additional powers including outright control of housing policy and funding and of the London Development Agency.

More important, after 10 years of devolution, the London-wide electorate supports such change.

As noted earlier with reference to Oates (2006), devolution does not just benefit a single local government. By providing an alternative to central government policies, it can give local electorates a voice and also serve as a sounding board for new ideas. More important, it can offer a place to put new ideas into practice and test them. If they fail, a region or locality might suffer, but a nation will not. As Lindblom (1965) noted when writing about what he called "the intelligence of democracy", more policy proposals can be a good thing. Different ideas are put into practice in different areas, and they compete for success. Frey and Eichenberger (1996) referred to this as Functional Overlapping Competing Jurisdictions (FOCJ), arguing that this was a healthy situation. The result is constant innovation in government and governance. Sometimes policies fail, sometimes they succeed, but successful ones are copied by other local governments. In the case of the United States, local governments compete to lower tax, to increase the number of services on offer, and to improve the quality of those services. They are accountable to local electorates, and if they fail to meet their demands, local councils are either voted out of office or as Hirschman (1970) observed, local residents leave an area in search of greener pastures. Either they exercise voice through the voting process to change local policies or they move to an area with policies they favour. Therefore, in addition, however, stronger, more decentralised local government with more administrative and financial powers can lead to greater residential satisfaction. They can better meet the demands of the local residents, who express themselves through local elections. As Tiebout (1956) pointed out, electorates can come to select locations based on the policies they support, which can in turn further fix those policies. This can be indirectly exclusionary since some social groups and economic activities may be banned from a given locality, but if there are large numbers of local governments with a wide range of different policies, finding an ideal location should become easier. Thus, rather than a one-size-fits-all system, local governments can be tailor made to meet the needs of individuals or small groups.

This is not to say that unbridled local control is always successful. Detroit and Newark in the United States are examples of failing cities. Due to a shrinking tax base, they have had to cut back on urban services and reduce the quality of those services. The result has been a spiral of decline: poor urban services lead to movement out of the cities, which in turn further reduces the tax base, which leads to still poorer services and further outmigration. Cities often become dependent on one single industry or industrial sector, as was the case with motorcar manufacture and Detroit. If and when that sector stumbles, the negative impacts can be severe: job losses, loss of tax income, and movement away from the region. Dominance by one single industry can stunt growth in other sectors, particularly if that sector is high paid. Local labour is sucked into that sector, leaving other sectors and businesses short of labour and raising labour costs throughout the local economy. Prud'homme (1994) wrote about the dangers of such a situation, in which tax revenue – and spending power – differs widely among different regions and localities, leading to social inequality and even social exclusion. Moreover, cities make mistakes. Perhaps, this can best be seen in the film “Roger and Me” (Moore 1989). General Motors closed down its factory in the city of Flint, Michigan, killing off the local economy. Unemployment soared, and the tax base shrivelled. In desperation, the city attempted to create a new tourist attraction, committing a substantial amount of city money to do so. Within a year, the attraction had failed, and the city was even more badly off.

Germany, particularly eastern Germany, offers an alternative vision of how to cope with decline. Since 1989, the population of the region has gone down by 12.5 %, or about two million people. In all, four million people have left the region but been replaced by two million newcomers. Those newcomers have moved mainly to suburbs, and the population of cities has declined proportionately more. There, the programme “Stadtumbau Ost” (Schader Stiftung 2010) has tried to manage downsizing cities in the former German Democratic Republic, offering grants and assistance to declining cities, helping them to refurbish some buildings and demolish others, and replacing them with parks and open space or lower-density housing.

Perhaps, then, while one might argue that devolution is good for local innovation and helps regions to become more adaptable, at the same time, carried too far, it can lead to exclusion from economic growth and even economic failure as in the case of Detroit. However, in the case of London, it seems to have worked and seems to have further to go. And as was noted in *Centre for Cities* (2006), a case can be made for giving similar powers to other UK cities.

References

Blitz R, Chung J (2003) Backslapping and handshakes turn to failure. Financial Times, 23 June 2003

Browne A (2010) More power to the Mayor. PS Publicser-vice.co.uk. Available at www.publicser-vice.co.uk/feature_story.asp?id=14486. Accessed 8 July 2010

Bundesministerium für Verkehr, Bau, und Stadtentwicklung (2006) Jahresbericht der Bundesregierung zum Stand der Deutschen Einheit 2006, pp 8–15

Carey JP, Carey C, Carey AG (1955) The South of Italy and the Cassa per il Mezzogiorno. West Polit Q 8(4):569–588

Carlino GA, Saiz A (2008) City beautiful. Working paper, no. 08–22, Federal Reserve Bank of Philadelphia, Philadelphia

Centre for Cities (2006) Greater Birmingham needs mayor with tax and spend powers. Available at http://www.centreforcities.org/253.html

Chung J, Roberts D (2003) Regional grants not delivering jobs. Financial Times, 23 June 2003

Clymer A (2010) Robert Byrd, respected voice of the Senate, dies at 92. New York Times, 28 June 2010. Available at http://www.nytimes.com/2010/06/29/us/politics/29byrd.html

Crossrail (2010a) Benefits of Crossrail: wider economic benefits. Available at www.crossrail.co.uk/the-railway/crossrail-wider-economic-benefits. Accessed 23 June 2010

Crossrail (2010b) Funding. Available at www.crossrail.co.uk/the-railway/why-crossrail/funding-for-crossrail. Accessed 23 June 2010

European Union (2015) Legislation summaries. Glossary. Subsidiarity. Available at http://europa.eu/legislation_summaries/glossary/subsidiarity_en.htm

Florida R (2002) The economic geography of talent. Ann Assoc Am Geogr 92(4):743–755

Frey BS, Eichenberger R (1996) FOCJ: competitive governments for Europe. Int Rev Law Econ 16:315–327

Groom B (2010a) Dynamic cities seen as solution to jobs deficit. Financial Times (London edition), 7 June 2010
Groom B (2010b) Cities look for an artful boost. Financial Times, 12 July 2010
Groom B, Tighe C (2010) Government scraps England regional offices. Financial Times, 23 July 2010
Hirschman AO (1970) Exit, voice and loyalty. Harvard University Press, Cambridge, MA
House of Commons Communities and Local Government Committee (2007) Local government finance: supplementary business rate, vol 1. Stationary Office Ltd (HC 719-1), London
Johnston RJ (1986) Individual freedom and the world economy. In: Johnston RJ, Taylor PJ (eds) A world in crisis? Blackwell, Oxford, pp 186–187
Kim B-J (2004a) Local government reforms in Korea: critical issues and prospects. In: Kim B-J, Cope GH (eds) Central – local relations and local government reform in Korea. Korea Association for Policy Studies/Kyonggi Resaerch Institute, Seoul, pp 11–54
Kim I-S (2004b) Devolution of central government functions to sub-national units in South Korea: progresses and prospects. In: Kim B-J, Cope GH (eds) Central – local relations and local government reform in Korea. Korea Association for Policy Studies/Kyonggi Research Institute, Seoul, pp 73–91
Lindblom C (1965) The intelligence of democracy. Free Press, Glencoe
London Assembly (2010) The London Assembly. Available at http://legacy.london.gov.uk/assembly/assembly_about.jsp. Accessed 12 July 2010
Mckinsey Global Institute (1999) Unlocking economic growth in Russia. McKinsey Global Institute, Moscow
Moore M (1989) Roger and me. Warner Brothers, Burbank, Ca.
Moyo D, Ferguson N (2009) Dead aid why aid is not working and how there is a better way for Africa. Farrar, Straus and Giroux, New York
Murphy J (2010) Cross rail backers threaten to axe funding if project is diluted. London Evening Standard, 4 June 2010. Available at http://www.thisislondon.co.uk/standard/article-23841397-crossrail-backers-threaten-to-axe-funding-if-project-is-diluted.do
New London Architecture (2005) Civilizing spaces: improving London's public realm (Exhibition and catalogue). New London Architecture, London
New London Architecture (2007) Public city: places for people (Exhibition and catalogue). New London Architecture, London
Oates WE (2006) On the theory and practice of fiscal decentralization. IFIR working paper, no. 2006–2005, Institute for Federalism and Intergovernmental Relations, Lexington
OECD (2005) OECD territorial reviews: Seoul. no. 33. OECD, Paris
Porter M (1990) The competitive advantage of nations. Harvard University Press, Cambridge, MA, pp 210–225
Prud'homme R (1994). On the dangers of decentralization. World Bank policy research working paper 1252. World Bank, Washington, DC
Schader Stiftung (2010) Stadtumbau-Ost. Available at www.schader-stiftung.de/wohn_wandel/86.php. Accessed 12 July 2010
Scott P (1997a) British regional policy 1945–51: a lost opportunity. Twent Cent Brit Hist 8:1–25
Scott P (1997b) Dispersion versus decentralisation: British location of industry policies and regional development 1945–60. Econ Soc 26(4):579–598
Scott P (2000) The audit of British regional policy: 1934–39. Reg Stud 34(1):55–65
Scott P, Rooth T (1999) Public policy and foreign-based enterprises in Britain prior to the second world war. Hist J 42(2):495–515
Seoul Metropolitan Government (2002) Vision Seoul. Available at www.english.seoul.go.kr
Seoul Metropolitan Government (2003) Cheong-gyecheon restoration. Available at www.english.seoul.go.kr
Sherwood B (2010) Johnson seeks more power for Mayor. Financial Times (London edition), 15 June 2010
Taaffe EJ, Morrill R, Gould P (1963) Transport expansion in underdeveloped countries: a comparative analysis. Geogr Rev 53:503–529

The Economist (2006) Local democracy: a funny thing happened on the way to the council. 25 Feb 2006. Available at www.economist.com/node/5555725

The Economist (2010) Scottish independence: one step back. 3 July 2010. Available at www.economist.com/node/16485590

Thrift N (1986) The geography of international economic disorder. In: Johnston RJ, Taylor PJ (eds) A World in Crisis? A Geography of World Problems. Blackwell, Oxford, pp 12–67

Tiebout C (1956) A pure theory of local expenditures. J Polit Econ 64(5):416–424

Tuppen JN, Thompson JI (1994) Industrial restructuring in contemporary France: spatial priorities and policies. Prog Plan 42(2):99–172

United Kingdom Government (2010) Regeneration projects in London. Available at www.businesslink.gov.uk/bdotg/action/detail?site=181&itemId=5001196529&type=RESOURCES. Accessed 23 July 2010

United States Bureau of the Census (2010) Statistical abstract of the United States. Table 416, US Bureau of the Census, Washington. Available at www.census.gov/compendia/statab/2010/tables/10s0416.pdf. Accessed 28 June 2010

Chapter 5
Theories of Metropolitan Government and the Post-socialist Experience: The Case of Poznań Metropolitan Area

Łukasz Mikuła

Abstract This paper, based on ongoing research in the Poznan metropolitan region, contrasts theoretical approaches to the issue of metropolitan governance in America and Western Europe with the practical experience of a post-socialist country – Poland – using examples from the Poznań Metropolitan Area. International comparative studies (e.g. Jouve and Lefèvre, Local power, territory and institutions in European metropolitan regions. Frank Cass, London, 2002, Herrschel and Newman, Governance of Europe's city regions: planning, policy and politics. Routledge, London, 2002, Salet et al., Metropolitan governance and spatial planning. Comparative case studies of European city-regions. Spon Press, London, 2003, Heinelt and Kübler, Metropolitan governance. Capacity, democracy and the dynamics of place. Routledge, London, 2005) have shown that "place matters". It is emphasised that the nation-specific "tradition" and character of administration are of importance to the metropolitan government model. The evolution of the territorial and administrative structures of the big cities and suburban areas in Poland is different from Western Europe and North America. Between 1950 and 1990, Poland has a monolithic state government system, in which the decentralisation principle did not apply. Reforms of 1990 and 1998 brought the model of territorial administration in Poland closer to European standards on the local and regional levels. However, these reforms took little consideration of issues related to the metropolitan, as opposed to the local or regional, scale in the country's territorial and administrative structure. As a consequence, few formal institutions for metropolitan governance have been created, and metropolitan governance has been neglected. This has resulted in local inequalities in the quality and provision of services, and competition among local governments for inward investment and development, often causing adverse impacts at the metropolitan scale. In Poland, in 1990, an almost instant democratic breakthrough in the political system resulted in embracing the ideals of local government and the autonomy of communes. In Poland, the model of governance in metropolitan areas has been moving towards one of public

Ł. Mikuła (✉)
Institute of Socio-Economic Geography and Spatial Management,
Adam Mickiewicz University, Poznań, Poland
e-mail: mikula@amu.edu.pl

© Springer Science+Business Media Dordrecht 2015

J. Buček, A. Ryder (eds.), *Governance in Transition*, Springer Geography,
DOI 10.1007/978-94-007-5503-1_5

choice, despite being limited by central government control which continues to be strong. The country's socioeconomic development and advancing suburbanisation have created problems which can only be resolved on a metropolitan scale. The position of local governments in Poland is strong enough to make radical metropolitan reform socially and politically unlikely. At present, the only feasible solution appears to be the intensification of voluntary collaboration stimulated, perhaps, by small legislative changes at the national level. Nevertheless, as compared to the situation 20 years ago, the trend towards fragmentation has reversed. After a period in which local governments gained and secured their independence, now we are facing a stage of integration.

Keywords Problems of suburbanisation • Metropolitan governance • Metropolitan management • Local government competition and development • Local inequality

Introduction

The aim of this chapter is to contrast the theoretical approaches to the issue of metropolitan governance adopted over the past several decades mainly by American and Western European scientists with the practical experience of a post-socialist country – Poland – using specific examples from the Poznań Metropolitan Area. International comparative studies (e.g. Jouve and Lefèvre 2002; Herrschel and Newman 2002; Salet et al. 2003; Heinelt and Kübler 2005) have shown that "place matters". It is emphasised that the nation-specific "tradition" and character of administration are of importance to the metropolitan government model. The evolution of the territorial and administrative structures of the big cities and suburban areas in Poland has been different from Western European and North American countries. The most significant difference between 1950 and 1990 was the existence of a uniform state government system, in which the decentralisation principle did not apply. The reforms of 1990 and 1998 brought the model of territorial administration in Poland much closer to European standards, both on local and regional levels. However, these reforms took little consideration of issues related to the position of metropolitan areas in the country's territorial and administrative structure.

Theoretical Framework

The issue of the political and organisational integration of metropolitan communities emerged to a wider extent in the scientific perspective of the 1930s, especially among American researchers. Since annexation and the expansion of city limits ceased to be used in response to the increasing range of the urbanised land, a growing spatial mismatch emerged between the mosaic of local administration units and the functional and economic metropolitan area (Brenner 2003). This change of approach in the political and administrative sphere coincided also with some important social and economic changes, such as the growing affluence of societies during the post-war economic boom and the resultant rapid development of low-density housing and individual motorisation. These had a direct effect on the dynamics of suburbanisation. In response to these political, administrative, social and economic challenges, two contrary theoretical "traditions" (Ostrom 1972) emerged: the idea of *metropolitan reform* and *the public choice* concept.

Metropolitan reform is, in a way, a modern version of the previous trend, i.e. the successive expansion of a big city's administrative limits. According to its proponents, the disadvantageous social and economic effects resulting from municipal fragmentation can be eliminated by radical changes in the territorial and administrative structure and the consolidation of governance in a metropolitan area. The major argument for a metropolitan reform is the thesis that a city's economic and social growth should be accompanied by appropriate transformations in the institutional sphere (Lefèvre 1998). Therefore, the main goals of metropolitan reform are aligning administrative structures with functional relations and establishing strong institutions that cover the whole metropolitan area, i.e. a metropolitan government. To justify this in more detail, three arguments are usually used:

1. The effectiveness and efficiency argument
2. The social equity argument
3. The democratisation and social participation argument

First of all, administrative consolidation should bring a greater efficiency and effectiveness of public service provision by achieving economies of scale related to a larger number of residents and an extended area of operations. According to this approach, the existence of a large number of independent municipal units is an obstacle to increasing efficiency and effectiveness in the public service sector. Fragmentation leads to the doubling of administration, infrastructure and institutions, whilst the lack of coordination in local policies leads to the dispersion and waste of resources. All these problems can be avoided through management integration, which enables better allocation of the resources by public authorities. A larger jurisdiction area also provides for more comprehensive and coherent spatial planning which leads to the optimised locating of public investments and a more effective distribution of specific land-use zones.

Secondly, metropolitan reform is conducive to raising social equity in the metropolitan areas. Fragmentation isolates the poorer communities of the core city from the often affluent suburbs. The administrative consolidation of a metropolitan area makes it possible to achieve a common standard of public services. It limits fiscal and income-related inequalities between local governments. Moreover, it eliminates the problem of "free riding", a situation in which the core cities bear the costs of maintaining the institutions and the infrastructure to the advantage of the suburban residents whose local taxes are used exclusively to sustain their municipalities (Hawley 1951; Neenan 1972; Green 1974). Administrative integration brings the metropolitan area much closer to an "ideal" situation, in which the public services are provided with the use of taxes paid by all the people who actually avail of them.

Thirdly, there are political arguments in favour of metropolitan reform. According to the advocates of this concept, larger-scale local government structures are more democratic and open than smaller units. This is because they substantially eliminate the possibility of being dominated by small but influential interest groups, which could easily take control in smaller communes. The establishment of a government unit which is clearly and directly responsible for public affairs on a metropolitan scale makes the authorities more independent from local pressure groups. The consolidation of governance provides for more constructive solving of supra-local problems because it makes an appropriate spatial and institutional scale available to deal with them. This also increases the residents' interest in metropolitan issues because they are finally aware of who is politically responsible for these issues and from whom they can demand specific action (Lyons and Lowery 1989).

Since the late 1950s and early 1960s, the public choice concept has been a competitive idea to the metropolitan reform (Tiebout 1956; Ostrom et al. 1961; Warren 1966; Bish 1971; Bish and Ostrom 1974). This approach decidedly rejects the idea of institutional consolidation in metropolitan areas and basically undermines all the major premises of metropolitan reform. This concept does, in fact, perceive administrative fragmentation as conducive to the effectiveness of public services, whilst any competition between local governments in this respect is, indeed, advantageous. The desired administrative structure of a metropolitan area consists of a large number of small local government units. The arguments used by the public choice theoreticians can be divided into two groups:

1. Economic and fiscal arguments
2. Social and political arguments

Arguments from the first group are based on the assumption that the metropolitan reform does not lead to a greater effectiveness of public services. On the contrary, it gives one institution a monopoly in the market, which does not favour innovation and cost reduction. In a situation of political and institutional fragmentation, when there are a large number of independent municipal units, a mechanism exists that is reminiscent of a market featured by a large number of manufacturers (local governments) who compete for consumers (the residents). Tiebout's model (1956) assumes that local authorities with extensive financial autonomy can freely determine their tax/service package. Owing to municipal fragmentation, residents of a metropolitan area can freely choose an appropriate "package" for themselves. It is also possible for them to appraise the management effectiveness by "voting with their feet", i.e. if they are dissatisfied with the quality of public services and the local governments' policy in their current place of residence, they can move to better-managed units within the same metropolitan area (Tiebout 1956; Ostrom et al. 1961).

The public choice proponents do not deny that there are some public services that require a larger spatial scale than a small local unit. The appropriate instrument for their organisation is not, however, a top-down reform of the basic administrative division and the establishment of new "heavy" institutional forms (of "metropolitan government" nature). The solution is the voluntary horizontal cooperation of the existing local government structures. Furthermore, the organisation and provision of public services, which are the responsibility of local authorities, can be easily separated from their actual "production" by entering into agreements with other actors, for example, with neighbouring local governments or private companies.

Social and political arguments are of equal significance in the public choice theory. A metropolitan government proposed by the advocates of the metropolitan reform is too remote from the ordinary citizens to give them effective political control over it. Real local democracy is only possible in small local units; hence, it is necessary to sustain their self-reliance and independence. A smaller community is capable of forcing a greater political responsibility on the authorities. This point of view emphasises the advantages of direct democracy and the importance of close contacts with local politicians (Parks and Oakerson 2000). Administrative fragmentation gives the residents of suburban communities a chance to separate themselves from the politics of the great cities where their interests could be lost among other problems or be outvoted. By holding a planning authority, the governments in the suburban zones can also indirectly determine the social and economic status of the prospective residents, for example, by determining the sizes of building plots, and exclude the undesirable forms of land usage (Pacione 2001).

Confronting both these classical models, i.e. metropolitan reform and public choice, is not an easy task, because both stem from fundamentally different values and principles (Lefèvre 1998). The core of the metropolitan reform concept is an integrated and global approach, the prevention of any form of social differentiation and segregation, as well as being guided by values such as solidarity and social equity. On the other hand, the public choice approach emphasises the advantages of individual freedom of choice and the proximity of local authorities and their direct responsibility to the residents, in addition to the free competition between local governments which should work to the advantage of improved management effectiveness.

What both approaches presented above have in common, despite their basic ideological differences, is the fact that they focused on analysing the functioning of traditionally perceived administration units – usually local governments. The essence of the dispute between their proponents was the possible need for introducing a new form of an integrated *metropolitan government*. More recent studies take into account the greater complexity of the modern models of governing metropolitan areas and the formation of their networking character involving a vast number of actors, including those which do not fit within the precise definition of the public administration sphere. To differentiate it from the traditional research perspectives, this new type of approach is often referred to as *metropolitan governance*.

In new economic conditions, the need to include metropolitan areas in the global economy has become more pressing. The central premise for the new wave of reforms is the belief that their implementation will improve a region's position in the worldwide metropolitan network and that this will be advantageous to all stakeholders within a metropolitan area (Swanstrom 2001). As a result, new important actors have emerged in the sphere of governing metropolitan areas. These include the institutions of the European Union, regional authorities, international corporations, nongovernment organisations and companies which provide public services which, until recently, were under public administration and excluded from market competition, for example, power and water supply, education, health care, transport and housing. In Europe, the gradual (or more radical as in the United Kingdom) withdrawal of central governments from many fields of social and economic life has led to a situation in which "instead of one pilot, a set of actors has emerged whose configuration and coordination is still forming" (Lefèvre 1998). This is also sometimes referred to as "multi-level multi-actor governance" (Marks 1993; Mitchell-Weaver et al. 2000; Hooghe and Marks 2001).

The changes in metropolitan management should be regarded as an element of a wider transformation process in the public sphere described as a transition from *government* to *governance* (Bailey 1993). The dominating feature of the government period was that of public administration dealing with the widest possible scope of public affairs, i.e. planning, building infrastructure, providing public services, etc. The government model was based on formal procedures and a vertical administrative hierarchy. In this context, governance means a transition from the vertical approach to a horizontal one (Barlow 1997) and from structures based on public authorities to a polycentric model featuring many actors. This also means that the boundaries between the public and private sectors are becoming indistinct and that the cooperation of administration with other actors is growing in significance. Moreover, some of the management techniques that were typically used in the private sector are now being taken over by public authorities to form the "new" public management. In the metropolitan governance system, the major actors are mutually dependent and complement one another. In order to achieve the potential that is necessary for effective operation, they establish policy networks in which the political legitimisation comes from public authorities chosen in general elections, whilst the financial and economic resources are mainly from the business sector (Jouve and Lefèvre 2002).

The achievements of the *metropolitan governance* concept so far are impressive. It has, primarily, expanded the scope of scientific interest in this field, placing greater emphasis on the role of key socio-administrative factors in contemporary public management, i.e. initiatives and structures based on agreements and collaboration, political leadership, incentives to cooperate from the central government, individual strategies of behaviour, ways in which decisions are made, the decision-makers' responsibilities and the relationships between the public and private (non-government) sectors.

To recapitulate the modern scientific dispute, one must conclude that, despite the differences in evaluating different theoretical approaches, there is reasonable agreement that the vision of the "perfect" metropolitan reform has been lost in the transition from the government to the governance type of analysis. The identification of an indisputable "best practice" or the complete replication of specific institutional experience seems impossible (Silva 2007). This leads to using a case study-type analysis, which takes into account the local and national contexts.

Public Governance in the Poznań Metropolitan Area: From Disintegration to Voluntary Cooperation

Poznań is the fifth largest city in Poland (556,022 residents) and the capital of the Wielkopolska Voivodeship (29,828 km^2; approx. 3,403,174 residents), the core region of the early Polish state which was established in the tenth century. From the Second Partition of Poland in 1793 until the reunion of the Polish state in 1919, the Wielkopolska region was under Prussian rule but with a majority of Polish population. Resistance against the Prussian political domination stimulated the rise of a civil society and dynamic economic development in the region, which were especially significant in comparison to the other parts of Poland that were under Russian and Austrian rule. Today, in terms of economic performance, Poznań is the second richest city in Poland (after the capital city, Warsaw). It is also, together with its suburban zone, a very attractive investment site for many transnational companies, including Volkswagen, GlaxoSmithKline, Bridgestone, Nestlé, Beiersdorf, Imperial Tobacco, MAN, SABMiller, Wrigley and Allied Irish Bank. The Poznań International Fair is one of the largest exhibition centres in Eastern and Central Europe.

The city of Poznań has a city-powiat status (similar to the former county boroughs or the current unitary districts in England or *kreisfreie Städte*, *Stadtkreise*, in Germany). It is surrounded by the powiat of Poznań which includes 17 communes (Fig. 5.1) and is the largest land powiat (county, *Landkreis*) in Poland. Together, they form the most important urban region in the western part of Poland (Table 5.1).

The scale of change in the Poznań Metropolitan Area since 1990 has been tremendous and probably more remarkable than in many other cities in Western Europe. The most important spatial outcome of the political, economic and social transition is the rapid growth of the suburban area in terms of population and employment (Tables 5.2 and 5.3, Fig. 5.2). It coincides with a decline in the number of residents and jobs in the core city. After 1990, the re-emergence of bid rent, with land prices gradually declining as the distance from the city centre increases, has significantly changed the spatial structure of the Poznań Metropolitan Area. Although market principles often work in favour of the pre-1990 residential districts (due to the existing technical infrastructure), new single-family houses in the suburban area are much cheaper than flats of the same size in the core city. This residential suburbanisation has been enabled by the increasing rate of car ownership. But traffic congestion has emerged as another result. Public transport is not well developed, although the Poznań Fast Tramway was opened in the late 1990s. Under these conditions, some of the most accessible sites can be found in the outer parts of the city or in suburban locations, where the traffic congestion is lower and the road network is relatively modern. This mainly affects the distribution of services and results in

Fig. 5.1 Poznań Metropolitan Area: city of Poznań and powiat of Poznań (Source: author)

Table 5.1 Basic data on Poznań Metropolitan Area (2009)

Territorial unit	Number of municipalities	Population	Area (km^2)	Population density (population per km^2)
City of Poznań	1	556,022	261	2,130
Powiat of Poznań	17	315,159	1,900	166
Poznań Metropolitan Area	18	871,181	2,161	403

Source: own compilation based on data from the GUS (Central Statistical Office) website, http://www.stat.gov.pl

Table 5.2 Population change in Poznań metropolitan area between 1995 and 2009

Territorial unit	Population 1995	Population 2009	Change	Change in %
City of Poznań	581,772	556,022	−25,750	−4.4
Powiat of Poznań	239,352	315,159	+75,807	+31.7
Poznań Metropolitan Area	821,124	871,181	+50,057	+6.1

Source: own compilation based on data from the GUS (Central Statistical Office) website, http://www.stat.gov.pl

Table 5.3 Employment change in Poznań Metropolitan Area between 2000 and 2008

Territorial unit	Jobs 2000	Jobs 2007	Change	Change in %
City of Poznań	221,830	227,846	+6,016	+2.7
Powiat of Poznań	60,267	89,457	+29,190	+48.4
Poznań Metropolitan Area	282,097	317,303	+35,206	+12.5

Source: own compilation based on data from the GUS (Central Statistical Office) website, http://www.stat.gov.pl

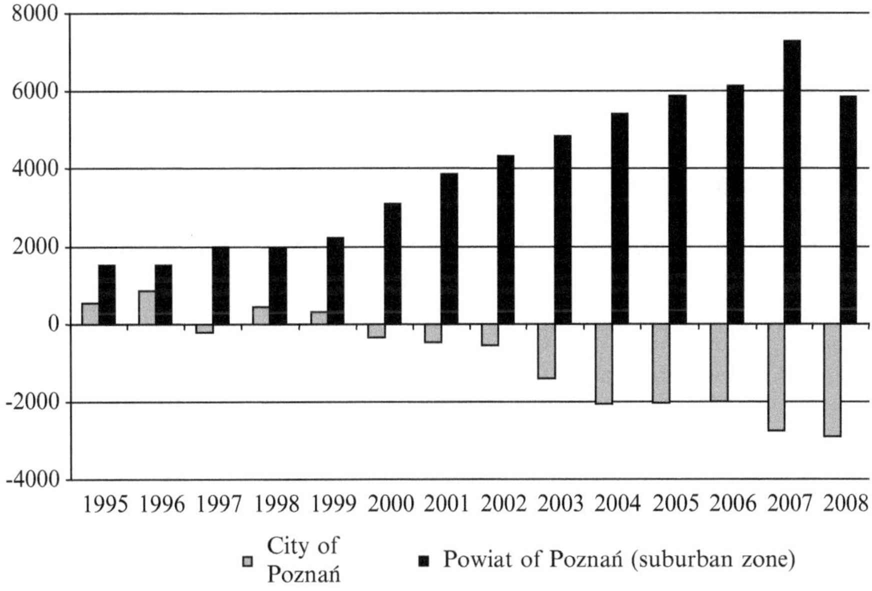

Fig. 5.2 Net migration in Poznań Metropolitan Area in 1995–2008 (Source: own work based on data from the GUS (Central Statistical Office) website, http://www.stat.gov.pl)

the construction of large out-of-the-city shopping centres. The lack of accessible and affordable greenfield sites within the core city has had an influence on the location of industry too. Most of the new industrial areas are situated in suburban communes (especially in the western part of the powiat of Poznań), and some enterprises that had been located in the city of Poznań also moved there. There is increasing inter-municipal competition to attract as many new residents and enterprises as possible. Some suburban communes (e.g. Tarnowo Podgórne, Suchy Las, Komorniki and Swarzędz) seem to be much more successful than the City of Po*znań*.

Institutional development in the Poznań Metropolitan Area has varied strongly over the past decades. After World War II, the issues of public governance in urban agglomerations were greatly affected by the general changes in the political system and the introduction of socialism. In 1950, local government and the dual system of public administration were dissolved and replaced with a uniform state power structure. The role of local government and local state administration was wholly taken over by "national councils" which were, in fact, state authority bodies. Territorial administration units (voivodeships and communes) did not have any decision-making or financial freedom. There were no democratic rules for electing political representations at these levels, and the centralised system of governing the state restricted the competence of local units in many important fields.

A new stage in the development of the urban management model in Poland began with a major administration reform in 1990. The reform is generally considered as one of the greatest successes of the Polish political transformation. It has contributed to a significant change in the approach to managing public affairs, has empowered local communities and encouraged them to take up civic activity. However, because local governments were only reinstated at the commune level, the changes of the early 1990s meant to a great extent that the big cities lost their administrative connections with their nearest surroundings. Obviously, these connections had not been, until then, based on the principle of local democracy and autonomy (state enterprises operated in the field of municipal services and public transport). They did, nonetheless, ensure a certain minimum of effective public service in the urbanised areas. Moreover, the changes in the political system did not cause any remodelling of the administrative division at the commune level, which, in fact, remains the same as that proposed between 1972 and 1975 and at a deeper level continues the traditions of the Act of 1933. With all due respect for the achievements of the local government reform of 1990, one must point out that it has contributed considerably to today's problems with administrative fragmentation in metropolitan areas. These are, however, not so much the result of communal reform itself but rather of delays in the subsequent stages of changes in the form of the higher administration and higher administrative divisions of government and the lack of a comprehensive vision of the whole local government system after 1990.

After 1990, the best opportunity to overcome the problem of administrative fragmentation in metropolitan areas and to introduce new structures of integrated governance was the territorial and administration reform of 1998. Introducing new tiers of local and regional self-government (powiats and new, larger voivodeships) could provide a chance to adjust the administrative status of big cities and metropolitan areas. In the end, however, the reform did not provide for any major shift towards metropolitan integration. Re-establishing the city-powiat category, which was traditional in the Polish administration system, made the large cities even more administratively isolated from their suburban areas. Furthermore, the concept of metropolitan areas as regional units was not approved by the authors of the reform. After the dissolution of the former small voivodeships and their inclusion in the larger regions, metropolitan areas were subject to administrative disintegration. The only links which have survived are the voluntary collaboration of communes and the cooperation of land powiats and city powiats. The three statutory forms of such a collaboration are joint boards, agreements and associations and more businesslike cooperation (joint-stock or limited liability companies).

So far, in the case of the Poznań Metropolitan Area, joint boards have had very limited significance. Out of the 233 such units in Poland, only five include communes from the Poznań Metropolitan Area. The relatively small role of this form of integration is due to the following features:

- The small number of actively functioning joint boards
- The limited scope of collaboration (only one or two issues)
- The small spatial range

The latter factor is especially significant. The existing joint boards usually include only some of the communes from the metropolitan area, and the city of Poznań is not a member of any of them. Similar problems, resulting to a great extent from the differences in size between the core city and the smaller suburban communes, are also an obstacle in this type of collaboration in many other metropolitan areas in Europe (see Kaczmarek 2005; Kaczmarek and Mikuła 2007b).

A municipal agreement between the city of Poznań and nine neighbouring communes plays a major role in the management of water supply and sewage disposal in the metropolitan area. It is executively operated by Aquanet – a registered company in which the city of Poznań holds the majority of shares. The municipal collaboration in this field dates, in principle, to the very beginning of the restored local government. In 1991, the hitherto state-owned water supply and sewage disposal company was divided. To continue its functioning, it was necessary for the communes it served to make an agreement and determine the conditions of taking over and administering the assets. The agreement specified the establishment of a limited company and the takeover of shares by the stakeholders of the agreement. This enabled the municipalisation of the assets administered by the state-owned company (1996). Aquanet is controlled by local government units; however, as a company subject to commercial law, it enjoys a certain degree of decision-making autonomy. Consequently, it is becoming one of the major actors in managing the development of the metropolitan area. Despite an analogous initial situation, a similar process was not implemented in the field of heat engineering. During the municipalisation of assets administered by the former state-owned heating provider, it was agreed that the particular towns and communes would take over exclusive ownership of these assets within their administrative limits.

So far, municipal agreement has been the most important instrument in providing integrated management in the Poznań Metropolitan Area; it is also used in fields other than water supply and sewage disposal. Inter-municipal agreements made between Poznań and the surrounding communes concern, for example, preschool education and day care in nurseries. During the past 3 years, a series of agreements was also made in the field of public transport. Within such an agreement, a suburban commune entrusts the city of Poznań to provide public transport services on a specified route. The city of Poznań delivers this service through its municipal unit – the City Transport Board. An agreement like this makes it possible to include the suburban lines it manages to the public transport system in Poznań and enables passengers to use a single ticket on their route to and from Poznań, as well as within the city.

The examples of municipal agreements at the commune level listed above aim to expand the territorial range of public services provided by the city of Poznań to neighbouring communes. Similar agreements are also made at the powiat level, mainly in the field of social services. Usually, the powiat of Poznań assigns to the city of Poznań public services delivered by specialist units, for example, adoption and child protection centres, as well as psychology and educational centres for children with special learning needs, which employ highly qualified staff. The number of children with such needs living within the land powiat is relatively small and establishing separate centres for them would be inexpedient.

Slightly different agreements are made in such areas as the employment market, business support, fire defence and emergency medical services. These concern the functioning of institutions that are clearly of a metropolitan character and whose operations cover both the city of Poznań and the powiat of Poznań. Taking into account the nature of the local employment market, it was considered reasonable to conduct a joint employment policy for both powiats. All the work in this field has been assigned to one employment centre which provides services to the whole metropolitan area.

Of major significance in the field of public security is the fact that the operations of the police and fire service units cover both the city of Poznań and the whole powiat of Poznań. Such territorial administrative organisation in these sectors is primarily the result of the regulations which introduced the territorial and administrative reform of 1998. It should be noted that the sphere of public security comes only partially within the scope of competence of local governments.

An interesting example of the wider metropolitan integration is the Poznań Local Tourism Organisation (PLOT), founded as an association in 2003. Among its members are the city of Poznań, the powiat of Poznań and ten communes from the metropolitan area. PLOT's statutory goal is to support the growth of the tourism market in Poznań and the Poznań Agglomeration, in addition to promoting them as attractive tourist sites. An important element of PLOT's operations is combining the public and private sectors. Apart from local government units, it also embraces tourism business chambers, cultural institutions, companies from the tourism sector (e.g. hotels) and actors that are of key importance to the local tourism market (e.g. the Poznań International Fair and the Poznań-Ławica Airport).

To date, areas of great importance to the development and functioning of a metropolitan area such as strategic programming, spatial planning and road infrastructure management have not been included in any form of institutional coordination. Hence, the signing, on 15 May 2007, of the agreement of cooperation between the local governments of the Poznań Agglomeration was a major step towards the stronger institutionalisation of the metropolitan area. The stakeholders included the city of Poznań, the powiat of Poznań and 17 suburban communes. They instituted the Agglomeration Council as a Permanent Conference of the Mayors of the Poznań Agglomeration. Initially, the Council, under a non-statutory institutional arrangement, was to act as a forum to exchange information between local governments and to determine the fields of cooperation which, in the future, could take on specific legal forms. However, as of April 2010, work on appointing inter-municipal joint boards had not been completed in some key areas within the metropolitan area, i.e. in waste management and public transport.

More promising than the results in establishing inter-municipal partnerships are the prospects for cooperation in creating the Metropolitan Area Development

Strategy. The members of the Agglomeration Council unanimously supported the idea of preparing such a document and assigned this task to a consortium of four universities in Poznań. The first stage of work consisted of some comprehensive diagnostic studies. An important aspect of this collaboration is the fact that an initiative undertaken with the Poznań Agglomeration Council, a non-formal body, resulted in operations within specific legal forms pertaining to fixing local government support for the project in proportion to their population. The powiat of Poznań provided financial assistance in the form of a special purpose grant for the city of Poznań, which performs the function of the Agglomeration Council's representative in their legal dealings with the academic consortium preparing the metropolitan strategy.

There is a great potential for further collaboration. So far, there is still no business and marketing public agency to develop and promote an integrated investment proposal and to provide a service for investors within the whole metropolitan area. A crucial factor which could strengthen the metropolitan consolidation in the Poznań area is the organisation in Poznań of mass events, for example, EURO 2012 and the cities competing for the title of the European Capital of Culture 2016. Organising such events requires strong collaboration between the city of Poznań and neighbouring local governments.

Western Theories in Post-socialist Reality: Opportunities and Limitations

The above empirical analysis of governance-related problems in the Poznań Metropolitan Area gives an opportunity to attempt to answer the question: how could Western theoretical concepts be applied in the conditions of a post-socialist country? The issue will be presented in relation to the three theoretical approaches presented in the first part of this paper:

1. The metropolitan reform concept
2. The public choice concept
3. The metropolitan governance concept

The main thesis of the metropolitan reform concept is the need to adapt the territorial and administrative structure of metropolitan areas to the expanding range of the big city's direct functional influence. Demographic and economic processes should, therefore, be accompanied by concurrent transformations in the institutional sphere. The dynamic population and economic growth of the suburban zone in the Poznań Metropolitan Area have been in recent decades very significant. In the functional sense, the city of Poznań has, indeed, already exceeded its administrative limits which have not changed since 1987.

The metropolitan reform concept was not put to practice during the general reconstruction of the country's territorial and administrative structure in 1998 at the powiat or voivodeship level. During the past several years, the significance of the metropolitan concept has been increasing in the subsequent drafts of legal acts aimed at formalising administrative structures on the metropolitan scale. But the final shape of the proposed solutions concerning the form of government has not been determined.

In Poland, a limitation to the usefulness of the metropolitan reform idea is the lack of some of the major challenges and problems that inspired the classical model. There is no ethnic (race) segregation in metropolitan areas (a problem that hardly exists in Poland or in Poznań), and there is a lack of strong fiscal imbalance between the poor core city and the affluent suburbs. Within the Poznań Metropolitan Area, the core city still has a relatively high budget revenue. This is not affected by the fact that some of the wealthier suburban communes have higher funds per capita in comparison to Poznań. There is, however, one considerable problem that is in the focus of the metropolitan reform proponents – the lack of integrated spatial planning on the metropolitan scale in the context of strong suburbanisation. The weak national regulations for metropolitan planning currently in force prevent any effective operations in this respect (Kaczmarek and Mikuła 2007a).

To date, the metropolitan reform idea has not been implemented in any way in Poland. But this does not automatically mean that the current model of governing metropolitan areas is related to the alternative concept of public choice. The restricted financial autonomy of communes and their lack of full freedom in determining tax rates make it very difficult to conduct a conscious policy of creating tax/service packages. The legally permitted property tax maximum rates are restricted to a low level, and local governments have no influence on the rates of personal and corporate income taxes – their main source of revenue. In Poland, there are also extensive mechanisms at the national level for equalising local government incomes at the national level (general subventions, including compensatory shares and special purpose grants) which, to some extent, moderate the differences in the affluence of the particular local units.

Another premise of the public choice concept is the competition between local governments within metropolitan areas. This is often perceived as an advantage. This type of competition, however, should include relatively equal actors, whereas within the Poznań Metropolitan Area the demographic supremacy of the core city over the suburban communes is still very obvious despite current suburbanisation trends. Furthermore, the relatively low mobility of the Polish people, in comparison to the residents of the United States, does not allow them to fully avail themselves of all the positive aspects of administrative fragmentation in metropolitan areas proposed by the public choice theory – for example, being able to freely choose the best commune to settle in.

The one field in which Polish communes have actual decision-making freedom is spatial policy, especially given the very weak planning tools at the metropolitan and regional levels. Nevertheless, the competition to win new residents and investors has resulted in the lack of planning coordination between neighbouring units, which is often characterised by the German term *Inselplannung* (island planning). This leads to unfavourable spatial effects (urban sprawl) in Polish metropolitan areas.

The political dimension of the public choice concept seems to be more relevant in Poland than the economic and fiscal factors. Local democracy and citizen participation in smaller municipalities seem to be very strong. Public opinion surveys and interviews with local politicians have confirmed that there is more direct contact between residents and mayors or councillors in smaller municipalities than in big cities (Mikuła 2010). This view is also supported by some of the experts who are involved in the development of a metropolitan bill (Kulesza 2006).

It is interesting to relate some of the elements of the *metropolitan governance* concept to Polish reality. These include the recognition of multiple actors involved in the governance of a metropolitan area; voluntary and flexible collaboration; the lack of hierarchic connections; the blurring of boundaries between the public, private and nongovernment sectors; and the establishment of more formalised metropolitan institutions from bottom-up initiatives. Despite the changes that are taking place in the public governance system in metropolitan areas, it seems that local authorities continue to play a fundamental role in the system, especially at the commune level. The integration of non-administration actors, for example, business and academic institutions, non-profit organisations and the media, into the governance model is still difficult in Poland, although there are examples showing that such attempts are being made. These are, of course, non-authority activities, like the organisation of the Poznań Agglomeration Economic Forum by the Wielkopolska Chamber of Commerce and Industry (WIPH) or support for metropolitan initiatives by the local and regional media. Activities also include collaboration among regional and local tourism organisations (Mikuła 2008). The role of scientific circles is clearly apparent in the formulation of a development strategy for the Poznań Agglomeration Council.

The institutionalised involvement of private investors in metropolitan area development poses an important dilemma. Some of these investors' operations on the local scale are controversial among the public. In some cases, private entrepreneurs can force local authorities to create specific policies which are advantageous to them in spatial planning (especially relating to site development conditions which are based on the principle of the so-called good neighbourhood) or with regard to real estate (obtaining attractive sites below their market value). Hence, some specific threats may emerge if metropolitan governance is strongly intertwined with the business environment. Furthermore, commercialised and privatised municipal companies are becoming relatively autonomous self-governing units and are becoming quite independent of the local authorities, in addition to gaining a capacity to determine the growth of the metropolitan area.

The *metropolitan governance* concept proposes a management model based mainly on the voluntary and flexible collaboration of numerous actors. It does, however, provide for the establishment of more formalised institutional forms but emphasises that this should primarily result from bottom-up initiatives. The latest drafts of the Metropolitan Act proposed by the central government lean towards such a solution, based mainly on "institutionalising" metropolitan areas. As far as the Poznań Metropolitan Area is concerned, there is a clear political will to ensure such a legal fixation for local governments' collaboration in the metropolitan area.

To conclude, as has been pointed out using the example of the Poznań Metropolitan Area, the experience of governing metropolitan areas in a post-socialist country is much shorter than in Western states. This is primarily due to historical conditions and the socialist heritage of centralisation. The transformations of the 1980s and 1990s in Western Europe led to a gradual withdrawal of the *welfare state* and the strengthening of the position of local governments and other important actors in metropolitan governance. Concurrently in Poland, in 1990, an almost instant democratic breakthrough in the political system resulted in, among other issues, embracing the ideals of local governments and the autonomy of communes. The model of public governance in metropolitan areas has been evolving towards one of public choice, despite being limited by central government control which continues to be strong. The country's socioeconomic development and advancing suburbanisation have created problems which can only be resolved on a metropolitan scale. The position of local governments in Poland is strong enough to make any radical metropolitan reform socially and politically unlikely. At present, the only feasible solution appears to be the intensification of voluntary collaboration stimulated, perhaps, by small legislative changes at the national level. Nevertheless, as compared to the situation observed 20 years ago, the general trend towards fragmentation has reversed. After a period in which local governments have gained and secured their independence, now we are facing a stage of integration.

References

Bailey S (1993) Public choice theory and the reform of local government in Britain: from government to governance. Public Policy Adm 8(2):7–24

Barlow M (1997) Administrative systems and metropolitan regions. Environ Plan C Gov Policy 15(4):399–412

Bish RL (1971) The public economy of metropolitan areas. Markham, Chicago

Bish RL, Ostrom V (1974) Understanding urban government: metropolitan reform reconsidered. American Enterprise Institute for Public Policy Research, Washington

Brenner N (2003) Metropolitan institutional reform and the rescaling of state space in contemporary Western Europe. Eur Urban Reg Stud 10(4):297–324

Green H (1974) Fiscal interactions in a metropolitan area. D.C. Heath, Lexington

Hawley A (1951) Metropolitan population and municipal government expenditures in central cities. J Soc Issues 7(1):100–108

Heinelt H, Kübler D (eds) (2005) Metropolitan governance. Capacity, democracy and the dynamics of place. Routledge, London

Herrschel T, Newman P (2002) Governance of Europe's city regions: planning, policy and politics. Routledge, London

Hooghe L, Marks G (2001) Multilevel governance and European integration. Rowman and Littlefield, Lanham

Jouve B, Lefèvre C (eds) (2002) Local power, territory and institutions in European metropolitan regions. Frank Cass, London

Kaczmarek T (2005) Struktury terytorialno-administracyjne i ich reformy w krajach europejskich (The territorial-administrative structures and their reforms in European states). Wydawnictwo Naukowe UAM, Poznan

Kaczmarek T, Mikuła Ł (2007a) Metropolitan areas in Poland – towards a new scale of urban governance? Quaestiones Geographicae 26B:97–105

Kaczmarek T, Mikuła Ł (2007b) Ustroje terytorialno-administracyjne obszarów metropolitalnych w Europie (The territorial-administrative systems of metropolitan areas in Europe). Bogucki Wydawnictwo Naukowe, Poznan

Kulesza M (2006) Konsolidacja zarządzania w samorządzie – wybrane zagadnienia. Uwagi do dyskusji (The consolidation of management in local government – chosen aspects. Points to discussion). Samorząd Terytorialny 7–8:5–17

Lefèvre C (1998) Metropolitan government and governance in western countries: a critical review. Int J Urban Reg Res 22(1):9–25

Lyons WE, Lowery D (1989) Governmental fragmentation versus consolidation: five public choice myths about how to create informed, involved and happy citizens. Public Adm Rev 49(10):533–543

Marks G (1993) Structural policy and multilevel governance in the EC. In: Cafruny A, Rosenthal G (eds) The state of the European community, vol 2. Lynne Rienner, Boulder, pp 391–410

Mikuła Ł (2008) Promocja i rozwój turystyki jako element współpracy metropolitalnej (The development and promotion of tourism as an element of metropolitan cooperation). In: Jażdżewska I (ed) Funkcja turystyczna miast. Wydawnictwo UŁ, Łódź, pp 25–36

Mikuła Ł (2010) Funkcjonowanie aglomeracji poznańskiej w opinii mieszkańców i polityków lokalnych (The Poznań Agglomeration in opinion of its residents and local politicians). Biblioteka Aglomeracji Poznańskiej, Bogucki Wydawnictwo Naukowe, Poznań

Mitchell-Weaver C, Miller D, Deal R Jr (2000) Multilevel governance and metropolitan regionalism in the USA. Urban Stud 37(5–6):851–876

Neenan W (1972) Political economy of urban areas. Markham, Chicago

Ostrom E (1972) Metropolitan reform: propositions derived from two traditions. Soc Sci Q 53:474–493

Ostrom V, Tiebout CM, Warren R (1961) The organisation of government in metropolitan areas: a theoretical inquiry. Am Polit Sci Rev 55:831–842

Pacione M (2001) Urban geography. Routledge, London

Parks RB, Oakerson RJ (2000) Regionalism, localism and metropolitan governance: suggestions from the research program on local public economies. State Local Gov Rev 32:169–179

Salet W, Thornley A, Kreukels A (eds) (2003) Metropolitan governance and spatial planning. Comparative case studies of European city-regions. Spon Press, London

Silva CN (2007) Reinventing local and metropolitan government. Public Adm Rev 67(5):954–957

Swanstrom T (2001) What we argue about when we argue about regionalism. J Urban Aff 23(5): 479–496

Tiebout C (1956) A pure theory of local expenditures. J Polit Econ 64:416–424

Warren R (1966) Government in metropolitan regions: a reappraisal of fractionated political organization. University of California, Institute of Governmental Affairs, Davis

Chapter 6
Regional Elites, Networks and the Beauty of Regionalism in Hungary

Ilona Pálné Kovács

Abstract The rescaling of administrative structures and hierarchies is a Europe-wide phenomenon, often associated with democratisation and increased accountability. This is particularly true within the European Union, where the principle of regionalisation, often referred to as subsidiarity, is embraced in the Copenhagen criteria. Those states which joined the Union after the collapse of communism had to reorganise their spatial administrative hierarchies and decentralise national administration, often creating two or more new levels of government. This was the case in Hungary, where, before 1990, the main regional actors were counties, which were responsible for directing the allocation of resources and managing development. After 1990, local government reforms strengthened local governments at the expense of the counties and left a vacuum at the regional level. However, in the mid-1990s, the Hungarian government attempted to create new NUTS2 meso-level regions, partly in response to demands from the European Union and partly to instil a new regional orientation among policymakers. This paper, based on research undertaken in 2002–2008, examines some of the problems in creating new regions. The research found that top-down regionalisation, demanded by the European Union, and imposed from above, failed to create a regional identity. Respondents, taken from among regional policymakers and business leaders, are identified with their locality and county, but not the new region. Even national politicians appeared ambivalent, supporters of the regionalisation legislation fearing that it would complicate administration by adding a new layer of government and opponents arguing that it would weaken the central government. Regionalisation remained a relatively hollow administrative exercise from which civic, business, and non-governmental organisations felt excluded despite their nominal participation. The research also found that bottom-up regionalism was potentially more successful, but it was unlikely to succeed unless there was broad-based political support at all levels of the government.

I. Pálné Kovács (✉)
Department of Political Studies, University of Pécs, Pécs, Hungary

Centre for Economic and Regional Studies of Hungarian Academy of Sciences,
Papnövelde utca 22, H 7621 Pécs, Hungary
e-mail: palne@rkk.hu

© Springer Science+Business Media Dordrecht 2015
J. Buček, A. Ryder (eds.), *Governance in Transition*, Springer Geography,
DOI 10.1007/978-94-007-5503-1_6

Keywords Regional governance • Regionalisation policy • Local-regional tensions • Intergovernmental relations • Region formation

Introduction

The rescaling of power structures is a Europe-wide phenomenon (Guilani 2006), but the timing and the measures differ by country. This paper focuses on changes in Hungary since the collapse of communist rule. The collapse led to systemic change, and as a result, the spatial division of power underwent deep and rapid changes. Under communist rule, Hungary had three tiers of government: the national government, county governments (equivalent to NUTS3 regions), and local governments (equivalent to NUTS4 regions). In the case of Budapest, the city functioned as a county and had 20 districts. Local governments were little more than state enterprises, managing housing, transport, and local services and relying mainly on central government for funding. County governments, analogous to local interior ministries under communism, controlled local governments, allocating resources to them, monitoring performance, and overseeing local policy formation and implementation. Between 1945 and 1990, the number of local governments was progressively reduced through a series of amalgamations, based on the idea of eliminating small units of government on efficiency grounds.

In 1990 for the first time in Hungarian history, municipalities gained self-government. The power of the formerly dominant counties decreased because they were held in deep distrust by local governments and their residents. Under communist rule, they had played an unpopular role in controlling municipalities and distributing public resources. The Act on Local Governance in 1990 delegated only peripheral competences and tasks to county assemblies. Therefore, it left them unable to play a role in integrating local governments and local government activities. Local governments also became responsible for financing their activities. At the same time, there was an explosion in the number of local governments, from 1,470 in 1990 to 3,158 and 3,175 in 2010, as communities joined together under communism reclaimed their administrative independence. In addition, the Act on Local Governance gave all local governments the same legal status, regardless of their place in the administrative hierarchy. As a result, county governments had the same status as small villages, and Budapest as a capital had the same legal status as districts in Budapest.

As a result of this process of administrative reform, the local government sector started to expand. The country had a large number of relatively small local governments with a limited financial base, which tended to be concerned with local management issues. The result was that a regional perspective disappeared from local government agendas. This created a democratic deficit at the local level, since regional problems were largely unaddressed by elected local governments. National government was left to manage relations among local governments, particularly regarding spatial issues.

From the middle of the 1990s, measures were introduced to reduce the democratic deficit and reduce malfunctions in territorial governance. However, a reallocation of powers among different levels of government appeared to be politically infeasible, and the national government had to find alternative methods to simplify the administrative system and re-instil a regional perspective into local government. Among other things, the Act on Local Government was amended to strengthen county assemblies. Later, a number of public administrative reforms were announced with the aim of limiting and coordinating the central administration's role in managing regional issues, but they had limited success.

The rescaling and shifting of the public power structure accelerated as Hungary approached accession to the European Union in May, 2004. A special act on regional development was passed, setting up special institutions at a new tier of government, consisting of groups of counties (generally three), analogous to NUTS2 regions. These were created in response to European Union demands for a multitier administrative structure within new EU member states and were also created with the aim of accessing European Structural Funds. Each of these macro-regions had a development council which was responsible for elaborating and implementing a plan for regional development throughout the region.

The need to create this meso-level administrative region and European regional policy together led to the formulation of a new governmental programme in 2002 under the leadership of the Hungarian Socialist Party and its parliamentary allies. This programme declared that regional governments with elected assemblies would replace county governments by 2006. This programme appeared to finally solve a decade-long dilemma: the counties would be replaced by elected regions which would, at the same time democratise meso-level governance, decentralise the strongly centralised Hungarian state and provide for the efficient management for the European Structural Funds. However, the reform programme was never implemented, even though the government set up an expert committee with the mission of preparing and implementing the reforms in order to demonstrate its commitment towards regionalism. Between 2002 and 2006, no single concrete measure was taken to introduce any elements of regional change. In 2006, the newly elected government, again, a coalition led by the Hungarian Socialist Party, confirmed the regional reforms, submitting a complex amendment package of acts almost as large as the national constitution itself, containing among other things a law on local governments and a law on national elections. However, the parliament did not accept it because the legislation had not been preceded by broader political, social consultation. The requirement that legislation be passed by a two-third majority vote was, of course, a big obstacle, because the opposition Fidesz and its allies, who had a conservative-nationalist stance, did not want to support any government reforms in what has become a permanent feature in Hungary's polarised political life.

During the above attempts to reform the territorial system of administration, new actors emerged in the public sphere in the form of decision-making and consultative appointed bodies, organised according to "corporative" models, as opposed to traditional elected and bureaucratic forms. In an effort to gain legitimacy, they have involved different institutions and social groups. The desire to create more democratic forms of governance could have been a sufficient explanation for the emergence of this jungle of actors, but in the eyes of many, the aim of the central government was to bypass and weaken county self-governments. An especially important step was the establishment of the so-called territorial development councils, based on the partnership principle found in European cohesion policy. These new territorial development councils have more decisive power and resources than the elected county assemblies. The institutionalised networks created from the top provide influence and informal power for the special elite who are appointed as members of these new types of organs and subvert the need for democratic decentralisation.

With hindsight, it appears that the comprehensive regional reform programme was too ambitious. Not only did it lack parliamentary support, but it also lacked professional support. The empirical analysis needed to support the reforms was also lacking. The most often-stated rationales for regional reforms were accession to the European Union and absorption of Structural Funds. However, these rationales no longer seemed convincing. The domestic management of the Structural Funds is strongly centralised, partly due to the pressure from Brussels, but also due to the centralising attitudes of the Hungarian government. This also suggests that the Hungarian national government places little trust in regional institutions and actors created by the government itself.

This does not mean that there is no need to strengthen the meso-level of public administration, but means that reform must be based on "just" our own needs and considerations. The reform of territorial public administration remains an issue among politicians and policymakers, and that is why there is a need to investigate the political conditions and processes between the top- and bottom-level power arenas and examine why the previous reform programmes failed. The question is whether top-down regional reform is realistic and to what extent it can expect local/regional support from below.

The paper deals with the content of the new Hungarian regions. It looks at how actors and their networks filled up these geographical voids which are still missing on the public administrative map of Hungary. The paper then attempts to draw lessons from the Hungarian experience to better understand aspects of "Europeanisation" and the role of culture and society in public administrative reforms.

Regionalism (Bottom-Up) vs. Regionalisation (Top-Down)

Regionalism might be defined as the construction of a regional identity from endogenous actors, that is, from within the region itself. Regionalisation might be defined as the construction of a region by exogenous forces, from above. In Hungary, the territorial reorganisation of public administration was not demand driven, but was initiated from the top-down by central government. Therefore, it makes sense to investigate differences between regional reforms initiated from the top and those initiated from the bottom due to local initiatives.

In the 1980s and 1990s, reforms were implemented at the meso-level of governance in many Western European countries. The strengthening or establishing of regional structures occurred, on one side, as consequences of the organic development of national public administrations. This arose from the need to improve technical performance, services, and infrastructure to better meet social needs and to cope with the effects of globalisation in general, putting more emphasis on the role of institutions at regional scale. On the other side the firm intention of the EU played a crucial role in this process. The Union favoured the creation of meso-level regions and the devolution of power to such regions. It was argued that such regions contributed not only to the efficiency of European cohesion policy but helped to advance the integration process by lessening or limiting the power of national governments.

The two main models of regional change implemented within the pre-2004 European Union were, therefore, bottom-up regionalisation and top-down regionalisation, depending on whether the regional power structures were created on the initiative of the local social, political, and economic actors or by the central government under the label of modernisation and administrative rationality (Keating 2004). In the so-called cohesion countries (Ireland, Greece, Portugal, Spain), as the primary beneficiaries of the Structural and Cohesion Funds, the most important motivation of region building was the externally imposed need to adapt to European regional policy regulations.

In many cases, bottom-up and top-down reforms undoubtedly supported each other. The emergence of the European multilevel governance model, cooperation among the European regions, and the formal institutionalisation of regional representation in the frame of the Committee of the Regions had a salutary effect on both types of regionalisation (Bache 1998; Bovaird et al. 2002).

When taking a closer look at reforms implemented in the EU member states and at the special development partnership institutions established in order to adapt to the cohesion policy, it appears that there were two approaches. Most countries preserved their traditional public administrative units, and the new regions were established over and not instead of them (like Greece, France), or regional development institutions were built outside the "normal" public administration hierarchy (Ireland, Portugal). The other important feature is that these reforms were gradually introduced and implemented after decades-long preparation (France, Italy) even if regional structures had historical traditions, antecedents, and cultural roots (Larsson et al. 1999).

Recent analyses on the consequences of changes in regional administrative structures suggest that reforms implemented or at least supported from below are more successful and less conflictive; new institutions are easier and more organically built; and new institutions are strongly socially embedded (Knodt 2002; Elcock 2003).

Regional public administrative reforms in Central and Eastern European transitional countries are particularly interesting (Surazska et al. 1997; Pickvance 1997; Emilewicz and Wolek 2002; Hughes et al. 2004; Pálné Kovács 2007). Where the reforms failed, they usually had not led to real decentralisation. However, in terms of success of or failure, there seem to be great differences between those reforms which were imposed by external agencies in the name of modernisation and those which were socially supported, "socialised", from below.

Therefore, it is obvious that the success of reforms depends to a large extent on multiple factors: the model chosen, the preparation before implementation, and the support or opposition of stakeholders, local society, and the political elite.

Theoretical Frames of the Research

The aim of this research, conducted at the University of Pécs and financed by the Hungarian National Research Fund (OTKA project No. 49453 in 2005–2008), was to investigate the internal cohesion of the NUTS2 regions, the extent to which they have developed the capacity to act as political units and the extent to which individual member governments and organisations had developed the ability to cooperate in forming and implementing policies. An additional aim was to determine the extent to which the regions had fostered regional identities. This chapter summarises the results of this research. They are based on two separate questionnaires administered in 2002 and 2007 to regional elites in the South Transdanubia region.

The definition of an action arena, following on the work of Elinor Ostrom, and the IAD (institutional analysis and development) model helped us to select the unit of our research and to assess interactions among actors by identifying the binding rules (position, boundaries, choice, aggregation, information, payoff, and scope) and to count the outcomes of the actors' actions (Ostrom 2005, 191).

We interpreted the region as an action arena. We assumed that particular actors and interactions determine a region, and there are common rules by which the region can be captured as "a part in the whole". Activities are determined by the context and where the efficiency of its actions depends on relationships among the actors of the arena, on driving forces, on knowledge of governance, and on internal cohesion.

Another pillar of our research was borrowed from the expanding scientific field dealing with networks which started in mathematics and technical and natural sciences, but which has reached many branches of the social sciences (sociology, psychology, economics) in the last decade.

Political science uses different approaches to analyse networks. A policy network approach was, for us, of special relevance. Policy network theory provides a lens through which to examine the interactions between groups or networks involved in policymaking and their influence on the policy process. Dowding (1995) and Marsh and Smith (2001) list four approaches to the study of policy networks: (1) the rational choice approach, (2) the personal interaction approach, (3) formal network analysis, and (4) the structural approach. In the governance model beyond hierarchical elements, horizontal network elements exist with different weights. As Kooiman (1993) noted, there are several definitions of governance. The British one was especially important for our research. This definition defines governance as a self-regulating network in an era when governmental decision extended beyond the walls of Westminster and Whitehall (Rhodes 1997, 49). This is based on the recognition that the British government, by creating various organisations outside the public sector (often referred to as quasi-autonomous non-governmental organisations, or QUANGOs), was able to bypass traditional public, state, and self-governmental institutions relocating more and more public tasks into this third sector. This method was useful not just for pluralising governance and decision-making and for introducing methods of New Public Management but also for creating a new governmental space, the region. Rhodes described this new type of governance as a self-regulating system of networks among institutions, where:

- There is mutual dependency among institutions involved.
- There are continuous interactions among the networks.
- These interactions are similar to games with defined rules and based on trust.
- The networks are independent from the state although, informally, it has a crucial governing impact on the networks (Rhodes 1997, 52).

A final approach to the study of networks, social capital theory, can also be regarded as relevant and is today found in numerous research projects analysing public policy decision-making. Little empirical research has been undertaken which lays the emphasis on civil society as a factor in region building, as opposed to those which analyse region building as a top-down or bottom-up process of institutionalisation. The value and the novelty of the latter approach are due to the fact that the very local scale is generally regarded as ideal for civil society participation, in contrast to the traditional approach focusing on public law and institutionalism. Putnam undertook a pioneering role (Putnam et al. 1993; Putnam 2000), directing attention to the role of civil society and social trust. Putnam distinguished three schools in researching institutional capacity and performance (Putnam et al. 1993, 9–10). The first one, based on the traditional formal *public law and institutionalist approach*, focuses on institutional design, declaring that the proper establishment of competences and the organisation of institutions is the most important condition for efficient and democratic governance. A number of works can be attached to this school, especially those by representatives of the *new institutional school*. Their research framework remains within the world of organisations and uses an organisational model to assess the efficiency of institutions. This approach has also emerged in the regional literature, emphasising the importance of development agencies, councils, etc. (Morgan 1999; Halkier et al. 1998; Danson et al 2000). The other school concentrates on the *socio-economic environment* of institutions. Dahl and Lipset represent this approach in modern political science. The third approach builds, according to Putnam, on the *sociocultural factors*, and his works follow this direction.

Putnam and his colleagues conducted empirical research in Italy two decades ago seeking to answer the question, "why do regional institutional systems produce different results and operate at different levels of efficiency in different parts of Italy?" In his work, he appears to have accepted the general opinion that "there is an old wine in the new bottle" – that is, that institutions alone are not able to change social solidarity (Putnam et al. 1993, 18). Putnam's research showed that even 20 years of regional reforms were not enough to diminish the 100-year-old social and cultural differences between the South and North. In an active social and economic environment, both the state and the market functioned better and more efficiently. The contract binding civil society was not legal, but moral, based on 100-year-old traditions. The civil environment in the South was stable but based on mutual distrust and vertical dependency. This strongly limited the performance of the regions established by the Italian national government in the 1970s. Putnam's research pointed out, however, that new institutions set off a learning process and that the formal change produced informal changes which become self-sustaining (Putnam et al. 1993, 184).

Social trust is the result or "by-product" of two forms of social capital: social barter and civil participatory networks (Paraskevopoulos et al. 2006). Together, these determine the efficiency of public/civic actions. Therefore, the kinds of social capital and trust which motivate the region-building process make a difference in the outcome of region building which can be driven or limited by networks of institutional actors.

Three kinds of human capital, material, knowledge, and social capital (Coleman 1990) are needed in region building. Each can influence the development of the other. Social network analysis (SNA) serves as a methodological mean to measure social capital and the strength and direction of cooperation among actors.

In Hungarian sociological and political science literature, the region is often mentioned but rarely investigated. There are works dealing with regional institutional settings (Kaiser et al. 2007), and there are some experiments to measure the extent of regional identity (Bugovics 2004), but our research investigating relations among regional actors can be regarded as a pioneer. This means that there are few studies to compare it with, both in terms of methodology and results.

One should also note that there is relatively little research on the formation of regional identity and politics even in Western regionalised countries. There are some excellent exceptions dealing with cultural and identity factors (Stiens 1987; Dirven et al. 1993; Deffner et al. 2003; Bukowski et al. 2003) or with the networking of regions (Kohler-Koch 1998), but the recognition within political anthropology that political spaces are constructed by different levels and scales of networks (Abélés 2007, 134) has not yet inspired that part of the scientific community studying regionalism.

The Process of Region Building in South Transdanubia

This research attempted to follow the process of region building in one of the NUTS2 regions of Hungary: South Transdanubia. We selected this region because the first steps for regional cooperation in Hungary were taken here, and we could base our work on earlier research conducted by the Centre for Regional Studies of the Hungarian Academy of Sciences, located in this region (Hajdú and Pálné Kovács 2003; Hajdú 2006).

The region, similarly to other Hungarian NUTS2 regions, has no traditions of institutional or political cooperation on this scale. The Hungarian government established the so-called territorial statistical regions in 1971. At that time, the South Transdanubian region consisted of four counties which were practically forced by the government to cooperate. Boundaries were drawn on the map, but no real cooperation evolved. The creation of a common development plan failed, and the redistribution of development resources remained within the competences of the counties.

The idea of regionalisation was struck from the agenda in the 1980s. Instead, the urban agglomeration became the geographical framework for regional integration motivated from the top.

Following this change, cooperation at the regional scale was even more neglected. However, in South Transdanubia the advantages of regional cooperation were recognised relatively early, possibly due to the role of the Centre of Regional Studies in promoting such cooperation. Within Hungary, regional-scale initiatives were born here, and this region is often referred to as the birthplace of Hungarian regionalisation and even regionalism. South Transdanubia was the first region to organise itself in the first half of the 1990s, when regionalisation was less popular. By the start of the 1990s, four county governments, cities, economic chambers, the research institute, and universities had established a regional development foundation. This organisation created the first regional development concept in the country, but the subsequent silent decline of this civil organisation highlighted the fact that lacking public power and resources, it was unable "to stimulate" regional cohesion. Subsequently, when regional development councils were established by the force of law, and as public actors took their seats in these institutions, the original informal organisation disappeared.

The amendment of the act on regional development in 1999 mandated the establishment of regional development councils in each NUTS2 region. The boundaries of the official NUTS2 region in South Transdanubia cover(ed) only three counties, although as we have mentioned, the bottom-up region was larger, which demonstrates that these new top-down imposed regions had no traditionally accepted boundaries. The regional development council is a so-called partnership organisation with members delegated by municipalities, county assemblies, and ministries. Its competences, resources, institutions, and processes are regulated by the law. In addition, as part of the reforms, regional development councils were created not just at the level of the NUTS2 region, but, in accordance with the act on regional development, at the micro-regional scale (174 NUTS4 units) and the county scale (19 NUTS3 units). The actors in this very fragmented system compete with each other, although they are partners at the same time, since they have seats at each council level. In approaching the EU accession, the macro-region (seven NUTS2 units) was strengthened vis-a-vis the lower units.

Within South Transdanubia, although the regional development council gradually became the dominant institutional setting for regional development activities, another bottom-up attempt was made to create an institution for regional cooperation. This was a pilot regional programme in South Transdanubia, initiated by the charismatic mayor of the city of Pécs. The programme seemed to accord with the central government's intention to demonstrate its commitment towards regionalism at a time when political regionalisation had already been put on the agenda. Three county self-governments and two county seat cities in the region (the third one did not initially join because its mayor belonged to the opposition Alliance of Young Democrats, Fidesz) decided to find a self-government association in 2006, aiming to foster cooperation in fields which were not among the competences of the regional development council, like public services. The pilot programme did not promise to be successful, since the national government did not provide any special legal framework to foster such cooperation. Therefore, it had to operate within the existing regulation, and extra resources for common development projects were limited. Thus the relatively weak counties and relatively poor cities, facing financial and administrative constraints, were not able to build a strong region. The regional self-governmental association provided little added value or positive evidence that could have contributed to the formation and evolution of legislation on regional governance. Later, after local elections, the new regional-level leadership belonged to the opposition Fidesz party. Therefore, the national government lost its enthusiasm to support this bottom-up regionalisation. Participants in the association were also unwilling to share their power and resources in real cooperation. Instead, it appeared that the hope of obtaining extra money motivated them to stay in the association established by their predecessors. As the financial situation worsened, local actors lost their willingness to cooperate. This is a bad omen for the chances of "bottom-up" regionalisation.

In sum, the selected region seemed to be ideal for our research for three reasons. It is a region where regional cooperation is not motivated exclusively by the government but is rooted in the efforts of local actors as well. Second, it is a lagging region and is therefore strongly interested in accessing European Union development funds and therefore in adapting to the requirements of the Structural Funds. Finally, at the same time, a strong regional identity is missing even in this region. Moreover, there are sharp conflicts between counties, cities, and micro-regions constituting the region. It is an open question whether regional-scale politics and regional policy networks are able to cope with these conflicts and substitute for the missing regional identity.

Regional Networks at the Beginning of the Twenty-First Century

Between 2001 and 2004, we had the opportunity to participate in a European Union's 5th Framework Programme of research (ADAPT 2001–2003; Pálné Kovács et al. 2004). The target of this research programme was to investigate how eligible cohesion regions were adapting to European regional policy. During this research a so-called social network analysis (SNA) was conducted to assess the process of regional institutionalisation. Thirty structured interviews were undertaken with actors in the region, selected from among members of the regional development councils (mayors, presidents of the county assemblies, delegates of ministries) and stakeholders involved some way in regional development decision-making, such as the representatives from business, universities, and other non-governmental groups and institutions. We investigated the positions of different institutions within the network by several methods and found that the regional development council and the regional development agency possessed the greatest level of centrality, justifying the assumption that actors from government, or appointed by central government, are the only ones with an opportunity to weave together the region. They are most endowed with competences and public resources, while civil and business actors are relatively peripheral (Table 6.1).

Table 6.1 Freeman's degree-based centrality measures

No	Actor	Valued graph (B)	Scaled graph (B_4)	Binary graph (B_2)
12	South Transdanubian Regional Development Council	70.5	78.2	93.1
22	South Transdanubian Regional Development Agency	65.3	69.0	89.7
14	Somogy County Development Council	54.7	57.5	58.6
8	Assembly of Baranya County (elected)	55.9	56.3	58.6
20	University of Pécs	55.0	56.3	58.6
9	Assembly of Somogy County (elected)	51.2	51.7	55.2
13	Baranya County Development Council	48.2	51.7	48.3
15	Tolna County Development Council	47.3	49.4	44.8
17	Chamber of Commerce and Industry of Baranya County	48.9	48.3	55.2
23	Local Government of the Town of Pécs with county rank (elected)	45.7	48.3	41.4
25	Local Government of the Town of Szekszárd with county rank (elected)	43.8	47.1	44.8
24	Local Government of the Town of Kaposvár with county rank (elected)	42.3	43.7	41.4
26	Micro-Regional Associations of Baranya County	41.2	41.4	41.4
16	South Transdanubian Regional Tourism Committee	40.6	41.4	44.8
3	National Development Centre	42.7	40.2	34.5
10	Regional Centre of Labour Force Training and Education	40.1	40.2	41.4
7	National Development Council	38.7	40.2	44.8
1	MARD regional and rural development divisions	37.5	37.9	41.4

(continued)

Table 6.1 (continued)

No	Actor	Valued graph (B)	Scaled graph (B_4)	Binary graph (B_2)
5	Prime Minister's Office, State Secretariat of Regional Policy	38.9	35.6	34.5
18	Chamber of Commerce and Industry of Somogy County	32.4	31.0	34.5
11	Hungarian Development Bank – regional unit	32.4	31.0	27.6
19	Chamber of Commerce and Industry of Tolna County	29.7	31.0	31.0
21	European Information and Development Ltd.	34.1	29.9	20.7
27	Micro-Regional Associations of Somogy County	30.0	26.4	20.7
30	Resource Centre Foundation	27.9	26.4	24.1
29	Industrial Park Pécs	26.9	25.3	20.7
2	MARD Sapard programme management	26.0	21.8	3.4
6	VÁTI – Territorial information system	23.4	21.8	17.2
28	Micro-Regional Associations of Tolna County	23.8	20.7	13.8
4	Ministry of Phare Affairs, Secretariat	24.4	18.4	13.8
	Mean	40.6	40.6	40.0
	Std. deviation	12.0	14.3	19.9

Source: ADAPT programme 2001–2003

Table 6.2 Connection densities between levels

	National	Regional	County	Micro-reg.
National	46.0	29.4	31.7	19.6
Regional	29.4	57.8	63.0	58.3
County	31.7	63.0	50.9	38.0
Micro-reg.	19.6	58.3	38.0	26.2

Source: ADAPT programme 2001–2003

The results showed that those institutions which were at the centre of the network and had the strongest integrating powers were those institutions based on the partnership principle (set up by the Act on regional development in 1996) and elected local and county governments.

Strikingly, as is shown in Table 6.2, the network was densest at the county scale, involving the largest number of actors investigated, although we found strong ties at the regional scale, due to regional development institutions (the council and the development agency) being able to contact organisations and actors in a wider environment. On the other hand, the larger cities were relatively isolated. Their desire to attract funding and development outweighed any desire to cooperate with smaller or less powerful jurisdictions in formulating regional development projects.

We used three models in approaching our research and interpreting the results. These were taken from the "development interest" constellation. The first was a public-dominated *developmental state*, where the public power designates the actors participating in the decision-making. Only constituted public institutions have the chance to influence development policy, and there is one single aim of growth (Korea, Japan). The second was the *neo-corporative state*, where representatives of leading employers and their employees participate in decision-making, seeking solidarity and equity (Sweden, Holland, Ireland). The last was the *pluralistic model* (USA), where there is a distance between the public and private sphere, and the latter pushes the state from the side of the playing field (Wilson 2001).

The model adapted by the Hungarian central government initially favoured strong state/dominance, representing a special type within the first model. The power structure has an hourglass form. The meso-level is the pinch point; the elected county governments are one bulge and state organisations the other. The elected county governments are very weak, and deconcentrated state organisations both depend on and reflect the central government. To legitimise meso-level decision-making, the central government introduced some so-called neo-corporative or partnership types of bodies related to employment, youth, and development policies, which seemed to be more vigorous in trying to undertake redistributive and planning activity than the county governments. However, both the public and neo-corporative spheres remained weakly embedded in civil society. The relatively weak public part and the stronger "partnership/corporative" part of the system of meso-level governance are strongly interrelated and complementary. The almost absent representative mechanisms were replaced by partnership institutions. However, the evolution of these could not compensate for the democratic deficit at the meso-level. The result was that, as a whole, they were rather dysfunctional. Our research suggests that:

- The fragmentation of the meso-governance increased. Alongside elected county governments, a number of decentralised offices and new partnership organisations emerged.
- The frames of interest representation were diffuse not just organisationally but geographically too. Beside the county-scale institutions, a series of macro- and micro-regional institutions emerged without establishing stronger contacts among each other and therefore unable to contribute to regional cohesion.
- Partnership organisations broadened the circle of actors participating in meso-level governance, but the actors, aspects, and interests channelled into the decision-making remained exclusively selected. Links with the broader public did not grow. Civil society was not strengthened, and even the business sector, mainly through a chamber of commerce, had also just a limited voice.
- Experts and technocrats took on an overwhelming role in the preparation of plans and policies, which paradoxically re-narrowed the partnership dimension.
- The functioning of partnership bodies was characterised by intensive exclusivity and an almost complete lack of publicity. The members had no willingness to involve a broader circle of stakeholders. They were satisfied with the limited transparency.
- During the decade after the creation of meso-levels, a new meso-level elite emerged through delegation. This elite possessed a set of cumulated personal positions, giving it a strong influence on territorial level politics. However, their activities in several partnership bodies were and are not controllable by local society or even by the organs delegating them (municipalities, county governments).

This type of networking reproduced all the disadvantages which were already recognised and reported on in Western democracies. In Hungary, this neo-corporative jungle portended problems for two reasons: first, because here there was no strong "traditional" elected self-government behind the partnership organisations and informal networks and, second, the traditionally weak civil society was not able to enforce transparency and thus exercise oversight over the networks and the elite.

These phenomena underline the necessity of change in the meso-level governance: the legitimacy of this subnational sphere has to be strengthened. This political aim was supported by development policy considerations driven by EU Structural Funds, but coupled with aspects of public administration – that is, stable, strong meso-regions were needed to fulfil the slogan of a "Europe of regions".

As mentioned in the introduction, the reforms were postponed or cancelled after the party (Fidesz) taking the power in April 2010 failed to support any regionalisation. This failure can also possibly be explained by processes occurring in the regions which we have analysed during the last few years. The question was could the official, top-down regionalisation agenda change the landscape of regional networks?

Regional Networks at the End of the Decade

The OTKA project mentioned above gave us the opportunity to repeat our empirical research in 2007–2008 to compare the then current networking process with that which had been emerging at the beginning of the 2000s. The sample, with 200 interviewees from the regional elite or leadership, was representative of the institutional setting of the region. This sample was bigger than the earlier one because we involved actors from outside the narrow institutional framework of regional development policy, assuming that elected and party politicians, the leaders of the state administration, civil organisations, and the media were also important stakeholders regarding the governmental programme of political regionalisation. The survey consisted of 47 questions, which are presented in Palne Kovacs (2009).

First, it is useful to examine Table 6.3, which summarises the overall frequency of contacts between institutions. The table shows that local-scale contacts were dominant and the national centres like the parliament and national political parties as institutions were distant from regional networks. The relatively isolated situation of the regional institutions can be explained by the narrow scope of competences exercised at this level. The low presence of political parties in the regional system of contacts is misleading: this reflects national politicians and political institutions, such as the parliament, government, and national political parties, and does not include local politicians who serve in various regional-level positions (members in the development council, local governments, etc.).

If we look at the strengths of contacts between the four institutional levels shown in Table 6.2 (national, regional, county, micro-region), it is apparent that relationships among units of self-government, politicians, the media, and development organisations are strongest. Those within the non-governmental civil sector have the

Table 6.3 The frequency of institutional contacts in decreasing order

	Standard
Municipalities	3.60
Local media	3.41
County assembly	3.34
County-scale media	3.34
Local-scale business	3.31
Universities, colleges	3.23
Local non-profit organisations	3.13
County-scale business	3.12
Ministry, national administrative organ	3.03
Professional chambers	2.95
County non-profit organisations	2.91
Regional development council and agency	2.87
Research institutes	2.86
Regional-scale business	2.84
County development council and agency	2.70
Central government	2.67
National media	2.54
Nationwide business	2.43
National non-profit organisation	2.34
Churches	2.29
Political parties	2.23
Trade unions	2.21
Parliament	1.97

Source: OTKA 2005–2008

On a scale of 1–4, all the time *4*; sometimes *3*; rarely *2*; never *1*

weakest weight, although they and the universities considered relations with development organisations to be very important. It is interesting that non-governmental organisations did not regard their contacts with regional decision-makers as important, which can be explained by their view that although they were formally involved in decision-making, in practice they were largely ignored or excluded. This is shown in Table 6.4.

We also tried to measure the geographical frames of relationships (Fig. 6.1). Our results showed that a number of organisations have stronger ties outside the region (e.g. universities, the state administration, and media), while contacts of the development organisations and self-governments were mostly within the region (and, as already mentioned, at a smaller scale, within the county or micro-region).

Investigation of the networks led us to conclude that some changes occurred in terms of cooperation. The integrative role of the regional development organisations remains strong. However, local governments (cities and micro-regional associations) are more open and active than before and have gained increasing knowledge in how to absorb EU Structural Funds; they no longer need the assistance of development agencies in the way that they did a decade earlier. At the same time,

Table 6.4 The strength of contacts among institutional sectors

Importance + frequency	Civil org.	University	Business	Media	Political party	Development inst.	Public adm.	Local gov.	Sum
Civil organisation	2.63	2.31	2.32	2.59	2.92	2.80	2.16	*3.02*	2.56
University	*2.96*	*3.95*	2.91	2.88	*3.04*	*3.07*	2.91	*3.25*	3.13
Business	2.75	2.88	*2.95*	*3.08*	*3.29*	*3.16*	2.84	*3.28*	2.97
Media	*3.04*	*3.03*	*2.96*	*3.31*	*3.46*	*3.32*	*3.05*	*3.44*	3.16
Political party	1.51	1.68	2.53	2.71	*3.38*	2.64	1.48	*3.03*	2.15
Development institutions	2.74	2.46	*3.03*	*3.06*	*3.34*	*3.50*	2.37	*3.24*	2.87
State administration	2.36	2.84	2.31	2.08	*3.32*	*2.95*	2.64	*3.05*	2.68
Local/county government	*3.42*	*3.07*	*3.20*	*3.42*	*3.88*	*3.82*	*3.37*	*3.85*	*3.47*
Sum	2.67	2.78	2.77	2.89	*3.33*	3.16	2.60	*3.27*	2.87

Source: OTKA 2005–2008

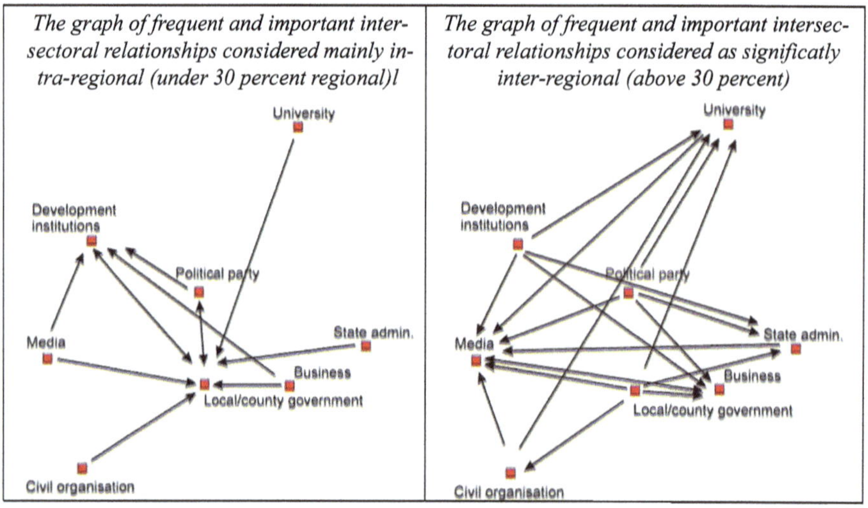

Fig. 6.1 Contacts within and beyond the region (Source: OTKA 2005–2008)

universities remain on the periphery of networks, in contrast to the official ambition of the regional development council of creating a knowledge-based economy.

Returning to the characteristics of the policy networks outlined by Rhodes (2000), it appears that, within Hungary, even as the central government wanted to change the system of regional administration by creating meso-level regions, changes in the last 6–8 years seem to have been necessary in any case. However, the policy networks formed as a result of regionalisation policies provide limited participation, where the public sector is still the most influential and civil and business actors are on the periphery. The regional development council sits in the centre, integrating the network at the regional scale. Local governments and their micro-regional associations play a stronger role in integrating local actors, but on a smaller, subregional scale. It is interesting that although mayors, county assembly presidents, and the technocrats working for the regional development agency are members of a variety of networks, elected politicians, like members of the European and the National Parliament and national-level party politicians, are not embedded. This phenomenon shows the centralised nature of Hungarian political life where many national-level politicians think that it is not worthwhile "to act locally" or regionally.

The Opinion of the Regional Elite About the Reform

The survey devoted a separate block to the question of the reforms which were proposed at the time. The failures of those territorial administrative reforms which have been announced reflect the fact that implementation is the "missing link" (Dunn et al. 2006). There is a lack of any real social or political support for the reforms. It is theoretically supposed that a regional identity is an important precondition for their success, but the lack of a regional identity does not always explain the failures. Hungary has no tradition of regional-scale governance. Its territory and ethnic composition are relatively homogenous. Therefore, regionalisation is often viewed as a top-down process, driven by economic considerations, and even a wish to "modernise" public administration hierarchies. The question is whether this process has been supported at the bottom. The attitudes and relation of the local, regional elite towards regionalisation are key factors during both the preparation and implementation phases. Nunberg (2003) considered the case of Hungary to be especially interesting. After 1989, the country was ranked first in the region regarding the implementation of political and economic reforms, and the national government formulated very ambitious reforms in the field of public administration. These ambitions, however, were not shaped into a firm plan. Instead, they were implemented in incremental small steps, under the prodding of the EU. The so-called reform programmes were more like spontaneous reflexes shaped by the strong traditions of centralisation rather detailing a long-term process of implementation. The few partial successes are due mostly to EU requirements or, more recently, due to fiscal pressures resulting from the current (post-2009) financial crisis rather than being due to a widespread recognition of the need for change (Nunberg 2003).

The government set out the reform proposals, and created some regional frameworks, but has done little to translate proposals into action. It was therefore argued by the researchers that the general public, and even the political elite, was not convinced of the prospects of success of the regionalisation programme. Consequently, the research aimed to identify what kinds of obstructions hindered the implementation of the reforms.

In one question of the block, we asked to what extent the respondents agreed with statements concerning regional reforms. As Table 6.5 shows, the elite are aware of the top-down nature of the reform and link the reform strongly to the demands of the EU. Most of the respondents thought that the aim of the reform was to gain access to the Structural Funds. They were also aware that regional traditions and cohesion were missing, and they rejected the idea of adopting a regional identity as opposed to a county one. Opinions differed, however, about the content and the consequences of the reforms. There was no agreement concerning the statement that the reforms would result in greater centralisation and increase their distance from decision-making levels. Nor was there agreement that they would contribute to the improvement of public services and decrease costs. Opinions were especially diverse regarding the boundaries and seats of the regions. While most respondents did not consider the region to be too big, they did not accept the idea that the seat of the region should be the city of Pécs (the largest city in the region where, of course, several respondents live). This disagreement over the location of the regional seat, or capital, contrasts with the overall view regarding the boundaries of the regions

Table 6.5 Statements on the content of the regional reform

	Standard	Deviation
Contacts with the EU are managed easier at the regional level	3.16	0.92
The regions have no traditions	3.15	1.06
The reforms cause too much difficulties and uncertainties	3.13	0.86
The region has no cohesion and solidarity	3.08	0.93
Decision-making will be too far removed from citizens	2.89	1.11
Reform causes centralisation	2.76	1.08
The political elite know very little about the region	2.69	1.04
Regionalisation can cause macro-political conflicts	2.68	1.04
The region is more efficient in public administration	2.66	1.08
Public services could be managed more efficiently in the region	2.64	1.00
Regionalisation means cheaper public administration	2.64	1.14
Development problems can be solved only on the regional geographical scale	2.48	1.00
With the emergence of the regions, the problems of peripheral areas can be more easily resolved	2.45	1.06
The boundaries of the regions are questionable	2.35	1.19
Regional reform gives a chance to replace the current elite	2.32	1.10
The location of the seat of the region is questionable	2.27	1.27
Smaller settlements will better find their role and opportunities to represent their interest in the region	1.97	1.00
Due to its bigger scale, the region is a crucial counterweight to the centre	1.79	0.86
The region is too large	1.68	0.97
People identify themselves more with the region than with the county	1.47	0.72

Source: OTKA 2005–2008
1 strongly disagree at all, *2* somewhat disagree, *3* somewhat agree, *4* strongly agree

themselves. The designation of the number and boundaries of the NUTS2 region was based on the existing county boundaries, leaving their long-standing boundaries unchanged. However, there was no consensus on the question of the seat of the regions. Within each NUTS region, several cities – the existing county seats – were competing to be the regional seats, and the government was not brave enough to make the decision. Cities, especially the largest ones, were not enthusiastic about regionalisation, as they were afraid to lose their privileged status within counties. If we assume that a region is determined mostly by networks of its cities, then this rivalry seems to be a crucial obstacle to reaching consensus and developing support from the bottom.

It is likely that different opinions can be explained by the party affiliation of respondents. These reflect the views of the two main parties, one, Fidesz, conservative, and opposed to the regionalisation reform, and, the other, the Hungarian Socialist Party, left wing, and in ostensibly in favour of it. However, to a large extent, both parties were reluctant to undertake regional reforms, and the socialists undertook them in order to meet EU demands for the creation of NUTS2 regions. Fidesz and its socialists argued that regionalisation would limit the powers of the central government, and the Fidesz ant its allies argued that regionalisation would cause too many difficulties and uncertainties by creating a new layer of administration. On the other hand, there was full agreement with the statement in the questionnaire that regionalisation is disadvantageous for the small villages and also that regionalisation is advantageous from the aspect of economic competitiveness.

A separate block of questions dealt with the management of Structural Funds, since this was one of the most important motivations or driving forces behind the creation of new regions. Responses to questions concerning the Structural Funds, shown in Table 6.6, indicate the disappointment which emerged after Hungary's accession to the EU. There was not just disappointment in EU institutions, but in the Hungarian management. This was due to the centralised and bureaucratic style of Hungarian management, far beyond the required level set by the EU. The elite interviewed in our survey were also critical of various lobbying groups which seem to be unavoidable in this field.

Table 6.6 Structural funds

	Standard deviation	
The management of SF is bureaucratic	3.40	0.76
The management of SF provides an opportunity to lobby	3.21	0.85
The management of SF is centralised, leaving no space for the regions	3.08	0.96
The management of SF is not transparent	3.01	0.95
The management of SF is as the EU requires	2.60	1.05
The management of SF is correct, efficient, close enough to the stakeholders	1.97	0.82

Source: OTKA 2005–2008

1 strongly disagree, *2* somewhat disagree, *3* somewhat agree, *4* strongly agree

We asked the sample whether they supported the establishment of regional self-governments in general. Interestingly, the majority supported regional reform (68 %), but this support is not linked to satisfaction with the reform process and the performance of the government. This means that the elite involved in regional-scale policy formation and implementation, such as regional development councils or the recipients of European money, accept the popular concept of regionalisation. On the other hand, many of the respondents did not support the government's reform proposals and particularly disagreed with the way the government tried to implement them.

In sum we can conclude that the support for the reform is relatively high. The elite feel the necessity of change at the meso-level, which has become a jungle of different levels and institutions, and are therefore too weak to counterbalance the overweight of the centre. Returning to the model of Ostrom, the region is not yet an identifiable action arena in Hungary since, based on its competences, regulated relationships, and networks of actors, it cannot yet be distinguished as a free-standing administrative unit. Its networks are characteristically institutionalised from the top, and mainly involve only public-sector actors, and neither reflect nor contribute to the existence of a coherent region. This is not surprising, since on the basis of the uncertainties of the reform and the paradoxes in governmental policy, local actors are unlikely to build regions from the bottom-up on their own.

Conclusion: Paradoxes of Regionalisation

Our assumption was that 10 years after regionalisation became an official government policy, the region would have become the main policy arena, where political decisions were taken by legitimised regional actors. The research aimed at measuring the density of networks in order to determine the strength of the reforms and the future of regional management. Our research could not cover every aspect of regionalisation, but the results are sufficient to draw conclusions about the success of local government reform and formulate an agenda for further research.

In Hungary, regionalisation is not occurring, not in the context of comprehensive reforms based on broad consensus, but instead is infiltrating the public arena through the backdoor. It is true that the reform faced many political and social obstacles, but reform by stealth involves a number of paradoxes, and it appears that the results have not always matched expectations.

The first paradox is an economic one. As we emphasised, regionalisation first of all was aimed at modernising the administrative system and enhancing economic competitiveness. When investigating the economic structure of the country and the regions, we do not find real economic cohesion or clusters. The spatial economic processes follow the location and expansion of foreign investments. The economic space remains fragmented, structured alongside urban agglomerations, rail and river axes, and highways. The gap between the regions is deepening. Only regions lagging behind show some homogeneity regarding their low GDP, high unemployment, and other negative indicators. It is possible to argue that regionalisation could not exploit advantages of agglomerative effects and economic development policy did not use regionalisation as a framework or lens through which to direct intervention and economic support for particular sectors or clusters. It appears that economic development was not and is not the focus of development policy. For this reason, economic actors – local chambers of commerce – were not able to influence regional development decisions because they were on the periphery of the networks. Public actors, like mayors, are interested mostly in the improvement of human and technical infrastructure. Therefore, there were no actors to represent economic issues. Thus, regional policies can be explained by the networks of decision-makers involved in regional policy formation.

The *second paradox is the regional development policy paradox.* European Union membership caused more losses than benefits for the regions. As the results of the survey demonstrated, expectations concerning accession to the European Union were closely based on the expectation that there would be some kind of regional decentralisation. This was due to the belief that regions would manage European Union Structural Funds. Consequently, it was a real shock when the regional actors were largely excluded from the domestic management of the Structural Funds. At the same time, the scope of action of domestic regional policy institutions, like development councils, was narrowed because domestic resources were used to complement project budgets financed by Structural Funds. So, the whole institutional setting, established in 1996, to prepare for EU accession, is essentially hollow. European cohesion policy created a one-off chance to modernise the country's administrative structure, but it could not result in the regionalisation of decision-making, and Hungarian regions were not able to become fully functional mid-level governments.

The *third paradox can be found in the realm of public administration.* The point of regionalisation should be decentralisation and consequently the more efficient representation of territorial interests than before. If we look at changes within the public administrative sector, political regions were not created. Instead, the emerging regions, without elected governments, and based on appointed bodies, represent the interests of the centre. This reform consequently cannot be regarded as a first step towards regional decentralisation. Instead, it is nothing else than the expansion of the central state at the regional level.

The last paradox is that of bottom-up regionalism. This refers to the process occurring in the region investigated. Although the regional reform was initiated from the top, its success depended to a large extent on support from below. There is no doubt that some kind of regional networking emerged, but the key actors of the networks were public quasi-governmental bodies and therefore dependent on the central government. In the real local institutions, actors have no dominant role. They just assist in the centralised distribution of the Structural Funds. It is no wonder, therefore, that local politicians and civil organisations identified themselves at a smaller geographical scale. They accepted the existence of a meso-level region, but considered it irrelevant as a framework for actions.

True regional decentralisation, the building of a political region, would require significantly more conscious and complex organisational activity. Local actors were able to accommodate and adapt to changing conditions and frameworks, but were not able to expand their competences and create denser networks. Our survey suggests that regionalisation in Hungary has failed due to the lack of complex preparation and consensus building. Further research could explain why this happened and examine whether the region building is only a question of strong will. We agree with the opinion that real power networks sometimes disfigure the official public boundaries (Abélés 2007, 134). The question is where is the real power, and who is able to shift it downwards?

References

Abélés M (2007) Az állam antropológiája. Századvég, Budapest. (Original edition: Anthropologie de l'État. 1990, Paris: Armand Colin)

ADAPT (2001–2003) EU enlargement and multilevel governance. 5'th Framework HPSE-CT-2001-00097. Panteion University-LSE, Leadpartners

Bache I (1998) The politics of European Union regional policy. Sheffield Academic Press, Sheffield

Bovaird T, Löffler E, Parrado-Díez S (eds) (2002) Developing local governance networks in Europe. Nomos, Baden-Baden

Bugovics Z (2004) A területi identitás és a társadalmi siker (Spatial identity and social success). In: Szakál G, Gergely A (eds) Társadalmi tőke, karrieresélyek, viselkedésminták. MTA Politikai Tudományok Intézete, Budapest

Bukowski J, Piattoni S, Smyrl M (eds) (2003) Between Europeanization and local societies. Rowman & Littlefield Publishers, Lanham

Coleman J (1990) Foundations of social theory. Harvard University Press, Cambridge

Danson M, Halkier H, Cameron G (eds) (2000) Governance, institutional change and regional development. Ashgate, Aldershot

Deffner A, Konstadakopulos D, Psycharis Y (eds) (2003) Culture and regional economic development in Europe: cultural, political and social perspectives. University of Thessaly Press, Volos

Dirven E, Groenewegen J, van Hoof S (eds) (1993) Stuck in the region? Changing scales for regional identity. Nederlandse Geografische Studies, Utrecht

Dowding K (1995) Model or metaphor? A critical review of the policy network approach. Polit Stud XLIII:136–158

Dunn W, Staronova K, Pushkarev S (eds) (2006) Implementation. The missing link in public administration reform in Central and Eastern Europe. NISPAcee, Bratislava

Elcock H et al (2003) Creating new regional regimes? In: Deffner A (ed) Culture and regional economic development in Europe: cultural, political and social perspectives. University of Thessaly Press, Volos, pp 87–107

Emilewicz J, Wolek A (2002) Reformers and politicians. SEAP-ELISPA, Białystok/Warszawa

Guilani E (2006) The rescaling of governance in Europe: new spatial and institutional rationales (Guest editorial). Eur Plan Stud 14:881–905

Hajdú Z (ed) (2006) Dél-Dunántúl (South Transdanubia). MTA RKK -Dialóg Campus, Pécs/ Budapest

Hajdú Z, Pálné Kovács I (eds) (2003) Portrait of South Transdanubia. Centre for Regional Studies HAS, Pécs

Halkier H, Danson M, Damborg C (eds) (1998) Regional development agencies in Europe. Jessica Kingsley, London

Hughes J, Sasse G, Gordon C (2004) Europeanization and regionalization in the EU's Enlargement to Central and Eastern Europe. Palgrave Macmillan, Basingstoke

Kaiser T, Ágh A, Kis-Varga J (eds) (2007) A régiók Magyarországa I. (Hungary of the regions). MTA-SZKI, Budapest
Keating M (ed) (2004) Regions and regionalism in Europe. Edward Elgar, Cheltenham
Knodt M (2002) Regions in multilevel governance arrangements: leadership versus partnership. In: Grote JR, Gbikpi B (eds) Participatory governance. Political and societal implications. Leske und Budrich, Opladen, pp 177–194
Kohler-Koch B (1998) Interaktive Politik in Europa. Regionen im Netzwerk der Integration. Leske und Budrich, Opladen
Kooiman J (ed) (1993) Modern governance: new government-society interactions. Sage, London
Larsson T, Nomden K, Petiteville F (eds) (1999) The intermediate level of government in European states. Complexity versus democracy? EIPA, Maastricht
Marsh D, Smith MJ (2001) There is more than one way to do political science: On different ways to study policy networks. Polit Stud 49:528–541
Morgan K (ed) (1999) Regional development agencies (Theme issue). Environ Plan C Govern Policy 17(6):669–719
Nunberg B (2003) The politics of administrative reform in post-socialist Hungary. In: Schneider BC, Heredia B (eds) Reinventing Leviathan: the politics of administrative reform in developing countries. Lynne Rienner Publishers, Boulder, pp 59–89
Ostrom E (2005) Understanding institutional diversity. Princeton University Press, Princeton
OTKA (Hungarian Scientific Research Fund) project N. 49453 (2005–2008) The region: a new scene of politics. Project leader: Ilona Pálné Kovács
Pálné Kovács I (ed) (2007) Regional decentralization in Central and Eastern Europe. University of Pécs, Pécs
Pálné Kovács I (ed) (2009) A politika új színtere a régió (The region: a new scene of politics). Századvég, Budapest
Pálné Kovács I, Paraskevopoulos CJ, Horváth GY (2004) Institutional 'Legacies' and the shaping of regional governance in Hungary. Reg Federal Stud 14:430–461
Paraskevopoulos CH, Getimis P, Rees N (eds) (2006) Adapting to EU multi-level governance. Ashgate, Aldershot
Pickvance CG (1997) Decentralization and democracy in Eastern Europe: a sceptical approach. Environ Plan C Gov Policy 15:129–142
Putnam RD (2000) Bowling alone: the collapse and revival of American community. Simon and Schuster, New York
Putnam RD, Leonardi R, Nanetti R (1993) Making democracy work. Civic traditions in modern Italy. Princeton University Press, Princeton
Rhodes RA (ed) (2000) Transforming British government. Changing institutions, vol 1. Macmillan, Houndmills
Rhodes RAW (1997) Understanding governance: policy networks, governance, reflexivity and accountability. Open University Press, Buckingham
Stiens G (1987) Regionalbewustsein und Regionalentwicklung. Informationen zur Raumentwicklung 7–8:377–502
Surazska W, Buček J, Malíková Ľ, Daněk P (1997) Towards regional government in Central Europe: territorial restructuring of postcommunist regimes. Environ Plan C Gov Policy 15:437–462
Wilson G (2001) The dual motives of interest group studies. Paper prepared for ECPR research conference in Canterbury

Chapter 7
Rural Governance in the New EU Member States: The Experience of the Polish LEADER+ Pilot Programme (2004–2008)

Marek Furmankiewicz, Wojciech Knieć, and Jane Atterton

Abstract The chapter examines the main features of local cross-sectoral partnerships in rural Poland and compares them with experiences in England, UK. These forms of local cooperation are considered a new form of rural governance, especially in post-socialist countries. Due to a European Union requirement regarding the operation of 'LEADER-type' programmes, partnership structures in Poland developed rapidly between 2004 and 2008. Approximately 210 partnerships existed in Poland in 2008. Most of them were created after 2004, within the framework of the LEADER+ Pilot Programme. Projects were focused on improving the quality of life and the development of natural and cultural resources. Local governments and local voluntary organisations were the most active partners. The main targets of activity were tourist promotion and infrastructure development, the restructuring of rural areas and local product promotion. Partnership projects were similar in Poland and the UK. However, the social, economic and institutional bases in the two countries are completely different. The challenges faced by the Polish partnerships (and not by those in England) were relatively low level of engagement by the business and third sectors, attempts by local governments to dominate partnerships, and additional rules and constraints, many due to central government controls, which were not applied to programmes in old EU member states.

Keywords Territorial rural governance • Civil society and economic development • Cross-sectoral partnerships • European Union LEADER Community initiative

M. Furmankiewicz (✉)
Department of Spatial Economy, Wrocław University of Environmental and Life Sciences, Wrocław, Poland
e-mail: marek.furmankiewicz@up.wroc.pl

W. Knieć
Rural Sociology Department, Institute of Sociology, Nicolaus Copernicus University, Toruń, Poland

J. Atterton
Rural Policy Centre, Scotland's Rural College (SRUC), Edinburgh, Scotland, UK

© Springer Science+Business Media Dordrecht 2015

J. Buček, A. Ryder (eds.), *Governance in Transition*, Springer Geography,
DOI 10.1007/978-94-007-5503-1_7

Introduction

In this paper, we examine the main features of the implementation of a new form of rural governance through territorial partnerships in rural Poland from 2004 to 2008 (connected with the utilisation of Structural Funds in the 2000–2006 EU programming period), drawing comparisons with the experiences of partnership governance processes in Western European countries, especially the UK. These forms of local cooperation have evolved over the last 20–30 years in advanced liberal democracies like the UK, but they are a new form of rural governance in post-communist countries.

The Common Agricultural Policy represents a top-down sectoral model of intervention in rural areas. It has been successful from an agricultural productivity point of view, but could be considered as having had a negative impact on the social development of rural areas (Van Depoele 2003). As a result, the European Commission decided to shift attention towards lagging regions and thus to limit the dominant sectoral approach in development policy, replacing it with a more territorially based approach. One tool in this approach was the LEADER Community Initiative. In 1991, the Directorate General for Agriculture launched LEADER I (1991–1995) as an experimental socio-economic initiative. After positive evaluation of this approach, the initiative was continued until 2006 in two further phases: LEADER II (1996–1999) and LEADER+ (2000–2006). LEADER represented a new rural development concept, based on territorial, bottom-up, cross-sectoral initiatives drawing on local resources and internal financial support and on local actions designed and undertaken by local communities (Ray 2000). The main aims were making the best use of natural and cultural resources, improving the quality of life in rural areas, adding value to local products and the use of new know-how and new technologies to make products and services in rural areas more competitive (European Union 2012). It was thought that a new rural policy oriented towards the socio-economic development and diversification of rural economies (tourism, small crafts, services, etc.) might be more effective than a sectoral agricultural policy alone in helping to solve the isolation and dysfunctions of rural areas (Saraceno 1999). The OECD 'New Rural Paradigm' report also confirms the limited impact of sectoral measures to support agriculture on broader rural economies (OECD 2006). Some authors describe these changes as a shift from agriculture to rural development (Van Depoele 2003, 81) or more precisely to neo-endogenous development (Ray 2005), whilst in the context of general policy processes, these changes were considered as representing a shift from 'governing' to 'governance' (Marsden and Murdoch 1998; Ray 2005).

Alongside the adoption of the LEADER Community Initiative in the EU, there were many other examples of territorial approaches developing in EU countries such as *contracts de pays* in France, Local Partnership Companies and Area-Based Partnerships in Ireland, PRODER in Spain, POMO in Finland, 'Regional Action' in Germany and others (Commins 1994; Westholm et al. 1999; Buller 2000; Ray 2005; Siebert and Dosch 2005; Macken-Walsh 2010). This new mode of rural governance has been the subject of much research and theoretical discourse in advanced western democracies and in 'old' EU countries from the beginning of the 1990s (Ray 2000; Moseley 2003). However, in new democracies in Central and Eastern Europe (CEE), these types of initiatives have only become common in the twenty-first century, and thus the literature is considerably more limited. After the EU's 'great broadening' in 2004 and 2006, LEADER+ type measures started to be implemented in five new member states (Poland, the Czech Republic, Hungary, Lithuania, Latvia and Estonia), whilst another three countries (Slovakia, Bulgaria and Romania) did not initially decide to adopt measures of this type (Wade and Rinne 2008). In Poland, partnership structures were adopted rapidly across the country from 2004 to 2006 as a result of the requirement to run an EU LEADER-type programme in the following programming period. In the 2007–2013 programming period, LEADER is no longer a discretionary rural development programme but rather is embedded in the Rural Development Programme of all member states with a requirement placed on delivery bodies to spend a proportion of funding according to the principles of LEADER through designated Local Action Groups.

In analysing the development of territorial, partnership modes of rural governance in the EU's new member states (i.e. those that joined in 2004 and 2007), it is important to take into consideration the different social and institutional environments in post-communist countries, which have shaped the central programmes supporting the partnerships and structures of Local Action Groups. According to Kovăch (2000), LEADER-type programmes are potentially a positive political force to break down bureaucratic rural policy regulations in CEE countries. However, there is evidence that local government remains the dominant player in LEADER partnerships in these countries and, in East Germany, that the voluntary sector remains a weak participant (Siebert and Dosch 2005; Dąbrowski 2008; Knieć 2009; Furmankiewicz et al. 2010). To understand the differences between the Polish LEADER-type pilot programmes and the EU LEADER Community Initiative, it is important to analyse the characteristics of the institutional and social environment in Poland.

Area-based partnerships are seen to enhance local societal activity, develop social capital and promote socio-economic development, based primarily on the use of local human, economic and natural resources. The Polish LEADER+ Pilot Programme had similar aims. However, the main rules and impacts were different from the EU LEADER Community Initiative. In this paper, we analyse the three aspects of partnership issues as distinguished by Yarwood (2002):

1. Partnership *programmes* which are initiatives instigated and organised by a central body (these include the LEADER Initiatives founded by the EU from 1991 to 2006 and the Polish LEADER+ Pilot Programme analysed in this paper).
2. Partnership *organisations* which often emerge in response to the above-mentioned partnership programmes. The LEADER-type Local Action Groups are one example of such initiatives.
3. Partnership *projects* which may occur as part of partnership programmes, implemented by a partnership organisation or based on an arrangement between two or more partners established to manage or implement a specific project.

In the context of these levels, we show that at the beginning of the twenty-first century, the institutional base of rural governance in Poland was very different from that of LEADER+ in the UK. At the end, we analyse the perspectives of rural governance in the 2007–2013 EU programming period in Poland, and we conclude with some actions which could improve the implementation of this new governance system in Poland and other post-communist countries.

We draw on material from our own research projects conducted in Poland (Furmankiewicz and Slee 2007; Knieć and Hałasiewicz 2008; Knieć 2009; Furmankiewicz et al. 2010; Knieć and Jarzębska 2011).[1] We analyse reports and evaluations dealing with issues of partnership rural governance implementation in Poland (Borek et al. 2006; Borek 2007) and the findings of other authors described in the Polish literature (Błąd and Kamiński 2005; Wasielewski 2005, 2009; Futymski 2007; Goszczyński 2008, 2009; Halamska 2009; Hanke-Zajda 2009).

From Governing to Governance in Post-communist EU Countries

Governance can be defined as 'the system of values, policies and institutions by which a society manages its economic, political and social affairs through interactions within and among the state, civil society and private sector' (Work 2002, 1). By participatory governance, we mean the existence of institutional arrangements that facilitate the participation of ordinary citizens in the public policy process within the realms of the local government (Andersson and van Laerhoven 2007). According to Little (2001), within the social sciences and human geography, this concept has been connected with broader theoretical debates on regulation theory, which resulted in the shift from a Fordist to a post-Fordist mode of accumulation. In the debate, researchers found that the distribution of governance responsibilities across multiple territorial jurisdictions allows for more flexibility in the provision and production of collective goods than does the centralisation of governance within a single central jurisdiction (Andersson and van Laerhoven 2007). It is connected with the concept of decentralised governance, which according to a UNDP report, 'carefully planned, effectively implemented and appropriately managed, can lead to significant improvement in the welfare of people at the local level, the cumulative effect of which can lead to enhanced human development' (Work 2002, 1). In effect, this concept supports the retreat of the state from a welfarist position as provider of support to one of coordinator and manager of the various participants in the process of governance. LEADER-type programmes are one tool to implement such an approach, as they aim to complement central state activities (government) using less institutionalised mechanisms of coordination (governance) (Siebert and Dosch 2005).

[1] Part of the data was collected in a project number NN 114171036 funded by the Polish Ministry of Science and Higher Education.

Changes in state institutions can be implemented more successfully when new patterns of behaviour are established amongst both the political elite and amongst citizens themselves. The key role is given to 'civic society' in which individuals or groups gathered in voluntary organisations take greater responsibility for processes of development (Siebert and Dosch 2005). In LEADER-type programmes, local partnership networks of actors from the voluntary, public and private sector are supported in producing a local development strategy through consensus building and based on the valorisation of local resources. Following the design of this strategy, the projects of local organisations (acting in compliance with these strategies) are implemented.

In post-communist countries, the processes of decentralisation are typically 'works-in-progress', but the implementation of participatory governance tackles several problems, such as passive citizens not engaging in voluntary organisations and cooperative actions to a great degree; low levels of trust in society, including trust in state institutions and voluntary actions; centralism and clientelism in the administrative system; and a lack of traditions of cooperation between the state and civil society (Siebert and Dosch 2005; Dąbrowski 2008). Local governments in Poland often do not appreciate the role of NGOs and see them rather as rivals and a potential threat to their influence, or at best as collaborators who should not be trusted (Grochowski and Regulska 2000). On the other hand, NGOs have already shown their power and potential as a source of 'new' local leaders, new types of initiatives and innovation for local and regional development (Lewenstein and Palska 2004). Almost 35 % of Polish adult society were members of voluntary associations in 2004 (Klon-Jawor 2005). However, many of these organisations – especially in rural areas – represent structures that were built-up in the top-down communist political system where they mainly played the role of actors legitimising the system. The activity of NGO 'alternatives' to the communist regime was thus limited or forbidden. Therefore, trust in voluntarism and non governmental organ isations decreased as a response to the 'collaboration' of 'official' NGOs with the oppressive state (Hubner 2002). Additionally the number of NGOs in rural areas is currently visibly lower than in urban areas (81 % of all NGOs in Poland were registered in towns and cities in 2006), making it even more difficult to create cross-sectoral partnerships.

In the context of the LEADER+ Pilot Programme (2004–2008), Błąd and Kamiński (2005) list five main obstacles to the implementation of the partnership mode of rural governance in Poland. These are a top-down way of thinking about rural development, administrative formalism (legal problems), special principles established additionally by the Polish central government, irregularities in the implementation of the programme (e.g. non-transparency of quality criteria used to evaluate applications, changes in programme rules) and a deficit of social capital and lack of tradition of cooperation in Poland.

Four of these five obstacles can be considered as exogenous factors, connected with the political system. In post-communist countries, this is often strongly centralised and hierarchical. The long-standing arrangements of top-town governance are deeply embedded through formal and informal norms and conventions. When reformers (e.g. local voluntary organisations) attempt to introduce new institutional frameworks – like cross-sectoral partnerships – they are faced with the problem of deinstitutionalising old ways of working, which are very bureaucratic and imposed by public authorities (Furmankiewicz et al. 2010). This is connected with the fact that politicians, as the main beneficiaries of the existing administrative system, are likely to strongly defend the status quo (Lowndes and Sullivan 2004).

According to Ray (2005), the partnership mode of rural governance often raises issues of democratic legitimacy (accountability) in the minds of civil servants and elected representatives. Rural partnerships often have no features of representative democracy between themselves and local residents, so different tensions can emerge between local authority members who often want to keep control of partnerships (as a primary administrative unit of action). These features are especially strong in post-communist countries with a tradition of strong public administration.

In the next parts of this paper, we consider the institutional frames of partnership governance development, showing the main rules of partnership support programmes, the typical structures of Local Action Groups and their actions. To highlight the main differences from West European modes of partnership rural governance, we compare the Polish cases to LEADER+ experiences in the UK (England).

The Partnership Programmes

The first organisation considered as a LEADER-type area-based partnership in Poland was the Strug Valley Association. This was created in 1994, and it remained probably the sole example of a partnership until 2000. In the mid-1990s, some ‘early bird’, bottom-up and regionally funded initiatives were established in rural areas in Poland, as a reflection of growing attention on a new type of governance amongst regional authorities. The Opolskie Region Rural Renewal Programme (Wilczyński 2000) was an example of a regional government’s funded and region-wide action to promote community actions for planning, preparing and implementing development strategies. Another interesting early initiative was Nakło County Renewal Programme, which was funded by the local government to promote small-scale, bottom-up village-based initiatives as a tool to activate rural communities (Kamiński 2005).

At the end of the 1990s, external programmes started to promote local cross-sectoral cooperation in Poland. From 1997 to 2004, the most important were the programmes of the Polish Environmental Partnership Foundation (in Polish – *Fundacja Partnerstwo dla Środowiska*) from Krakow (initially financed by the Polish-American Freedom Foundation and followed by other external grants), which enhanced the creation of ‘partnership groups’ aimed at local sustainable development. The main target of the programme was to overcome passiveness in rural areas, to give local communities a chance to generate their own innovative ideas and put them into practice and to help gather experience in local initiative implementation by creating cooperative networks on a regional and national basis. The programme provided support to organise local area cross-sectoral coalitions. These could act to facilitate programmes by obtaining funding from local and external resources. This supported a variety of local initiatives, including organisation building, advisory services, vocational training, study visits to the US and Western Europe, local conferences and meetings. Through this programme, 14 partnerships had been created by 2004, all with an important role granted to the voluntary sector. At the same time in Poland, approximately 40 other local networks were also initiated with the support of other institutions, e.g. with EU PHARE programmes, British Know-How Fund and Dutch Ministry of Economic Affairs funds. However, many of them did not survive until the end of the grant support period.

After joining the EU in 2004 (towards the end of the 2000–2006 programme period), financial support available to beneficiaries in the rural sector in Poland amounted to EUR 5.4 billion. This included EUR 3.6 billion under the Rural Development Programme (direct payments, technical assistance and rule 1268/1999 implementation) and EUR 1.8 billion under the SOP (Sectoral Operational Programme). SOP comprised support for changes and adjustments in the agro-food sector, sustainable rural development and technical assistance (Wigier 2006). The centrally implemented LEADER+ Pilot Programme (LPP) was measure 2.7 of the SOP and eventually had a budget of EUR 30.4 million. The Managing Authority was the Ministry of Agriculture and Rural Development, the Department of Rural Development, and the Implementing Authority was the Foundation of Assistance Programmes for Agriculture (FAPA).

The first phase of LPP (2004–2006) was divided into two main budget lines: *Acquisition of skills* and *Pilot integrated strategies*. The maximum level of aid amounted to 100 % of eligible costs (up to EUR 37 000). It aimed at establishing Local Action Groups, analysing the development potential of territories and writing integrated development strategies. Training, information and consultancy activities were supported to get local populations involved in the process of building public–private partnerships and creating local development strategies. The beneficiaries could be local governments or NGOs (foundations, associations, unions of associations and other non-governmental organisations) located in rural and urban–rural municipalities. 249 representatives of local rural communities (also part of the old partnerships) sent applications to the LPP in December 2004. The formal applicants were mostly local authorities (67 %), voluntary organisations (32 %) and other institutions (1.2 %). After qualification, 167 partnerships were given financial support for organisational affairs and strategy building in the years 2005–2006.

In 2006, the second part of the LPP, Scheme II *Implementation of a development strategy*, was announced to all created partnerships. It was an open competition. Partnerships created in Scheme I and those created independently outside the Pilot Programme could apply for a grant. Through LPP Measure 2, 162 LAGs were created and 149 LAGs received funding for the implementation of their integrated local strategies which included the development of rural territories and cooperation between local partners. The projects had to be realised by early 2008. Only soft (non-investment) projects were financed from the programme. The LAGs in the

LPP had to follow specific national requirements which are specified in Table 7.1 and which are compared to the rules of the LEADER+ Community Initiative (2000–2006) implemented by the England Rural Development Plan (UK).

Comparing the LEADER programme in England (UK) and Poland, we observe a higher level of restrictions in the Polish LPP, for example, the strict exclusion of 'urban municipalities' (some of which have 5,000–10,000 inhabitants, with agriculture important in employment terms) and the need to be formally registered as an NGO. Additionally, and very important in terms of civil participation, was the exclusion of 'global grants' (small grant funds managed by

Table 7.1 The eligibility criteria of LEADER+ in England (UK) and in LPP (Poland)

<table>
<tr><th>UK, England – LEADER+ Community Initiative 2000–2006</th><th>Poland – LEADER+ Pilot Programme, Measure II, 2006–2008</th></tr>
<tr><td>The economic and social partners (e.g. voluntary, community and private sector) must comprise at least 50 % of the decision-making body</td><td>The same as England</td></tr>
<tr><td>The group must have either an administrative and financial leader with the ability to administer public funds or a legally constituted common structure which will guarantee the ability to administer public funds</td><td>The LAG has to be a formally registered unit, registered as one of the following, (a) a foundation, (b) an association of individuals or (c) a union of associations, whose statutory aim has to be rural development. This unit administers the public funds</td></tr>
<tr><td>The territory of a given LAG could not cover the area of another LAG</td><td>The same as England</td></tr>
<tr><td>LEADER+ areas (either wholly or partly) cannot be part of areas selected for the URBAN programme</td><td rowspan="2">All municipalities classified as 'urban' are excluded from the Pilot Programme; in 'urban–rural' classified municipalities, towns above 20,000 inhabitants are completely excluded (the URBAN Community Initiative was not available in Poland)</td></tr>
<tr><td>A local action group area should not have more than 40 % of its total population in settlements over 20,000 inhabitants</td></tr>
<tr><td>Local action group areas should not exceed the population threshold and/or the population density by more than 25 % or fall below the minimum population threshold by more than 5 %</td><td>The population of the partnership area has to contain between 10,000 and 100,000 inhabitants; the density of population for the LEADER area is a maximum of 150 inhabitants per km^2</td></tr>
<tr><td>The strategy must be built around one or more of the four themes listed in LEADER+ Programme</td><td>The same as England</td></tr>
<tr><td>The development plan and its programme of proposed activities must not benefit simultaneously from funding from other Structural Funds, England Rural Development Programme or other Common Agricultural Policy (EAGGF guarantee) schemes</td><td>LAGs cannot combine LEADER funds in Scheme II LEADER project realisation with any other public funds</td></tr>
</table>

(continued)

Table 7.1 (continued)

UK, England – LEADER+ Community Initiative 2000–2006	Poland – LEADER+ Pilot Programme, Measure II, 2006–2008
The approaches to LEADER in England are differentiated on the basis of their business plan and its delivery and the project approval processes. These are the Strategic Plan and Action Plan approaches. The Action Plan groups are responsible for all aspects of delivery, decision-making, administration and reporting in relation to their business plan. The partnership includes a locally accountable body that is responsible for the administrative and financial compliance of the programme. Strategic Plan groups are differentiated by their requirement to submit annual plans to the relevant Government Office (GO), which acts as the accountable body. LAG responsibilities are similar, other than that the ultimate responsibility for the approval of projects rests with the GO. In this case, the GO is also responsible for the processing of claims. LAGs have funds in the form of 'global grants' so the local organisation could apply for small grants	LAGs as a legal unit realise one large project based on the created strategy, having full responsibility for its realisation and the settlement of accounts. There is no 'global grant', so LAGs cannot implement the small grant funds (from LEADER resources) and the local organisation cannot apply independently for funds and realise individual projects (even in concordance with strategy); however, they could propose actions for LAGs in the strategy building process

Source: *England LEADER+ Programme 2000–2006*: Department for Environment, Food and Rural Affairs 2001; *Uzupełnienie Sektorowego Programu Operacyjnego Restrukturyzacja i modernizacja sektora żywnościowego oraz rozwój obszarów wiejskich, 2004–2006*. Załącznik do rozporządzenia Ministra Rolnictwa i Rozwoju Wsi z dnia 8 września 2004 r. (poz. 2117), Warszawa 2004

LAGs). In such cases, 'regranting' would have strengthened the role and position of newly created partnerships as important players on the community development scene. The central state administration was aiming for greater control of funds and did not allow the transfer of financial responsibility to the LAG. The project could be realised only by the LAG. It was not possible for local organisations in partnership territories to apply independently for small grants (however, their needs were to be taken into consideration in larger LAG projects). As a result, most commonly, small elites within the LAG management boards decided on project planning and implementation. This was not conducive to the aims of transparency and greater engagement of local actors in project implementation. This resulted in a decreasing level of enthusiasm for LEADER amongst independent, local organisations and the business sector and had a critical influence on the style of governance represented by LAGs (Knieć 2009). On the other hand, the LPP in Poland created fertile ground for the re-establishment of a new rural elite on the basis of strategies of local development, as an alternative to the existing community development strategies prepared by local governments in the late 1990s. The LAG strategies effectively created a new paradigm of rural development, which had previously not been known at such a local level (Kaleta 2004).

The Partnership Organisations

According to Knieć and Hałasiewicz (2008), amongst the 162 LAGs established in 2005–2006 in Poland, 99 had the legal form of 'associations of individual persons', 50 were foundations, 11 were 'unions of associations' and 2 were associations in the form of so-called Local Tourist Organisations. In the authors' own questionnaire research in 2007 on partnerships across Poland, on average, representatives of 22 different legal units were involved in cooperation within these partnership structures (regardless of their legal forms). These included voluntary organisations (on average 7 per partnership), municipal councils (including municipalities gathered in associations) (5), powiat (county) (average less than 1), business entities (5 and less), 2 representatives of institutions run by local councils (community centres, museums, schools and other cultural and educational or sports and recreation centres) as well as 2 farm owners. When we take into consideration that the areas within partnerships have between 10,000 and 100,000 inhabitants, the average participation of two farmers and five entrepreneurs testifies to their very low engagement in LAGs. On the other hand, it is evident that local authorities were over-represented in Polish LAGs. Regardless of the overall membership structure, all the surveyed organisations formally met the requirement of the minimum 50 % participation of social and private sectors in the managing bodies of LAGs as legal entities. However, Goszczyński (2008), in his analysis of forms of social capital in LEADER partnerships in Łódź region (Voivodship) in central Poland, found that this rule on participation did not effectively prevent government domination. The authors also found that some local organisations or individual members formally classified as 'independent' were closely related to local government or were under the influence of government, which reinforced the dominant role of local government in partnership structures. In such cases, the structure of the management board was legitimate from the point of view of the programme rules, but strong connections were observed between the local authorities, which showed features of Putnam's bonding social capital (Putnam 2000). In several surveys (see, e.g. Goszczyński 2009; Knieć 2009), the phenomenon of the over-representation of local authorities in LAGs was even deeper. In many cases, individual LAG members who officially represented NGOs, actually represented the interests of the public sector, for example, being employed as local government clerks. Although fundamentally this was an instrument to meet parity standards, in reality this often led to the abandonment of internal democracy within the partnership and so-called colonisation of the partnership by local government (Knieć 2009). The dominance of elites (often political elites) in the creation and participation of LAGs confirms research undertaken by Halamska (2009) in the Łódź region. Similarly, Hanke-Zajda (2009, 207) found that the main members of LAGs in this region were local government officials, some of whom stated that they participated in the LAG *to control* it and *to use the financial resources* available for LAGs.

In comparing the Middle Odra River Partnership in Lower Silesia (Poland) and Somerset Levels and Moors LEADER+ Partnership in England (UK), we found the

Polish partnership was based on a network of representatives of local institutions rather than on individual citizens. The UK organisational model, with its embeddedness in local Parish Groups and the election of civil representatives to a Partnership Council Forum, had stronger participatory features (Furmankiewicz and Slee 2007). Overall, we can say that Polish partnerships probably had more elitist structures than English partnerships.

The Partnership Projects

According to the final reports produced by LAGs in the LPP, between 2006 and 2008, at the national scale, about 43 % of the funds were spent on making the best use of natural and cultural resources, 34 % on improving the local quality of life and about 11 % on each of two categories: 'adding value to local products' and 'the use of new know-how and new technologies to make products and services in rural areas more competitive'. The authors found that partnerships typically concentrated their activities on tourist promotion and cultural issues rather than on more innovative actions. Detailed analysis of a sample of 40 local strategies created by Polish Local Action Groups was undertaken by Futymski (2007). He found the main activities planned in the strategies were:

1. Tourism (most commonly agrotourism or ecotourism). Measures in this field were defined in all of the 40 strategies regardless of the leading theme and the geographical area they pertained to.
2. Encouraging local residents to create and develop small and medium enterprises. One in every two strategies envisaged measures in this area.
3. Developing renewable energy, including energy crops. Approximately 40 % of the strategies analysed provided measures in this area.
4. Developing local, traditional and organic food (ecological) products. These types of projects occurred in one in three strategies. Approximately 16 % of LAGs planned to promote and register these types of products as a local brand.

Thus tourism seems to be the expected driving force for local development for most LAGs. The authors agree with Futymski's conclusion that this expectation is too optimistic and that the implementation of the plans in the partnership areas would not always give the expected results. It is difficult to imagine that all LAG areas in Poland would achieve significant income generation for local communities from tourism. In Futymski's opinion, very few strategies paid adequate attention to the active participation of the local community in strategy and projects. He argues that the authors of the strategies perceived their local communities more as passive recipients of measures taken by LAGs, rather than as active participants in implementing policies. Often the projects did not involve local society at all. Grants were often spent on preparing different investment documents, local spatial plans, etc. which were more connected with local authority plans.

The authors even have evidence of dominant local authorities excluding local inhabitants' projects from the LEADER strategy, considering infrastructure development as the most important aim (Furmankiewicz et al. 2010). Hanke-Zajda (2009, 210) also found that members of partnerships rarely had any influence on partnership management body decisions and generally did not participate in the realisation of projects.

Comparing the main features of projects realised by one LEADER+ LAG project in South West England, UK (Somerset Levels and Moors) and in Kujawsko-Pomorskie region, Poland (Toruń Rural Boroughs Association), we can see many similarities in the scope of action (Tables 7.2 and 7.3). However, in the UK, there appears to have been a more 'authentic' response to local needs, through projects prepared and realised by local organisations in a Small Grant Scheme. In the Polish LPP, this was impossible, and decisions about the actions taken were most commonly made by the LAG management board. Even when local needs were taken into consideration in LAG strategy building, the responsibility for implementing new ideas was not transferred to local organisations, so it is not possible to foresee great effects in terms of real 'capacity building' or 'civil society' creation. This shows the weaknesses of institutional support for bottom-up initiatives in Poland in the first years of membership of the EU. Some of the best organised partnerships in Poland (like the Middle Odra River Partnership) gained additional money from other resources (e.g. the 'Act Locally' programme financed by the Polish-American Freedom Foundation). However, in the authors' research in 2006–2007, only 2 out of 100 investigated partnerships implemented these kinds of local funds.

Nevertheless, it must be acknowledged that even in conditions that were so unfavourable to real bottom-up actions in Poland with an absence of complete democracy and the dominance of local authority representatives, it is still possible to observe many interesting and innovative initiatives, for example, those presented by 'Toruń Rural Boroughs' LAG. This organisation prepared and implemented actions that represented a real alternative to the more usual infrastructure development focused strategies. The lack of regranting possibility in the LPP definitely lowered the value of projects implemented, as was stated above. However, the innovative character of actions taken in the environment, where ideas of sustainable and integrated rural development were previously unknown, led to a real – albeit not a widespread or holistic – shift in the definition of local community development goals. Therefore, it can be said that the LPP in Poland established the conditions for the development of liberal governance in rural areas in terms of the content of actions, rather than in terms of the implementation methods for this content. In reality, the Programme started to disseminate the concept of partnership governance all over the country: at the end of the 1990s, there was only one area-based rural partnership in Poland, in 2004 there were 20–40 such partnerships, but in 2007 there were nearly 210 and by 2010 (according to the Ministry of Agriculture and Rural Development), 336 partnerships covered almost all rural regions in Poland.

Table 7.2 The main projects realised in years 2000–2006 by Somerset Levels & Moors LEADER+ (England, UK)

	The name of the action	Main aims/objectives	Intended beneficiaries	Actions	Results
1	Somerset Levels & Moors LEADER+ Small Grants Scheme	Provide small grants to enable locally based projects to launch or add value to larger projects. Examples include grants for illustrations for a book about local bird species, promotion of locally distinct brands of food, production of leaflets for walking and to promote local areas	Local community groups, small businesses, individuals, tourism suppliers, producers	Advice on small grants. Feasibility studies, leaflet design and printing, public infrastructure created. Festivals supported and launched	Festivals, walking routes, book publications, leaflets, school visits organised, branding of local products. A wide variety of small and large projects have been supported using the Small Grants Scheme that has been used throughout the Levels and Moors area
2	Selling The Levels	Focus on opportunities for adding value to local products, developing marketing initiatives and creating sustainable market outlets through links with other small businesses in the area, which can use those products	Local producers and outlets, farmers, tourism providers, rural businesses	Feasibility studies for brand and market development. Creating promotional materials, map existing production in the area, facilitate collaboration and cooperative working	Database of local producers and stakeholders. Brand of 'Levels Best' launched. Marketing groups and coops established, display material produced, a number of publicity events and markets hosted linking in with food festivals. National recognition through mainstream media interest

(continued)

Table 7.2 (continued)

	The name of the action	Main aims/objectives	Intended beneficiaries	Actions	Results
3	Avalon Marshes Heritage Network	To develop literature and interpretation boards for tourists celebrating and representing the Avalon Marshes area of the Levels	Local communities, tourists, tourism outlets	Historical research and documentation. Scripting for boards and book, publication process, consultation about format and style, local interviews	Interpretation boards erected. Leaflets printed and widely available. Books published in print, Braille and on CD-ROM
4	River Parrett Development Programme	Develop and promote walks and trails of interest on and around the River Parrett Trail. Promote local places of interest and boost tourism in nearby villages and for local businesses	Local communities, walking groups, tourists, tourism providers	Risk analysis of walks, engaging local landowners. Mapping of routes, creating leaflets and distributing information. National promotion and involvement of regional and national tourism agencies	Production of leaflets to promote the area. Identify and publicise specific walks and trails. Engage local community groups and individuals in process of developing walks. Raising profile of the River Parrett Trail and addressing landowner issues

Source: *Own research*

Table 7.3 The main projects implemented or planned in years 2006–2008 by Toruń Rural Boroughs Association (Kujawsko-Pomorskie region, Poland)

	The name of action	Main aims/objectives	Intended beneficiaries	Actions	Results
1	Healthy Life – Healthy Style	Development of active tourism activities on rural boroughs near to Toruń city	Local tourism entities, agrotourism farms, local governments	Promotional activities, surveys, training, new tourist trails promotion	Creation of 'eco-cultural trail of Toruń Rural Boroughs'
					Establishment of action plan for promotion and development of tourism activities on the trail
					Growing attention to tourism development amongst local authorities
2	Green fairs	Establishment of month-to-month fairs – with wide offer of local food products and crafts	Farmers, food processors, retailers from local communities	Organisation of 'green fairs' – an event gathering approx. 150 participants	Integration of farmers and food processors in informal networks
		Establishment and development of alternative food network amongst farmers, local processors and consumers from Torun to diversify local economy		Promotion of 'green fairs' in local media and the internet	Creation of alternative food network
				Further actions to develop 'box-scheme' offer of local food products for urban and suburban consumers	Growing attention to food quality and the role of diversification of farm businesses for rural development
					Establishment of close links between rural entrepreneurs and urban consumers

(continued)

Table 7.3 (continued)

	The name of action	Main aims/objectives	Intended beneficiaries	Actions	Results
3	Sport and recreation for community integration	Promotion of group sport and recreation activities through sporting events, festivals, feasts and other events	Local sport associations	Organisation of football cup	Integration of different social groups mixed in rural boroughs and isolated from each other – through sport and recreation
		Promotion of sport associations	Informal sport and recreation groups	Organisation of half-marathons	Putting more attention to the development of sport and recreation infrastructure
		Increase in quality of life through creation of attractive offer for filling free-time for children and adults	Local communities	Promotion of healthy life through study visits, training and events	
		Integration of generations through sport activities			
4	Small grants (only planned)	Promotion of community-based, bottom-up actions on the basis of one village	Informal and semiformal groups of inhabitants and local voluntary organisations	**Failed** – due to lack of possibility of financing regranting instruments from LEADER+ Pilot Programme	**Failed** – due to lack of possibility of financing regranting instruments from LEADER+ Pilot Programme

Source: Own research

The Partnership Perspectives

Under the current Rural Development Plan (programming period 2007–2013), the rules underlying rural partnership support in Poland are now different in the current when compared to the LEADER+ Pilot Programme. The new principles state that the LAG territory can be inhabited by 10,000–150,000 people, so the upper limit has been increased. LAGs have to prepare a new strategy. The area of the partnership must be spatially coherent and it may cover urban–rural municipalities (excluding towns above 20,000 inhabitants) and urban municipalities with the exclusion of the towns above 5,000 inhabitants. An example of a LAG for which this change has been useful is the Kaczawskie Partnership in Lower Silesia. In this LAG, the local government of the town of Wojcieszów, inhabited by 3,964 people with 16 % of them employed in agriculture, applied to participate in the LAG. They were excluded according to the rules of the Pilot Programme. However, they can participate in the partnership in the current programming period. The LAGs created in the LPP can be supported, but in 2007 a new legal form was created for all new LAGs in Poland. This new form enables membership by legal units representing different sectors, as well as local inhabitants, within the framework of one association (to date it has not been possible to create an organisation with formal membership by legal units representing different sectors). However, in the authors' research, respondents doubted the benefits of giving equal rights to individuals and legal units within one organisation. The new associations can engage in not-for-profit activity listed in the statute, aimed at realising the local strategy. The financial resources available to LAGs are now wider than in the Pilot Programme, and at the same time investment projects (such as the renewal of traditional buildings, infrastructure reconstruction, etc.) can be implemented. However, local government tensions can be exacerbated by efforts to dominate local partnerships. As was revealed in the nationwide partnership survey in 2009, LAGs declared numerous different attempts by local government to take control over them, as well as growing pressure from LAGs internal bureaucracy to influence decision-making processes within the organisation (Knieć and Jarzębska 2011).

The duty of a LAG is to form a Partnership Council, which has a significant role in encouraging cooperation and an appropriate balance of power between different sectors. For each measure (implementation of the local development strategy, enhancing nonagricultural activity, rural renewal and development, funding and developing microbusiness enterprises, small rural renewal projects), a LAG should plan at least 10 % of the partnership budget in the LEADER axis, so the local grant fund is obligatory. It is now possible to include voluntary and unpaid work as a part of the budget. However, that has been limited to 10 % of eligible costs. It is also possible to apply for all types of projects to be undertaken under the LEADER axis. These features are more conductive to supporting the voluntary sector than in the Pilot Programme. However, the support level should not exceed 70 % of whole project costs, so given the 10 % allowance, in reality voluntary organisations should engage 20 % of their own financial resources. In the 'small rural renewal projects' scheme, both NGOs and local government can apply and compete for grants, which creates unclear divisions of interests between municipalities and NGOs in LAGs. These conditions can eliminate local voluntary organisations from participating in current LEADER activities. In our opinion, the voluntary sector needs extra support to participate in this programme and indeed in other similar programmes. A designated support fund only for NGOs could be provided to help fill this financial gap.

Concluding Remarks

Some authors suggest the features of rural areas in Europe vary from place to place, and the programmes that address them need to be locally sensitive (Moseley 2003). On the other hand, the opportunity to create extra rules in the Polish LPP resulted in some additional constraints that did not exist in the LEADER+ Community Initiative in western liberal democracies. These restrictions made the fund spending safe from the public administration point of view, but it did not sufficiently enhance local capacity or build civil society. Additional constraints were added by the Polish central government as a result of their distrust of voluntary organisations and cross-sectoral cooperation and their resistance to full decentralisation. Such decisions made it more difficult to develop cross-sectoral partnerships and to achieve the engagement of local communities and the third sector.

Polish society has no experience of cross-sectoral cooperation or the realisation of LEADER-type projects. This was bound to be evident in the implementation of the pilot projects of LAGs. The partnership idea – according to Lewenstein and Palska (2004) – still seems to be absent from the commonly produced vision of voluntary work and civil society in Poland. This vision is based upon the assumption that the third sector still plays the role of a 'necessary add-on' to local democracy, but it does not define this situation. Polish Pilot LEADER case studies show that the corporate governance of rural areas is still impeded by a very limited understanding of partnership principles and/or limited perception of its effects at the local level, but also at the regional and – perhaps surprisingly – the national (or central) level. The LEADER methodology as a tool for rural governance comes from a Western European cultural and political milieu. It is not clear if this model can be easily transposed to CEE conditions, where different historical experiences have shaped specific attitudes towards the voluntary sector, partnerships and governance processes.

In conclusion, we suggest the following recommendations for future programmes supporting cross-sectoral partnerships in post-communist, new member states:

- The rule of a maximum 50 % share by public bodies is not effective in LAGs in post-communist society with their strong public administration and weak voluntary sector. Thus we propose the use of the Finnish model of partnership in the post-communist countries in which the public sector share in the management body is limited to one third (see Rizzo 2009, 207).

- Combating the phenomenon of the 'colonisation of partnerships' by the local public sector is possible only when LAGs become financially independent from municipalities and develop their own independent administrative structures, which are open to new members.
- The possibilities of prefinancing should be explored in countries with a weak voluntary sector (e.g. payment on account in small grant funds for voluntary organisations) or additional support should be established, especially for voluntary organisations (e.g. special loans with very low bank costs, national funds to cover bank costs, etc.).
- In particular, innovative projects should be promoted (they can be adapted from Western European partnerships, so international networking between west and east is very important). Local rural society often lacks innovative ideas, so some projects could be also prepared by experts and promoted to rural partnerships for adoption with advisory help made available.
- Local farmers and entrepreneurs should be involved and greater entrepreneurship encouraged, as this is the main way to secure income for local communities. Help based on social aid was largely responsible for creating the passive, aid-dependent groups.

In Grochowski and Regulska's (2000) opinion, the associations, unions, foundations and other institutions that support local governance are beginning to assume the same or even greater representation of social interests as formal political institutions, such as political parties. The programmes supporting partnership governance should pay greater attention to enhancing the development and role of the voluntary sector and real local participation in partnership management, which in post-communist countries are typically weaker than in Western European democratic countries. Monitoring should disclose cases of domination, decision-making in small, exclusive groups, etc. and address these situations. This is the most important area for further research on power relations in rural partnerships in post-communist countries to find the model through which local community (or voluntary sector) participation will be substantially enhanced.

References

Andersson K, van Laerhoven F (2007) From local strongman to facilitator: institutional incentives for participatory municipal governance in Latin America. Comp Political Stud 40(9):1085–1111

Błąd M, Kamiński R (2005) Social capital enhancement in the Polish countryside: experiences from the implementation of LEADER-type programmes. In: Zawalińska K (ed) Rural development in the enlarged European Union. Institute of Rural and Agricultural Development, Polish Academy of Sciences, Warszawa, pp 235–247

Borek T (2007) Realizacja Pilotażowego Programu LEADER+ w Polsce (broszura informacyjna). Fundacja Programów Pomocy dla Rolnictwa FAPA, Warszawa

Borek T, Fałkowski J, Giejbowicz E, Janiak K, Pośledniak A, Zielińska M (2006) Inicjatywa LEADER – pierwsze doświadczenia i szanse rozwoju. Fundacja Programów Pomocy dla Rolnictwa FAPA, Warszawa

Buller H (2000) Re-creating rural territories: LEADER in France. Sociol Rural 40(2):191–199
Commins P (1994) Rural development in Ireland. Economic and Social Research Institute, Dublin
Dąbrowski M (2008) Structural funds as a driver for institutional change in Poland. Eur-Asia Stud 60(2):227–248
European Union (2012) Leader plus programme. Available at: http://ec.europa.eu/agriculture/rur/leaderplus/intro_en.htm. Viewed June 26 2012
Furmankiewicz M, Slee B (2007) Cross-sectoral cooperation for rural development in old and new EU member states: a comparative case study of English and Polish area-based partnerships. Agric Econ Rural Dev 4(3–4):5–20
Furmankiewicz M, Thompson N, Zielińska M (2010) Area-based partnerships in rural Poland: the post-accession experience. J Rural Stud 26(1):52–62
Futymski A (2007) Local action groups' objectives (conclusions drawn from the analysis of IRDS). LEADER+ Quart 1:6–9
Goszczyński W (2008) Kształtowanie synergicznych form kapitału społecznego jako warunek prawidłowej aktywizacji społeczności lokalnych na przykładzie Pilotażowego Programu LEADER+. In: Podedworna H, Ruszkowski P (eds) Społeczne aspekty zrównoważonego rozwoju wsi w Polsce. Partycypacja społeczna i kapitał społeczny. Wydawnictwo Naukowe SCHOLAR, Warszawa, pp 102–122
Goszczyński W (2009) Program LEADER jako mechanizm kształtowania kapitału społecznego na obszarach wiejskich. In: Wasielewski K (ed) Tworzenie partnerstw lokalnych i ich sieci na obszarach wiejskich. Doświadczenia z funkcjonowania programu LEADER w Polsce we latach 2004–2009. Wydawnictwo Uczelniane Wyższej Szkoły Gospodarki w Bydgoszczy, Bydgoszcz, pp 15–28
Grochowski M, Regulska J (2000) New partnership and collaboration: local government and its supporting institutions – the case of Poland. In: Amnå E, Montin S (eds) Towards a new concept of local self-government? Recent local government legislation in comparative perspective. Fagbokforlaget, Bergen, pp 73–100
Halamska M (2009) Zasoby własne wsi a zewnętrzne programy rozwoju na przykładzie zależności kapitał społeczny – PP LEADER+. Wieś i Rolnictwo 3(144):9–28
Hanke-Zajda K (2009) Potencjał kapitału społecznego lokalnych grup działania województwa łódzkiego. Wieś i Rolnictwo 4(145):199–211
Hubner P (2002) Pojęcia i tradycje samoorganizacji społecznej w Polsce. In: Gliński P, Lewenstein B, Siciński A (eds) Samoorganizacja społeczeństwa polskiego: trzeci sektor. Wydawnictwo IFiS PAN, Warszawa, pp 72–75
Kaleta A (2004) Cultural activity of rural inhabitants during the system transformation. Polish Soc Rev 3:331–337
Kamiński R (2005) Sami o sobie, Odnowa wsi w powiecie nakielskim. The brochure issued by Nakło County Partnership
Klon-Jawor (Association) (2005) Badania III sektora w Polsce. Klon-Jawor Association, Warszawa
Knieć W (2009) Władza w lokalnych grupach działania. In: Wasielewski K (ed) Tworzenie partnerstw lokalnych i ich sieci na obszarach wiejskich. Doświadczenia z funkcjonowania programu LEADER w Polsce w latach 2004–2009. Wydawnictwo Uczelniane Wyższej Szkoły Gospodarki w Bydgoszczy, Bydgoszcz, pp 61–72
Knieć W, Hałasiewicz A (2008) Leader: Budowanie potencjału społecznego. Fundacja Programów pomocy dla Rolnictwa, Warszawa
Knieć W, Jarzębska A (2011) Local partnerships in Poland – towards citizenship in rural society. [unpublished report]
Kovách I (2000) LEADER, a new social order, and the Central-and East-European countries. Sociol Rural 40(2):181–189
Lewenstein B, Palska H (2004) Organizacje pozarządowe na scenie publicznej Polski okresu transformacji: dynamika rozwojowa i relacje z władzą – analiza badań jakościowych. In: Gliński P, Lewenstein B, Siciński A (eds) Samoorganizacja społeczeństwa polskiego: III sektor i wspólnoty lokalne w jednoczącej się Europie. IFiS PAN, Warszawa, pp 80–105
Little J (2001) New rural governance? Prog Hum Geogr 25(1):97–102

Lowndes V, Sullivan H (2004) Like a horse and carriage or a fish on a bicycle: how well do local partnerships and public participation go together? Local Gov Stud 30(1):51–73

Macken-Walsh Á (2010) Towards a 'transverse inter-sectoral debate'? A case study of the rural partnership programme (RPP) in post-socialist Lithuania. East Eur Contrus 16:45–64

Marsden T, Murdoch J (1998) Editorial: the shifting nature of rural governance and community participation. J Rural Stud 14(1):1–4

Moseley MJ (ed) (2003) Local partnerships for rural development. The European experience. CABI Publishing, Wallingford

OECD (2006) The new rural paradigm: policies and governance. Organisation for Economic Co-operation and Development, Paris

Putnam RD (2000) Bowling alone – the collapse and revival of American community. Touchstone Books, New York

Ray C (2000) The EU LEADER programme: rural development laboratory. Sociol Rural 40(2):163–171

Ray C (2005) Neo-endogenous rural development in the EU. In: Cloke P, Marsden T, Mooney P (eds) Handbook of rural studies. Sage, London, pp 278–291

Rizzo F (2009) Local institutions and agrarian structures matter in LEADER: case studies from Finland and Italy. Finn J Rural Res Policy, Special Issue 2:71–84

Saraceno E (1999) The evaluation of local policy making in Europe: learning from the LEADER Community Initiative. Evaluation 5(4):439–457

Siebert R, Dosch M (2005) Cooperative networks in rural development experiences from the pilot programme "Regional action". East Eur Countrys 11:111–126

van Depoele L (2003) From sectoral to territorial-based policies: the case of LEADER. In: The future of rural policy. From sectoral to place-based policies in rural areas. OECD, Paris, pp 79–87

Wade P, Rinne P (2008) A LEADER dissemination guide book based on programme experience in Finland, Ireland and the Czech Republic. Final report of the transnational LEADER dissemination project for the Finnish Rural Policy Committee. Rural Policy Committee, Vammala

Wasielewski K (2005) Program LEADER+. Szansa czy tylko nadzieja na aktywizację polskiej wsi? In: Fedyszak-Radziejowska B (ed) Proces demarginalizacji polskiej wsi. Programy pomocowe, liderzy, elity i organizacje pozarządowe. Instytut Spraw Publicznych, Warszawa, pp 69–96

Wasielewski K (ed) (2009) Tworzenie partnerstw lokalnych i ich sieci na obszarach wiejskich. Doświadczenia z funkcjonowania programu LEADER w Polsce w latach 2004–2009. Wydawnictwo Uczelniane Wyższej Szkoły Gospodarki w Bydgoszczy, Bydgoszcz

Westholm E, Moseley M, Stenlås N (eds) (1999) Local partnerships and rural development in Europe. A literature review of practice and theory. Dalarna Research Institute, Countryside & Community Research Unit, Cheltenham and Gloucester College of Higher Education, Falun

Wigier M (2006) Structural policy for agriculture and rural areas before and after Poland's accession to the EU. Agric Econ Rural Dev 3(9–10):12–21

Wilczyński R (2000) Odnowa wsi perspektywą rozwoju obszarów wiejskich w Polsce – doświadczenia województwa opolskiego. Fundacja Fundusz Współpracy, Poznań

Work R (ed) (2002) The role of participation and partnership in decentralised governance: a brief synthesis of policy lessons and recommendations of nine country case studies on service delivery for the poor. UNDP, New York

Yarwood R (2002) Parish councils, partnership and governance: the development of 'exceptions' housing in the Malvern Hills district, England. J Rural Stud 18(3):275–291

Part II
Crossborder and International Co-operation in Development

Chapter 8
Strategic Cooperation Between Regions: Building and Utilising Transnational Relations

David Rylander and Pontus Tallberg

Abstract Although cross-border cooperation is often thought of as involving adjacent regions, transnational cooperation is increasingly important, particularly within the European Union and among its European neighbours. Transnational cooperation often involves networks of several participating regions, at a variety of levels: local governments (NUTS 4 and 5 regions), regional governments (NUTS 3) and supra-regional governments (NUTS 2). Looking at NUTS 3 regions in five countries, this paper examines different forms of transnational cooperation and examines their success. Exchange of best practice, networking, learning about the experience of other regions in a similar situation, learning about mistakes to avoid and greater clout in European decision-making are all reasons for transnational cooperation. In some cases, they have started as a way of delivering aid, in the form of goods or training, or sharing experience with new EU member states. However, most of these have evolved into mutual exchanges, offering both sides a chance to learn from each other and share experiences. Some have a narrow focus and are organised in the form of discrete projects with a limited term. Others take the form of ongoing collaboration across a range of fields. As well being mediated through the European Union, some partnerships operate through other European networks, particularly the Assembly of European Regions (AER), the Conference of Peripheral Maritime Regions of Europe (CPMR) and the Association of European Border Regions (AEBR). Often, transnational partnerships were originally created by local authorities which were later subsumed into larger regions. These regions have in turn adapted and extended the agreements. Not all partnerships are within Europe: one region has links with Guangdong Province in China, aimed at promoting the development of herbal medicine. Regional propinquity is not a requirement for successful cooperation. Success

D. Rylander (✉)
School of Life Sciences, Södertörn University, Huddinge, Sweden
e-mail: david.rylander@cultid.eu

P. Tallberg
Region of Scania, Kristianstad, Sweden
e-mail: Pontus.tallberg@skane.se

© Springer Science+Business Media Dordrecht 2015

J. Buček, A. Ryder (eds.), *Governance in Transition*, Springer Geography,
DOI 10.1007/978-94-007-5503-1_8

depends more on the focus of the programme and how projects are defined. Successful cooperation often requires a champion or set of champions to push forward cooperation and engagement. They work better if regional authorities have some control over finance and over policy-making.

Keywords Transnational networks • Intergovernmental partnerships • Regionalisation • Decentralisation of decision-making • Strategic cooperation

Building and Utilising Transnational Relations

Incentives

In an era, which is characterised by flexibility, fluidity and variation, transnational networks stand in stark contrast to hierarchical government organisations from which region builders wish to distance themselves (Rylander 2004). Before the end of the cold war, national states, alongside international organisations, were the main players in international cooperation. Under pressure originating from the challenges of globalisation and from the development of the European Union, regions have taken on an increased freedom of manoeuvre. The questions are how are the regions making use of these opportunities to build transnational relations in practice and to what extent do they build transnational relations at all. As Beck puts it in his dichotomy of the first and second modernity, in the first nation state modernity, the world consists of national players, and there are only two ways to stability: either equilibrium (terror balance) or hegemony. During the second modernity, globalisation, the apparent alternatives are the loss of national sovereignty or transnational cooperation. The latter implies, among other things, the creation of what are effectively transnational states and transnational regional states (Beck 1997). Transnationality looks at alternatives to a somewhat parochial national outlook and seeks ways to think and act that penetrate the walls around states (Beck 2004). Therefore, in this context, we are not talking about international cooperation between states but about transnational cooperation between regions. Here, there are no players on the central national level but only local and regional players.

Transnational interaction has become increasingly important with the expansion of the European Union. Cooperation with adjacent areas in Russia has intensified, and north-south cooperation has been consolidated within the union. Regions link themselves together to create better opportunities for development and to better position themselves on the global-local axis. European integration gives transnational cooperation political legitimacy and provides growing resources to support transnational cooperation on interregional and interlocal levels. Today a growing number of regions are members of an organisation of European regions. The most important of these are the Assembly of European Regions (AER), the Conference of Peripheral Maritime Regions of Europe (CPMR) and the Association of European Border Regions (AEBR).

Decentralisation and Regionalisation

One clear trend in the process of regional development within the European Union is the delegation of certain areas of responsibility or competences from national to regional and municipal levels. Another is the forming of partnerships between local authorities and the business sector, institutions and associations. In some cases, this leads to a process of integration and the formation of a stronger regional organisation, i.e. to regionalisation. In several countries, some of the national authorities have been moved from the capital city to other regions. This minimises the concentration of advanced skills in the capital and enables these to be spread to other regions. The delegation of responsibilities for an increasing number of tasks and the decentralisation of decision-making processes to local and regional levels mean that municipalities and regions face the prospect of making decisions that carry more weight than previously. This situation has arisen primarily within the areas of primary health care, elderly care and education.

One of the reasons behind regionalisation is efficiency. A large public sector is easier to run if it is divided into units. This allows local adjustments and cooperation with concerned parties. The delegation of responsibility for different operations has fostered a new attitude towards decision-making processes and delivery of services. Local and regional authorities change their work methods. This is partly in response to changes in people's values and expectations and partly in response to technical changes. Local forces work to gain greater influence over those decisions that concern their local communities. Information and communication technology (ICT) infrastructure supports this decentralisation process by reducing the costs of transactions needed for people to be kept informed and reach agreements. Increasing importance is being placed on municipalities' abilities to develop strategies for using ICT in a manner, which improves their ability to provide services to citizens, businesses and tourists. In this regard, the development of e-administration and e-democracy is a necessary measure to obtain increased interaction.

Decentralisation can be achieved in a variety of ways. These include staff autonomy, policy autonomy and economic autonomy. Staff autonomy implies that the municipality has the right to select civil servants within its administration. Policy autonomy means that there is a broad-based existing framework for decision-making processes. The more the state governs through laws and directives which are characterised by goals and frameworks, the greater is the freedom enjoyed by lower administrative and governmental bodies. Economic autonomy refers to the lower bodies' authority to collect taxes. Those who pay for the various organisations also govern them (Hadenius 2001).

The decentralisation of decision-making processes may result in a strengthening of the democratic process if the local authorities make use of the opportunities to involve a greater segment of the local community in political processes. Local adaptation is an important argument in favour of decentralisation and small-scale development, as the problems and solutions are not the same in all areas. People with roots in the local environment, who are familiar with the prevailing conditions there, are in a better position to formulate the public policies which should be applied to that area than officials in a higher level of government. Decision-makers can put their knowledge of various problems and local attitudes to good use. Furthermore, as local representatives of their communities, they will have a popularity and legitimacy which will make it easier to implement public policy. The existence of many decision-making centres which are independent of each other also provides a great scope for initiative. New work methods can be tested and spread throughout the rest of the country if they are successful. Decentralisation may also have a conflict-reducing effect, since the political system then becomes more open and the political arguments less polarised. Minority groups will find it easier to influence and to engage with decision-making processes when political offices at the local and regional levels are at stake. This may be of great significance to segregated societies. Nevertheless, a decentralisation process may be accompanied by disadvantages in the form of duplicate administrative structures and a lower degree of professionalism on the part of the decision-makers. This may lead to a "leaking" of resources as a result of irregularities and mismanagement lower down in the system.

How to Build and Utilise Transnational Relations

A region's transnational cooperation can include friendship agreements, participation in transnational organisations and participation in different projects. An important step is to create contacts and confidence between authorities, organisations and businesses, thus creating a common ground for long-term development. Creating networks in a variety of fields improves the possibility that common activities will achieve the intended results. The main barriers to joint cooperation are the different national structures of authority, administrative structures and legal systems (Rylander 2004).

Border regions have often had more or less cooperation with regions on the other side of the border. This type of cooperation has increased and is supported by the EU through different kinds of project funding. Juridical opportunities to formalise cross-border cooperation came into force through an EC regulation (EC regulation 1082/2006). However, in this article, we do not deal with cross-border cooperation. Instead, we focus on cooperation that takes place between regions without physical border between them, i.e. where cooperation is not expected due to geography. Instead, it builds upon features and ideas in common and the goal of reaching some kind of shared decision.

A clear trend is that business life continues to internationalise with consequences for branches of the economy and businesses. This influences how regions work and how they relate to the broader world. If the regional level has a responsibility for development, it should support, investigate and consider different future scenarios. Political mobility, social interplay, cultural attraction and an ability to function nationally and transnationally are some crucial factors for successful regions (Lindeborg 2003; Green-Tallberg 2006). The opportunities for a region to develop are related to its geography, political circumstances and national policy. An important question is whether the regions have that space of manoeuvre which creates the ability to function efficiently in a national and transnational context. What mandate do the regions have in their national context?

Strategic Cooperation

Strategic cooperation consists of network-based cooperation between industries, public sectors, universities and NGOs with the aim of facilitating and advancing regional development. Strategic cooperation focuses on such activities as business, culture, environment, health, infrastructure, research, development, innovation and social issues. Strategic cooperation includes long-lasting cooperation agreements, participation in organisations and contributions to projects. A strategic aim is of more profound importance for a region's development. Figure 8.1 illustrates how structural conditions form the preconditions for cooperation activities and the attitudes to cooperation. The actors' activities and attitudes interact, and as an outcome, we get different sets of effects for companies, organisations, cities and regions.

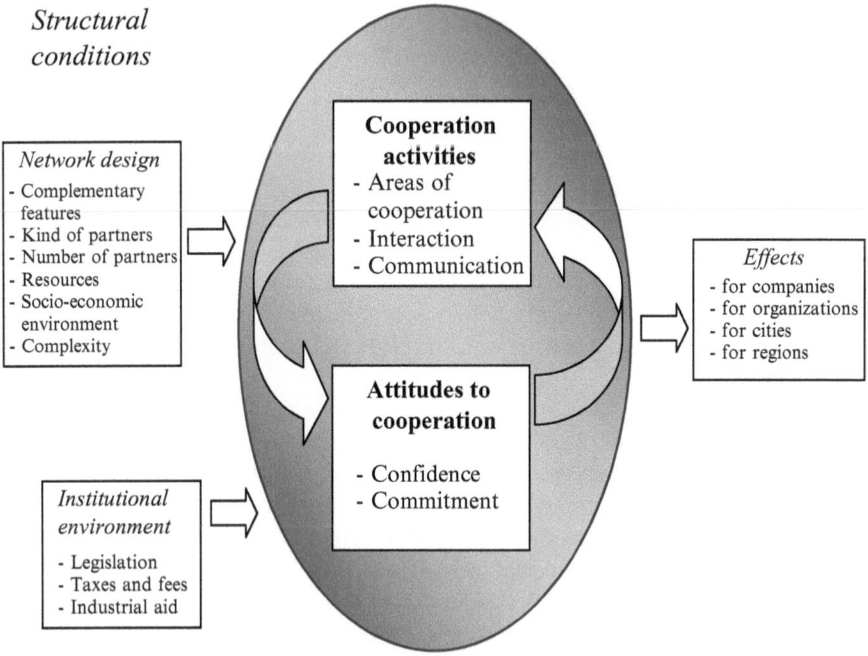

Fig. 8.1 Model for analysis of regional strategic networks (Adapted from Hallén et al. 2009)

Five Regions and Their Strategic Transnational Cooperation

In this section, we examine the character of transnational cooperation practised by five regions. How did it come about and what does it include? A comparative analysis is carried out and followed with conclusions.

Brandenburg

Germany is a federation with 16 states (Bundesländer). Each state has its own constitution, parliament and government, which means that the states are strong compared to regions in the other cases in this study. However, the federal system could be characterised as a cooperative federal system and not a dual system (Lindeborg 2008) as is the case, for example, in the USA.

As a Bundesland, Brandenburg has comprehensive decision-making powers. The state has a population of about 2.5 million. Brandenburg surrounds Berlin and therefore cooperates to some extent with the city. The state's capital is Potsdam, located just outside Berlin. Economically, Brandenburg and Berlin constitute a single region totalling approximately six million inhabitants. Strong industries in the area include biotechnology and media IT. In cooperation with Berlin, a new, extensive airport is being built, which will increase the connectivity of the region. Brandenburg has an international border with Poland on the Oder River, measuring 250 km, and even though cross-border cooperation mostly takes place between the individual cities on the border, this certainly affects the state.

The state of Brandenburg distinguishes three types of transnational cooperation. The first type is cooperation in clear-cut border areas. The second type consists of cooperation with other regions surrounding a national capital, and in this category is included cooperation with the region surrounding Moscow, with Saitama surrounding Tokyo, with the region surrounding Warsaw and with Île-de-France surrounding Paris. The third type consists of cooperation with a choice of regions and areas such as Kaliningrad, Byelorussia and Dnepropetrovsk region in Ukraine. This type of cooperation is characterised by subsidies from Brandenburg to support training and development and an effort to overcome differences.

The means of cooperation are often joint declarations, and these entail different forms of means for action. In some cases, there are yearly programmes and yearly summits on an executive level. Areas of cooperation typically include certain business areas, tourism, exhibitions, education, artisanship and fishing.

Brandenburg has a trilateral cooperation with Île-de-France in France and Mazowieckie region, surrounding Warsaw, in Poland. The aim is to pursue a deeper level of cooperation. Brandenburg has an office for representation in Brussels and offers cooperation with other regions in Brussels (Hoerschelmann 2004).

The Jönköping Region

Constitutionally Sweden has two levels, the central level and the local level. The local government consists of two tiers: the municipal level (*kommun*) and the regional level the county (*lan*), with a local legislative council (*landsting*).

In the Jönköping Region, the responsibility for regional development is shared between *the county council* (Landstinget) and *the regional council* which was formed in 2005 (region Jönköping). After being founded, the region Jönköping took over some of the tasks from Landstinget concerning business support (Johansson and Rylander 2012).

The county council has a department for regional development with activities divided into four parts: education, culture, international relations and international affairs and development. The international department works across the county council and offers a supportive function for the politicians. Transnational cooperation consists of two pillars: political cooperation within the Assembly of European Regions (AER) and cooperation and joint activities with Alytus County (*apskritis*) in Lithuania.

Connected to these two pillars, the County Council of Jönköping participates in EU projects regarding prevention of alcohol abuse, rational use of energy and preservation of cultural heritage.

The County Council of Jönköping became member of AER in 1994, as did other the Swedish county councils in preparation for Sweden's accession to the EU. After 1995, the counties gained increased freedom of manoeuvre within the European Union. The "Europe of regions" was a European byword by the end of the 1980s and was agreed on in the Maastricht Treaty, formally known as the Treaty of the European Union that was signed in 1992. Today, the Assembly of European Regions (AER) consists of 270 member regions from all parts of Europe, including regions in Russia and Ukraine. AER is a process-orientated organisation and has initiated several EU-founded projects.

The engagement of key personnel in the county council has led Jönköping to take a leading role in the AER committee for economic and regional development. The chairman of County Council of Jönköping, Håkan Sandgren, was president during 2009–2010. Jönköping is also active in a second committee which focuses on health care and medical treatment. For Jönköping, the benefit of AER cooperation is the ability to lobby. Membership serves as a springboard to EU projects and helps to make the region's voice heard in the context of the EU.

Cooperation with Alytus County in Lithuania started in 1996. During the first 10 years, it was to a large degree aid oriented, with a focus on transferring used medical equipment and sharing and transferring of work methods. The exchange has since 2007 become more two-sided. It now focuses on learning from each partner's development processes and exchanging experiences. Cooperation consists mainly of cooperation between medical personnel in Alytus and Jönköping, focused mainly on areas of professional development and personal development, in particular to meet gaps in professional skills and experience in Alytus. A two-way learning exchange has been experienced by medical personal in Jönköping, such as when older but still useful work methods have been rediscovered during visits in Alytus.

Since the mid-2000s, after Lithuania joined the European Union in 2004, Alytus has taken a more active role in cooperation with Jönköping. In 2007, Alytus was ready for the first time to partially finance a project with their own resources, aiming at developing skills through an exchange with the medical care and health service in Jönköping. However, the severe setback caused by the onset of the financial crisis in the Baltic countries put a stop to that turn for the time being, and Jönköping still finances the exchange.

After the elections in Jönköping in 2006 and the change of government, a new letter of intent was agreed upon between the two regions. The new text makes it easier to cancel the cooperation agreement, without a major delay. Alytus County became a member of AER in 2008. Before that Alytus was a stakeholder in the EU project PRESERVE, a project whose initial purpose was to develop methods for preserving cultural heritage. Jönköping County had hesitated to take part in PRESERVE due to its large involvement in the alcohol prevention project. Thanks to Alytus's engagement in PRESERVE and the cooperation between Alytus and Jönköping, Jönköping finally decided to join.

The Region of Southern Denmark (Region Syddanmark)

In 2007, a major administrative reform took place in Denmark, which led to the country being divided into five regions. At the same time, the number of communes was significantly reduced to 98 from over 200. Danish communes are, among other things, good at transnational cooperation, a strength not shared by Swedish communes to the same extent. The Danish regions have no taxation mandate. Instead, they depend on block financing, mainly from the central government but also from their communes. Their main tasks are health care and regional development. The regions may not themselves reallocate funds between the different competences, and thus do not have the same ability to determine their priorities as those in Sweden.

The main task of the region is health care. In addition, the region has a broader responsibility for regional development. The region has approximately 1.2 million inhabitants.

The region of Southern Denmark cooperates with three regions:

Guangdong Province in Southeast China. This is characterised by high economic growth. The cooperation, among other things, aims to facilitate exchanges between research institutions regarding the development of herbal medicine. The cooperation between the region of Southern Denmark and Guangdong is an extension and expansion of an agreement which the county of Fyn concluded with the province in 2004. It included a very broad spectrum of areas of cooperation and was taken over by the region of Southern Denmark. A new agreement was signed in 2007, including first and foremost health care, but also the development and commercialisation of herbal-based medicine and as well as cooperation related to culture and the environment.

Olomouc Region (*kraj*) in the eastern central part of Czech Republic. This is a transport node and an area of economic growth, especially in industry. The cooperation facilitates a number of initiatives under regional development, e.g. culture, education and tourism. In June, 2008, an agreement was signed, renewing an earlier agreement with one of the counties which is now a part of the region. Cooperation resulted in a range of projects within the environment, tourism, education, culture and business development.

The Małopolska Region (region) in southern Poland. With its capital Krakow, this is a centre for education and growth. Cooperation involves regional development, structural funds and health care. The cooperation with Małopolska started in 1991 as a cooperation between the Danish county of Fyn and the city of Krakow (itself an administrative region) and the surrounding Krakow voivodeship. The Małopolska Region was formed following a larger structural reform in Poland in 1998, and the new, much larger region took over cooperation with the county of Fyn. The structural reform of Denmark came into effect in 2007, and the merger of the counties of Fyn, Sønderjylland, Ribe and Vejle resulted in the region of Southern Denmark. Therefore, a renewal of the cooperation agreement was necessary. In March 2008, a south Danish delegation visited Małopolska's capital, Krakow, to sign the agreement. The agreement with Małopolska is mainly a framework for cooperation through projects, most of which are of limited duration, between actors from both regions, incorporating health care, regional development and the use of EU structural funds. The region of Southern Denmark has advocated expansion of the agreement to include more local actors, allowing communes and other local actors within the regions to reap the benefits of cooperation.

In 2007, the transnational strategy of the region of Southern Denmark was approved by the local legislature. It aims to promote a transnational orientation and meet the challenges of globalisation by increasing the transnational connection within public and private partnerships in a variety of ways. The strategy is part of an effort to increase the level of knowledge among citizens, communes, enterprises and institutions for higher education about transnational cooperation. It aims to give the region increased readiness to address the challenges of the future; to increase the awareness of politicians and civil servants in the region; to make a contribution to the increase, on behalf of these actors, of transnational cooperation; and thus to ensure that the transnational possibilities of the region are maintained and developed.

To implement this strategy, clear areas of action have been defined, such as participation in transnational networks, support to enterprises acting internationally or within an EU context, internationalisation of partnerships, internationalisation of the region, internationalisation of the public sector, opening and building cross-border links and using structural funds in a cross-border context (The Region of Southern Denmark 2007).

Usti Region (Kraj)

The administrative reorganisation of the Czech Republic occurred after the Velvet Revolution in 1989 and the splitting up of what was the Czech and Slovak Federal Republic in 1993. Following discussions and compromises, an agreement was reached on creating 14 regions. The first regional elections were held in 2000. The Czech administrative system comprises, apart from the regions, a large number of communes (6,243). The relationship between the regions and the central government is described as integral, and the Czech Republic can be described as a centralistic state (Österberg 2008). Usti region lies in the Bohemian part of Czech Republic, between Prague, the Czech capital and Dresden in Saxony. The region has a long border with Germany. The region is heavily industrialised and has a long tradition of coal mining but is striving to modernise its industry and make use of its strategic location between Prague and Dresden. The region has about 800,000 inhabitants and contains 46 towns and 354 villages.

Czech regions are responsible for education, culture, planning and environmental management and planning. In addition, they are responsible for regional development and run hospitals. As well as these roles, they are also obliged to execute some tasks given by government. They have no mandate for taxation. Because the region is located on the border to Germany, there is some border-related cooperation.

Usti has made some attempts at and developed some strategies regarding transnational cooperation. In 2010, Usti region started cooperation with the Banat district of the Vojvodina region of Serbia, specifically with the municipalities of Bela Crkva and Vrsac. A project was developed and implanted, and in 2011, a contract was signed with Bela Crkva and Vrsac, as well as the regional development agency in Zrenjanin and the central employment agency of Serbia. In 2010, the topics were focused on structural funds and the capacity of institutions, especially in rural areas. In 2012, the focus was on employment and investments and trying to get enterprises from the Usti region involved in business in Serbia.

There are also contacts with Republika Srpska in Bosnia and Herzegovina. These instances of cooperation have no clear form. They are at an early stage and can be described as having the character of support in topics like tourism and regional development. Usti region is also trying to start cooperation with Croatia, but the negotiations are just beginning. Usti region has in the past not actively sought transnational cooperation but increasingly sees value in such cooperation and exchanges, which lead to more substantial projects of joint interest (interviews with Lucie Kuželová and Josef Svoboda, 22 January 2010, and Lucie Kuželová, 19 June 2012).

The region is represented in Brussels with an office. Its basic mission is to strengthen the ability of Usti region to create and maintain an effective dialogue with the EU institutions and partner regions in order to fully realise the potential of EU membership for the needs of Usti region (Usti Region 2008).

The Uppsala Region

Compared to Jönköping, the model for regional development is slightly different in Uppsala County (*lan*), even though both regions are within Sweden. In 2003, it became possible to create indirectly elected regional development councils in Sweden. It resulted in an asymmetric system. Uppsala Regional Council was created in 2003, but the Uppsala County Council with health care as its main task still exists. Thus, the region parallels the county government. The main actor for transnational cooperation is the regional council. The regional council is not so much a superior administrative body, as one which brings together county and local governments in one single forum and has its own, separate responsibilities.

Uppsala Regional Council is a coordinating body consisting of Uppsala County Council and the county's eight municipalities. It is a political organ, and members are selected by the county council and the municipalities. The county has 331,000 inhabitants and is one of the fastest-growing counties in Sweden. The county is affected by its proximity to Stockholm (60 km) and the fact that its main city, Uppsala, is one of the most important centres for education in Sweden. Under Swedish legislation, the regional council is responsible for, among other things, questions concerning regional growth, which requires that it develop and implement a regional development programme. It is this responsibility which is of interest for strategic transnational cooperation and gives the region its mandate to pursue such cooperation. Still, there exists no obligation for the region to pursue transnational cooperation. The governmental mandate also includes responsibility for the county's infrastructure planning and enables it to participate in the preparation of partnerships for structural funds. Other tasks are given to the regional council by its members.

The regional council lacks a clear strategy but pursues transnational cooperation in the form of cooperation by agreement with Małopolska Region, Poland, and Wallonia, Belgium. The regional council is a member of AER and participates in their work. The regional council is also involved in European cooperation with other regions on ecological innovations. This means that the regional council pursues cooperation in three forms.

Cooperation between the regional council and Małopolska is based on an agreement signed in 2009 and includes the promotion of a beneficial business climate, cooperation between universities and industry, cooperation on renewable energy sources, cooperation on climate and environmental issues, cooperation on education and cooperation on medical and social services. The intention is to pursue common projects within these areas. The current focus is on concretizing the agreement, but one area already in operation is to arrange thematic study trips to examine certain aspects of the corresponding region, for example, innovation.

Cooperation with Wallonia is largely built on a historical link to the so-called Wallonian mills in Uppsala County. A dialogue existed for many years but was only formalised in 2009. The cooperation between Uppsala County and Wallonia is structured in a way similar way to that between Uppsala County and Małopolska (i.e. exchange of experiences, thematic study trips, Brussels' events, common projects, deepened relationships among sister cities). Wallonia is not represented at KIC InnoEnergy, but cooperation between universities is instead maintained by a Wallonian research attaché at Uppsala University. Together with Małopolska and Wallonia, Uppsala is developing a three-part cooperation in order to, among other things, identify common projects.

Uppsala Regional Council cooperates with 12 other regions in nine countries as part of the ECREIN+ (European Clusters and Regions for Eco-Innovation Network Plus). The aim of this cooperation is to learn from each other and together create a manual for developing practical energy solutions and innovations.

Comparative Analysis

When comparing the five regions, the first item to note is that the main tasks and degree of independence from the national government vary. Brandenburg, as a Bundesland in a federal state, has far broader independence than some of the other regions discussed in this chapter. Jönköping and Uppsala have the right to tax, despite the fact that the two regions are divided among three authorities. On the other hand, Southern Denmark and Usti present more unified organisations with one regional council but lack the right to tax. All five regions have direct elected regional assemblies (see Table 8.1).

Table 8.1 Characterisation of the regions' main tasks and independence

	Regional authorities	Main tasks	Election of regional assembly	Right to tax
Jönköping	County council	Medical care and health service	Direct	Yes
	Regional council	Regional development	–	–
	County governor	Regional development	–	–
Usti	Regional council	Regional development	Direct	No
		Medical care and health service		
Uppsala	County council	Medical care and health service	Direct	Yes
	Regional council	Regional development	–	–
	County governor	Regional development	–	–
South Denmark	Regional council	Medical care and health service	Direct	No
		Regional development		
Brandenburg	Federal state government	Culture, schools, higher education, broadcasting, building code, police, medical care and health service	Direct	Shared
	Federal state parliament			

Table 8.2 Features of transnational cooperation

Region	Areas being in focus for transnational cooperation	Degree of activity	Regulated cooperation
Jönköping	Medical care and health service	Medium	Agreement
Usti	Environment	Low	Development aid
Uppsala	Industry, R&D, medical care and health service	Medium	Agreement
Southern Denmark	Health service, culture, environment, business	High	Agreement
Brandenburg	Comparative experience exchange in many areas	High	Agreement

Table 8.3 Transnational cooperation within cooperation organisations

Region	Member of CPMR	Member of EARALL	Member of AER	Office in Brussels
Jönköping			X	Yes, together with regions in South Sweden – SydSam
Usti				Yes
Uppsala			X	Yes, together with Stockholm region
Southern Denmark	X	X		Yes, together with the municipalities in the region
Brandenburg				Yes

All of these regions engage in transnational cooperation, which is summarised in Table 8.2. Cooperation often consists of exchanging of experiences, sharing learning processes and comparing development strategies and administration. All the regions have some kind of regional representation in Brussels: an office of their own like Brandenburg and Usti or an office shared with partner regions. Other European regional organisations are important to some of the regions, i.e. Jönköping and Uppsala are engaged in the Assembly of European Regions (see Table 8.3).

The most extensive forms of cooperation are those practised by *Brandenburg*. Being a German state – a Bundesland – Brandenburg is in a strong position, and cooperation borders on quasi-diplomacy. Brandenburg's transnational cooperation is strategic in the sense that the region has chosen to cooperate with other regions surrounding a capital, which are in the same type of situation planning wise as Brandenburg. The purpose of the cooperation is to learn from each other.

Usti is a relatively newly formed region and has a weaker national position compared to other more well-established regions. The region does not participate in any comprehensive cooperation agreements but has shown an interest, even if the regional authorities remain passive rather than proactive, and mainly reply to invitations.

Southern Denmark engages in many forms of cooperation, including several which have the character of binding agreements. The actors' task is to fill these agreements with content in the form of concrete projects. The region is also active in several different cooperative projects. Southern Denmark is the region which, together with Brandenburg, has the most clearly defined strategy through a politically adopted strategy. This, in particular, highlights the need for transnational cooperation to strengthen the region's capacity to address future challenges. This gives transnational cooperation strong legitimacy and priority, a fact which is reflected in the extensive amount of transnational cooperation the region is involved in.

Jönköping has cooperated with Alytus in Lithuania for some time. Cooperation started as a form of development aid but has broadened and deepened, fostering many exchanges and affecting many in the region. Jönköping also participates in cooperation related to the main task of the region's county council, namely, health care. Jönköping is also an active member of the AER and views this as an important strategic element in its development. Within this and within the wider European cooperation, Jönköping uses its contacts with Alytus in Lithuania to exchange information about new EU transnational projects, in some cases joining together to develop and implement projects.

Uppsala has cooperation agreements with the Małopolska Region in Poland and with Wallonia, Belgium. The agreement with Małopolska was created by the former county governor and has, so to speak, been inherited from the earlier main Swedish contracting party. The agreement was signed in 2009. Uppsala also pursues cooperation within a greater European network. However, Uppsala lacks a specific transnational strategy when engaging in transnational cooperation. One of the driving forces behind their cooperation is benchmarking. Uppsala has identified on a strategic level other European regions with conditions similar to the Uppsala region (having a university, being close to a larger city, similar GDP etc.). Both Małopolska and Wallonia are considered to be of strategic importance to Uppsala.

Conclusions

All the regions examined here engage in some form of transnational cooperation. Even so, there are differences as to the extent, the aims and the prospects of cooperation. In Sweden, approximately half of the communes and three fourths of the counties pursue transnational cooperation, with a focus on regions within the EU (Montin 2010). However, Sweden lacks a basis for assessing the depth of cooperation with other regions and communes. The most clearly developed form of cooperation is that of the state (*Land*) Brandenburg. To a certain extent, it follows logically that a state (or a unit) within a federal system has a stronger position than a region in a centralised state. Even so, the cooperation engaged in by Southern Denmark might be described as well-developed because of the strong and clear political commitment.

All five regions studied here recognise the value of transnational cooperation since it offers an exchange of experiences without involving central government.

However, the nature of cooperation is not always well thought out. They have engaged in transnational cooperation more or less by chance, and often due to the interests of a single actor, or because an actor simply has come into contact with a region and found it interesting to start cooperating. Departments for transnational cooperation have no given role in the organisation of regions. They also lack an obvious aim in their work. Often, they make no decisions regarding strategy. Means of guidance, as well as appropriate competences, have often been lacking when it comes to making good use of the possibilities that exist in a transnational context.

The types of cooperation that can be described as more strategic include several of the projects the regions are involved in. This is especially true for Southern Denmark. Cooperation often consists of limited projects concerning a particular field of practice. The choice of Brandenburg to concentrate on regions in a similar situation is the result of a strategic way of thinking, in order to create circumstances and partners from which they can learn. However, the possibilities that exist for a region to make use of its transnational cooperation for strategic lobbying on a European level seem to be undeveloped. There is considerable future potential for this, and there are already some tendencies towards this. A region gains strong legitimacy through a clear political standpoint.

We are of the opinion that the different types of cooperation which do exist have generally come about by chance, without a clear strategic purpose. Sometimes cooperation is formalised in more or less comprehensive agreements, which subsequently are supposed to be filled in with concrete programmes, aims and objectives by the regions. To pursue cooperation with other regions is advantageous for a region, and depending on its aim and purpose, may take many forms. Formalising and creating institutional support for cooperation may be advantageous, since it may result in regions learning from each other. This gives an opportunity for regions to learn from better and worse examples, and requires that the contracting parties have a good level of knowledge about each other. It demands long-term enduring cooperation, exemplified by the cooperation which has been pursued for a long time by Jönköping. That cooperation started as development aid, and has over the years developed into a more mutual type of cooperation which expanded to a broader European level after the admission of Lithuania into the EU. In this article, we do not discuss to what extent civil society (including enterprises) are involved in transnational cooperation. Still, there is an assumption that these actors ought to be included to achieve the goals often stipulated by formal cooperation. To be able to involve these actors, it is necessary that the region has the legitimacy at home to pursue transnational cooperation. It is noticeable that this is true for the region of Southern Denmark and that there are vague hints of getting at least the communal level involved in Uppsala.

Another form of cooperation, which we found was relatively undeveloped yet of great potential, is more purely strategic cooperation. What we mean by this is a type of cooperation where the region actively seeks to identify the thematic area within which the region (1) wishes to cooperate and (2) actively seeks out a region with which to cooperate. This type of cooperation should, reasonably (at least initially), be limited with regard to its contents and timeframe. The region of Southern

Denmark appears to use this strategy to a rather large extent. This form of cooperation may produce strong learning effects.

A third form of cooperation we have identified as rather undeveloped is to use cooperation with other regions as a means of networking within the EU. This type of cooperation can be developed through either long-term cooperation or be more strategically limited but can be developed from the outset with the sole purpose of working together in different ways at the EU level. In such cases, cooperation would require the region to have an office of representation in Brussels and would depend on how that office searched for cooperative partners among other regions, also represented in Brussels, as well as on the interests of the region.

Recommendations

Regions should perhaps work with several strategies for transnational cooperation. Long-term cooperation between and among regions can provide deeper understanding and constitute a kind of learning process. Besides such long-term cooperation, there should be strategic and limited cooperation on special matters. This can also develop from long-term cooperation. The third type of transnational cooperation is to be involved in a transnational organisation like AER. From this, cooperation can foster and enhance long-term cooperation and networking at the EU level. We suggest that there are sound reasons for developing long-term cooperation with a region at some geographical distance, focusing on a themes at heart of development for both regions. If the involved regions also differs in several ways it gives fuel to the learning processes.

References

Beck U (1997) Was ist globalisierung? Irrtümer des Globalismus Antworten auf Globalisierung. Suhrkamp Verlag, Frankfurt am Main

Beck U (2004) Der kosmopolitische Blick – oder: Krieg ist Frieden. Suhrkamp Verlag, Frankfurt am Main

Green E, Tallberg P (2006) Regionalisering och Globalisering – exemplet Skåne. In: Conference paper Nuuk, Greenland

Hadenius A (2001) Demokrati – En jämförande analys. Liber, Malmö

Hallén L, Johanson M, Roxenhall T (eds) (2009) Regionala strategiska nätverk i praktiken. Studentlitteratur, Lund

Hoerschelmann A (2004) Regionpartnerschaften del Landes Brandenburg. Landes- und Kommunalverwaltung, Heft 10. Available at http://beck-online.beck.de

Johansson M, Rylander D (2012) Swedish regions: network, knowledge and change. In: Halkier H, Danson M, Bellini N (eds) Regional development agencies – the next generation? Routledge, London

Kickert WJM, Klijn E-H, Koppenjan JFM (eds) (1997) Managing complex networks: strategies for the public sector. Sage, London

Lindeborg L (2003) Framgångsrika regioner. In: Regionernas Europa. Region Skåne och Västra Götalandsregionen, pp 67–75. Available at www.skane.se/upload/Webbplatser/RU/Dokument/RegionernasEuropa.pdf

Lindeborg L (2008) Fler kompetenser för tyska förbundsländer. In: Tallberg P (ed) Regioner i Europa. Utgiven av Region Skåne, Västra Götalandsregionen och Regionplane- och trafikkontoret, pp 70–96

Montin S (2010) Kommunerna och flernivåstyrningen inom EU Inflytande, anpassning eller inbäddning. In: Tallberg P, von Bergmann-Winberg M-L (eds) Flernivåstyrning framgångsfaktor för kommuner, regioner och staten. Region Skåne, Västra Götalandsregionen, Regionplanekontoret vid Stockholms läns Landsting och Sveriges Kommuner och Landsting, pp 121–147

Official Journal of the European Union (2006) EC-regulation 1082/2006 on a European grouping of territorial cooperation

Österberg O (2008) Den centralböhmiska regionen. In: Tallberg P (ed) Regioner i Europa. Utgiven av Region Skåne, Västra Götalandsregionen och Regionplane- och trafikkontoret 2008, pp 151–164

Rylander D (2004) Nätverkan och regional utveckling. Om gränsöverskridande samarbete i södra Östersjöregionen och nätverkstjänsternas roll i den regionala utvecklingen. Doctoral dissertation, Göteborg University, Göteborg

Sande Lysa J, Ogard M (2011) Fylkeskommunalt intenasjonalt engasjement: Hvordan ser det fasktiskt ut? Eksemplene Telemark og Aust-Agder. In: Baldersheim H, Vegard Haug A, Øgård M (eds) The rise of the networking region: The challenges of regional collaboration in a globalized world. Nordregio, Stockholm, pp 117–145

Tallberg P, von Bergmann-Winberg M-L (eds) (2010) Perspektiv på flernivåstyrning (forthcoming)

The Region of Southern Denmark (2007) Internationalt samarbejde. Available at http://www.regionsyddanmark.dk/wm211398

Usti Region (2008) Representation of Usti region in Brussels – our mission. Available at http://www.kr-ustecky.cz/EN/vismo/dokumenty2.asp?id_org=450021&id=1039&p1=86254

Interviews

AbildtrupKS, Region Syddanmark, 6 April 2010
Didier M, Region Uppsala 20 March and 25 March 2010
Kuželová L, Regional council of the Usti region, 22 January 2010 and 19 June 2012
Österdahl A, International Department, County Council of Jönköping, 24 February 2010
Svoboda J, Regional council of the Usti region, 22 January 2010

Web Sites

www.regionsyddanmark.dk
www.kr-ustecky.cz/EN/
http://www.regionuppsala.se/
http://www.lj.se/
http://www.brandenburg.de

Chapter 9
Changing Patterns of City-Hinterland Relations in Central and East European Borderlands: Szczecin on the Edge of Poland and Germany

Péter Balogh

Abstract This paper examines cross-border cooperation in a broader European context, looking in particular at the experiences of the Szczecin region since the end of the communist rule in 1990. Political reform allowed local and regional governments to create direct linkages with regions in neighbouring countries, as well as with regions further afield. However, as well as administrative and geographical factors, historic and political factors can play a role in the development of cross-border relations and the ability of a city or region to fulfil its broader locational potential. This is particularly true in the case of Szczecin, which, as the city of Stettin, was part of Germany until 1945. After becoming part of Poland, it turned its back on that part of its former hinterland in Germany, and rather than being orientated towards Berlin and the West bank of the Oder, it became orientated towards Warsaw. After 1990, Berlin and Mecklenburg-Vorpommern in Germany favoured closer economic and political collaboration, both to promote development on their territories and to restore the region's relative centrality within Central Europe. However, the Polish government is far more centralised than the German one, and cross-border cooperation was straitened because of this. In addition, local and national politics also influenced the strength and breadth of cooperation. Despite this, cooperation between Szczecin and the neighbouring German regions has grown and continues to grow, both formally, through joint agreements and plans, and informally through bottom-up cooperation, such as cross-border migration in search of housing, cross-border trade in goods and services and, more recently, the development of Szczecin as a low-cost flight destination. Such cooperation has been enhanced by Szczecin's participation in broader European networks. Additional national actors have to some extent removed some of the political sensibility attached to cooperation with Germany alone.

Keywords Re-scaling • City-hinterland relations • Cross-border urban development • Networks • Szczecin

P. Balogh (✉)
Institute for Regional Studies, Centre for Economic and Regional Studies, Hungarian Academy of Sciences, Békéscsaba, Hungary
e-mail: baloghp@rkk.hu

© Springer Science+Business Media Dordrecht 2015
J. Buček, A. Ryder (eds.), *Governance in Transition*, Springer Geography,
DOI 10.1007/978-94-007-5503-1_9

Introduction

In the past two decades, many scholars, including geographers, have talked of the decreased importance of physical space, place and distance. The idea is that, with information and communication technologies (ICTs), geography and place will matter little (e.g. O'Brien 1992; Mitchell 1996; Flint and Taylor 2007). However, more recent research is showing how geographic identities, places and relations are being reproduced even with the usage of ICTs. With this comes an increasing awareness that many processes of globalisation are taking shape within and are thus limited to state-territorial and local frameworks. Another macro-level example is the recurring reference by states and other actors of "spheres of influence", "economic backyards", etc. One can say that after the Cold War, space, place and geopolitics have received renewed attention (e.g. Götz et al. 2008). Another related new(-old) field is geo-economics, which "analyses economic strategies – notably commercial – decided upon by states in a political setting, aiming to protect their own economies…" (Lorot 1997, 29). In response, one may ask two things. One, must geopolitics and geo-economics only analyse strategies decided by states, rather than political entities on other geographic level (cities, regions)? Second, must these strategies be self-interested, instead of aiming at a win-win situation? One positive example is the Luleå-Narvik railway line that – albeit being built when the two countries were still unified, to ship Swedish iron ore via the Norwegian coast – has led to gains for the economies of both countries, despite their choosing separate paths (Lundén 2004).

While urban areas are widely regarded as the main drivers of economies and innovation, their geographic and economic orientation is of crucial importance. The importance of profiling is related to increased globalisation and emergence of a new-old world order (instead of entering this discussion here, let us just say that we are referring to the past approximately three decades). It is also related to cities' awareness of the importance of geographic location and distance for their development. A Scandinavian example is the fairly strong commitment by several cities to the European corridor project, a broad link for physical transportation between Stockholm and Hamburg, including other places along the way (Europakorridoren AB 2009). The perceived need of cities to profile themselves in a system increasingly based on the division of labour requires strategic thinking and clear policies of what their respective roles ought to be, regionally and internationally. Even though the system is not static, certain path dependencies are sometimes recognised. Thus, urban managers are busy defining and redefining potential or already existing development axes for the territories under their responsibility: Stockholm focuses on the Mälar Valley, Copenhagen and Malmö on the Öresund region, while Vienna focuses on "Centrope".

In regions where major historical events and territorial disputes are more distant or accepted, such strategies are less controversial. In many places of East-Central Europe, the ice between long-standing "archenemies" has only recently begun to melt. In light of this, it is particularly interesting to look at what happens to a rather large city, Szczecin, located at the border between two strong and well-integrated states, Poland and Germany. In this paper, the German name Stettin will be used in pre-1945 contexts (with the obvious exception of quotations), and elsewhere the Polish (and today single official) name Szczecin is used.

The paper's aim is threefold. One is to increase the understanding of the historical and contemporary geo-orientations of the city of Szczecin. By geo-orientations, I refer to physical spaces towards which a place – in this case a city – unconsciously or deliberately tries to or indeed does develop. The scale can be small (e.g. suburban planning) or large (e.g. directions for international trade and flows), but these can sometimes be interrelated. Such a study has to our knowledge not been done before, arguably for reasons we will return to.

One main reason for doing this is to test empirical observations against more abstract approaches, mostly rescaling theory and central place theory: to what extent can these be applied to the case of Szczecin?

Another aim is to analyse statements and policies on Szczecin's orientations by a number of actors and institutions from Germany and Poland, the two countries which embrace its geographical and historical hinterland. Here, an interesting question is how geography and history are used in support of various theses regarding Szczecin's geo-orientations.

Cities in the Rescaling Literature

As van der Heiden (2007, 4) argues, in the past,

> the ... state took care of foreign relations in general and the economic well-being of the country, local authorities were traditionally in the position to manage social policies... The cities were tightly under control of the ... state, regulating market and fiscal transfers, tax law and tariff policy. The tight fit between urban dynamism and national economic growth (Sassen 1991) was unquestioned as long as GDP growth rates allowed large redistribution programmes. City governments could concentrate on the management of the cities and on distributing public goods ... (Brenner 2003, 299).

Now, with a global market, countries compete both at the national scale and below the national scale, defending themselves by offering the lowest taxes, unrestrictive zoning laws, large subsidies and low- or no-tariff policy. Therefore, the importance of the national scale in policy-making has decreased dramatically (Brenner 2006). However, the question discussed in the rescaling literature is whether other scales, that is, actions at lower levels of government, can compensate for losses at the national scale.

One example of a city that is increasingly aware of the need to have its own "foreign policy" is Berlin, whose senate recognised that although constitutionally foreign policy lies in the hands of the federal government, it is also the concern of the *Länder* to maintain close relations with other regions. There is even a law defining the legal framework for cooperation between the federal government and the Länder in EU affairs (Land Berlin 2003). The senate explicitly proposes that "Berlin has to be attuned to increased locational competition with regions like Öresund and Centrope"; thus, "identifying and taking measures for common German-Polish interests must be done without delay" (Land Berlin 2005). Another example from the Central European post-socialist region is Riga, which "has been separating itself more and more from Latvia's image, in turn creating its own popularity", leading some to wonder which one of them is more significant and has more to say (Murashova 2010, 5).

The rescaling literature further recognises that

> Although capital is said to be completely mobile nowadays, companies still need an infrastructure to operate and this infrastructure is still mostly immobile, and production processes are also locally bound. Swyngedouw (1997) called this process glocalisation, meaning that although globalisation leads to heavier competition and the erosion of the national scale, economic processes are still rooted in local places, especially in metropolitan areas. Thus, city regions are nodal points for globalisation processes, as they link the national economy to the international market place (Savitch and Kantor 2002).

Therefore, the question is not if some scales have become unimportant in the age of globalisation (deterritorialisation) but rather which territories are the ones where political decision-making is still possible (reterritorialisation) (see Brenner 1999). Cities and city regions, as places where human activities in general concentrate, are nodal points in these economic and political processes (Sassen 1991). Therefore, they do not need to be "leaves in the wind" of globalisation (Savitch and Kantor 2002, 346). Often, the economic development of these city regions seems more and more disconnected from the development of the rest of their country (Brenner 2003, 298).

> In an era of globalisation, there is global competition between large metropolitan areas, competing for the location of businesses on a global scale. It is within metropolitan areas where economic prosperity is still possible, or as Brenner (1999, 298 and 437) puts it, the geoeconomic power of cities is increasingly disarticulated from the territorial matrices of the interstate system. Rescaling can thus be defined as decentralisation of the national scale of accumulations, urbanisation and state regulation in favour of new sub- and supranational territorial configurations (Brenner 1999, 435) (van der Heiden 2007, 4–5).

Perhaps most importantly here, "rescaling theory … predicts a shift of political steering capacities from the … state to urban areas, whereby increased international activities are one of the most prominent possibilities for such shifts to take place. International activities of urban areas are currently mushrooming at a global and especially European level" (van der Heiden 2007, 2).

To what extent were and are these observations valid for Szczecin? It is with this question in mind that we shall review the city's historic and contemporary development axes.

Szczecin's Historical Orientations

Szczecin has been a meeting point of various tribes and peoples for centuries. This is true of many settlements in Central Europe and elsewhere, but the uniqueness of this city lies in its location at the mouth of the Odra River, making it a gateway between important Central European regions and the Baltic Sea and beyond. Various

Germanic and Slavic tribes have resided here, including the Wends and the Lusatian culture. The Germanness or Slavicness of some of them has been a subject of debate for some time (Piskorski 1999, 21). Within the region, towns – which in time became cities – are widely regarded to have been the main drivers of economic development, with many constituting autonomous political entities throughout much of the Middle Ages (which, as geographers and others are now suggesting, they are becoming again). Up until the nineteenth century, trade and wealth mostly depended on access to the sea and/or a river. Thus, many larger cities developed around river mouths. So did Stettin, which owes its growth to trade (first in agricultural products and later natural resources) between the areas around the Oder with its tributaries and the Baltic Sea and beyond.

In relation to its direct neighbourhood, by the twelfth century, Stettin had become the regional centre of the area. Settlements directly located on the Baltic coast (e.g. Wolin, which once saw itself as the most important Baltic town) soon saw their population surpassed by Stettin. In the south, trading vessels from towns on the upper Oder (Greifenhagen, Gartz, etc.) had to pay tolls in Stettin in order to pass up or downstream and were thus outcompeted (Piskorski et al. 1994).

During the eleventh century, a lively trade can be detected between the forerunners of the Germans, Danes, Poles, Russians as well as other Slavic tribes, and the city was part of the Hanseatic League of trading cities. Like Venice, it was less orientated to its immediate hinterland than to more distant markets to which it was linked by trade. From 1310 onwards, Stettin was the centre of a herring trade, and its merchants had established contacts in Skåne (Kirchner 2009). By the fifteenth century, Stettin's significance was more limited to its closer neighbourhood in comparison with Danzig or Lübeck, which served much larger hinterlands (much of Poland-Lithuania and the German states, respectively) (Piskorski 1999). In the seventeenth century, the city's growth potentials beyond the provincial borders were further limited by extensive damage to the whole region's economy due to several wars, protectionist policies and customs by Brandenburg (which controlled parts of the lower Oder), difficult conditions for shipping (exits to the Baltic were becoming too narrow) and lack of investments during the era of Swedish rule (Piskorski 1999).

In addition, during the period of Swedish rule, the city had a peripheral location within the Swedish realm. After Gustav Adolph's landing in Pomerania, Sweden's aim with Stettin was to establish it as a centre for its military operations against Germany rather than as a centre of trade between German lands and the Baltic (Kirchner 2009).

Nevertheless, the conditions for Stettin to become a production and trade centre started to improve after its incorporation into the Prussian state in the eighteenth century, as the borders dividing the city from its hinterland were abolished (Piskorski 1999). This was even more the case after 1740, when Prussia annexed the economically important region of Silesia. However, the Prussian state's trade and customs policies now hampered imports from Poland. Commerce with Western Europe was important, as before, but was restricted by the Danish straits customs (Piskorski 1999). The main driver of Stettin's economic recovery in the nineteenth century was instead its transit role in Central European imports of Swedish iron and other metals and in coal exports from Silesia. In 1843, a direct Berlin-Stettin railway was opened, in order to link the (then) "rich, industrial and commercially significant capital with the kingdom's number one trading port… The transport of people and goods was soon flourishing" (Buchweitz et al. 2007, 8). The railway lines between Pasewalk and Stettin (west of Stettin) and between Ducherow and Swinemünde (northwest) were opened in 1863 and 1876, respectively (Buchweitz et al. 2007). A direct road was completed between Berlin and Stettin in 1828 (Deutsche Städte 2009), and a new water corridor linking the two cities was finished by 1914, more straight than the one first constructed in 1605 (Wahl 2009). Stettin's harbour became the largest on the German Baltic coast (Kirchner 2009), and it was in these years that Usedom, a nearby island, received the name "Berlin's bathtub". The recreational functions in the area became so popular at the turn of the century that tourists came from as far as the Czech lands, at that time part of Austria-Hungary. Immediately before World War I, the city had a population of 250,000 (a tenfold increase since 1816) and a large port turnover (Mieszkowska 1996). Regarding its extension in relation to the Oder River, it was characterised as a one-bank dominated agglomeration situated on the higher western bank, with only some port functions on the low eastern bank (De Geer 1912).

In 1920, Swinemünde became Germany's maritime link to East Prussia, a province that became cut off from Germany proper by the new Versailles borders. Furthermore, with a part of Eastern Germany transferred to Poland, Stettin lost important suppliers of export coal from Upper Silesia and of crops from the areas around Posen/Poznań (Kirchner 2009). Consequently, during the interwar period, Stettin fell into serious economic decline. The competition of Hamburg and Bremen reduced the port turnover by half, ruining the shipbuilding industry and creating high rates of unemployment. But the city retained its position as a trading centre, adding entire new neighbourhoods and vast parks to its urban fabric. On the eve of the Second World War, its population has grown to 381,000 inhabitants (Mieszkowska 1996). Links to the German capital were further strengthened by the new Reichsautobahn highway that was opened by 1937 (Meynier 1937). During the Second World War, just as during the First, Stettin became the main port for traffic with Scandinavia (Kirchner 2009). Nevertheless, "World War II was a disaster for the city. Allied bombing, mainly during 1944, turned its oldest parts and industrial sites into heaps of rubble. The harbour was 80 % destroyed … By the decision of the victorious powers in Potsdam, Stettin, with the major part of … Pomerania, was passed to Poland" (Mieszkowska 1996, 110).

The entire German population of about 400,000 had to leave Stettin, henceforth Szczecin. By 1947, Szczecin was an almost entirely Polish frontier city (Musekamp 2009). The border towards occupied Germany was quickly closed off. On the other hand, contacts with Sweden were quickly restored, despite Szczecin being transferred to a state under Soviet influence. Swedish trade vessels resumed coal and food imports and viscous cellulose exports to Wrocław via Szczecin from as early as the summer of 1946 (Kirchner 2009).

> Port turnover grew steadily, and in the 1960s the united Szczecin-Świnoujście harbours came to outdistance other Baltic ports. Along with the port, other branches of the maritime economy developed (shipyards, fisheries) as well as several industries based on the flow of materials and products from the hinterland and overseas. … Szczecin grew as a cultural

> centre of the north-western part of Poland, housing several academic schools and various cultural institutions. Its seemingly free contact with the world was, however, severely limited by tight state borders and the Iron Curtain... (Mieszkowska 1996, 111).

The new Polish regime legitimised the Polishness of the "recovered territories" by emphasising its Slavic history (Musekamp 2009). During the Cold War, local contacts with neighbouring East German regions were not maintained, despite the official rhetoric of brotherly relations between the GDR and Poland. The border was closed, with an extremely limited level of permeability (with the exception of the 1970s, when some cross-border mobility was allowed). The entire Polish-German borderland, including Szczecin and its surrounding areas, became a geographically and economically peripheral lagging region. This is particularly true for Szczecin, into which investments were limited up until 1970 (Dutkowski 2009) and which lost its function as a port for Berlin. During the time of the People's Republic, between 1945 and 1989, Szczecin was almost entirely ignored by regional development policy, even though the city went through a strict reorientation eastwards (Dorsch 2003). One illustrative example is the construction of a wide road in the 1960s that crosses the pre-1945 old town and the Odra River, creating an important corridor towards the central parts of Poland. This special treatment of the city had to do with its relatively uncertain status within Poland until the border's recognition by West Germany in the early 1970s, but even so in 1989–1990, some concerns were (re) raised in Poland until reunified Germany finally reconfirmed recognition of the border in late 1990.

From the overview, we can see that Szczecin's historical orientations are the north-south axes between Scandinavia and Central Europe (mostly Silesia and Berlin) and connections with its direct vicinity (in its role as a regional centre), especially to the west, north and east (the areas to the south have tended to have closer ties to Berlin).

Szczecin's Orientation After 1989

Szczecin has received less scholarly attention than much smaller but divided cities along the German-Polish border such as Görlitz-Zgorzelec, Frankfurt-on-Oder-Słubice and Guben-Gubin (e.g. Stokłosa 2003, Jańczak 2009). This may be because in many divided cities, a sort of balance existed where each half of the city was orientated towards its respective country. It is perhaps because of their pre-1945 history as one unit that such cities have received more attention, with many projects aimed at overcoming the divisions. This is less the case with agglomerations located on a state border (such as Trieste, Strasbourg, Oradea, etc.) but – at least until recently – clearly in one country. Nevertheless, as these metropolitan areas grow, urban managers need to consider the direction and location of future growth. Such discussions are not always uncontroversial, as is seen from the following overview of the main hypotheses around Szczecin's geo-orientations.

"Relocating and Leapfrogging" or "Integrating and Restructuring"?

Two hypotheses have dominated the discussions on Szczecin's orientation in the recent years. One predicts that a dynamic economic development of the city will concentrate on the axis Berlin-Szczecin-Skåne/Öresund, while no substantial and lasting economic linkages will develop with the directly adjacent German border area. The other one envisages an emergence of an intensive cross-border economy and labour market, so that a positive restructuring process will start through the utilisation of synergies and the division of labour in strategic sectors and political areas (Maack et al. 2005).

Consequently, studies on the future development of the cross-border region have reached different conclusions. On the one hand, some studies have placed the German-Polish border space in a larger developmental context and reached negative conclusions regarding the area's growth perspectives. These mostly emphasise the particular problems and weaknesses of the German side, which manifest themselves in negative demographic development and a weak economy. At the same time, other studies have focussed on the growth opportunities of border regions in the eve of European integration. According to these, such regions have significant advantages, since they can benefit from diverging conditions in neighbouring states, including wage costs taxation and costs of living. By their nature, arguments in favour of foreign direct investments (FDI) support engagement with border regions. These include opening up new markets through foreign export and production, the reduction of costs by taking advantage of salary and labour cost differences, as well as of differences in environmental quality and environmental legislation, and the activation of innovation potential by utilising local technological competences (Maack et al. 2005).

Both hypotheses find support in spatial theory, where the study of distance and location aspects has an almost two-century-old tradition. Market-hinterland relations were analysed by von Thünen (1826), Alfred Weber (1909), Christaller (1933), Lösch (1940) and others. The latter two developed central place theory, hypothesising spatial patterns of settlement structures in an ideal, homogenous space – based on their market functions – and finding that settlements in a free-market system would occupy different hierarchical ranks. As they move up the hierarchical order, they offer rarer and more specialised services to their surrounding areas. Thus, unless major physical barriers (large waters or mountains) or non-open political borders are present, towns and settlements on various levels of a hierarchical system would be distributed more or less equally. One should note that Lösch paid special attention to border areas, using the relatively open US-Canadian border as an example. He argued that they truncated hinterlands and stunted economic growth. One can find works inspired by this thinking up to this day, not least in Germany (e.g. Güssefeldt 2005, Maack et al. 2005) and East-Central Europe. A recent study from Poznań (Ciok 2009) has identified the hypothetic catchment areas of cities in the German-Polish borderland along this principle, with Szczecin's area of influence clearly located on both sides of the boundary. Another contribution by an international group of scholars analysed Szczecin's cooperation and conflicts with its direct German neighbourhood, finding that the city's role as a market centre of the theoretical hinterland is unevenly influenced by the state boundary (Lundén et al. 2009).

The past two decades have seen the rise of new approaches that have changed the way geographers and others think about (spatial) distance. For instance, today the perspectives of the New Economic Geography as developed by Krugman (1991) have come to the fore. Krugman, Porter (1990) and others have focussed on factors that contribute to the development of clusters and agglomerations. Another groundbreaking author is Castells (1996), whose works shifted the focus from place and location towards flows and networks. According to this line of thinking, in the information society, networks and flows between actors and cities are more significant than direct physical contact and environment.

The City's Geopolitical Orientations Since 1989

Compared to Gdańsk, Szczecin is weakly represented in various transnational urban networks. However, association with the Euroregion Pomerania (an association of German, Polish and Swedish municipalities located near the border and involved in cross-border cooperation) has played an important role since the mid-nineties (Kommunalgemeinschaft Europaregion Pomerania 2007). Moreover, numerous bilateral partnerships exist with other European cities (such as Rostock, Lübeck, Malmö, some districts of Berlin, etc) (Dorsch 2003).

The process of establishing the Euroregion is itself illustrative of Poland's initial cautiousness towards cooperation with Germany. The initiative came from the German side. The Polish side was fairly enthusiastic as long as the cooperation would also include other parties from Denmark or Sweden. In December 1995, the Euroregion was finally established (later than all the other ones along the German-Polish border), and in February 1998, the Swedish county of Skåne joined in (Kommunalgemeinschaft Europaregion Pomerania 2008). The role of the Swedish side thus seems at least have been to act as a kind of mediator or balance and lessen Polish fears of German dominance.

Overall, however, in the 1990s, Szczecin concentrated on the Baltic space and only more recently on the German-Polish borderland. A transborder orientation towards the Polish-German border region and the Baltic was considered something natural: the spatial hinterland of Szczecin is located on both sides of the German-Polish border, and as a port (and until December 2011, the location of a shipyard), the Baltic connection is inevitable – even if external links were curbed between 1945 and 1990 (Dorsch 2003).

The city experienced a very strong cross-border policy while Bartlomiej Sochanski was mayor, from 1994 to 1998. He concentrated on a strengthened "urban foreign policy" focussing on the borderland. Instead of focussing on the old shipyard as a motor of growth, he saw Szczecin's long-term future in becoming a regional and even European centre based on services, culture, education and recreation. Despite leaving some imprints on the city's 2002 development strategy (City of Szczecin 2002), after he was voted out of office, cross-border activities were reduced. Subsequent developments reflect the mixed level of support for cross-border cooperation among the city's leadership. In the current (Michalski 2009), no explicit reference is made to the German part of Szczecin's hinterland.

This can be compared to Berlin's "foreign policy" position papers, in which direct references are made to Szczecin and the West Pomeranian Voivodeship in strengthening "Berlin as a part and centre of a cross-border macro-region", particularly through deepening city partnerships and networks, improving rail, road and water connections, as well as common spatial planning strategies (Land Berlin 2003). Explicitly,

> It is the declared goal of Berlin's Senate and economy to develop the space on both sides of the Oder for the long term into an infrastructurally closely linked region, with cooperation in many areas, creating a dynamic economic space with urban pulse generators, or growth poles, consisting of Berlin, Szczecin, Poznań and Wrocław. All of these were once part of Berlin's larger economic and political hinterland. With regard to developing regional economic spaces, the Senate does not question whether there is a common Oder region but asks how opportunities and development possibilities can be shaped and used to the advantage of the entire region. Crucial for cooperation in the Oder region are the related interests of politics and economy, since the strengthening of Berlin's economic power and growth in employment is thought to strongly depend on how the cross-border regional economy develops. The nearby western Polish voivodeships constitute a natural 'cooperation space', especially Szczecin, Poznań and Wrocław" (Land Berlin 2005, 1)
> [the original document refers to them by their German names: Stettin, Posen and Breslau].

A similar rhetoric can be found in Szczecin. For instance, the proclamation of the city as a "Europolis", as a "centre of European integration and a supra-regional centre of the Southern Baltic space", as formulated in various documents as well as statements by the leader of the city's office for international cooperation, makes clear that more recently there has been a reorientation from a Pan-European or even a Baltic orientation to a stronger orientation in the direction of the borderland. Szczecin wants to brand itself as the "centre of a multinational region". On the historical basis of the Pomeranian province, a multinational, cross-border space is to be created, with Szczecin as the urban centre. This historical reference to Pomerania and German Stettin is not without controversies, particularly on the Polish side. However, economic constraints seem to make closer cooperation inevitable for both sides of the border – if they want to escape the peripherality of the past fifty years. With Szczecin being the only large city in the northern border region, it is also the sole bearer of this potential. This stronger orientation is reflected in increasing enthusiasm for concerted action and collaboration within the Polish-German border region as well as by efforts to create a so-called Pommern Forum (Dorsch 2003) (which could not be traced online in late 2009). The Oder partnership is a similarly uneasy undertaking but nevertheless shows some concrete results such as an improved rail connection between Szczecin and Berlin (Rada 2008), although in

2012, it was still necessary to change trains and the total travel time was approximately 2 h, almost twice the pre-1945 time (Deutsche Bahn 2012).

Currently, Szczecin is a leading power on the Polish side of the Euroregion Pomerania. However, initially, the city was not even member of the union of participating municipalities. Dorsch (2003) notes the general lack of support from national governments as a further problem.

Berlin plays an important role for Szczecin and its development into a "Europolis". The Szczecin region does not just aspire to retake its earlier role as the port and recreational area of Berlin but also to create tight economic connections to the German capital. However, concrete steps towards this direction so far are few, as Berlin's interest in Szczecin seems rather limited compared to its interest in Poznań or Warsaw. According to Dorsch (2003), one further problem is that Szczecin lacks engaged experts with sufficient knowledge and skills (arguably, this is valid on both sides of the borderland). Neither is there a greater general willingness than elsewhere in Poland or Germany to learn the language on the opposite side of the border.

Nevertheless, some interesting cross-border developments are taking place. One is a new group of Polish settlers buying family houses on the German side where, thanks in part to inner-German emigration, prices for dwellings are already lower than those in central Szczecin. The majority of these dwellers commute to Szczecin on a daily basis, but some have opened up SMEs in their new hometowns. During 1991–2003, the annual average of Polish citizens in Mecklenburg-Vorpommern was stable at about 1,650; by December 2007, the figure had risen to 3,631. Growth was particularly high in Uecker-Randow, the German district closest to Szczecin, which had 109 registered Polish citizens in 1995, 283 in 2003 but 992 in late 2007 (Deutsches Polen-Institut 2009, 9), and by the end of 2010, the population of foreign nationals had reached 2,746 (Statistisches Bundesamt 2012). Clearly, Poland's admission to the EU has contributed to this development, which accelerated after it joined the Schengen Area in 2007. From a temporal perspective, the growth of cross-border commuting is easy to explain, as control-free crossing points allow individuals to travel freely back and forth across the border.

However, the recent wave of Polish immigration has triggered a rise of far-right political extremism, targeted especially against new residents, in a region which is already a stronghold of such movements.

Here, one should note an interesting comment by a former Polish ambassador to Germany. He noted that "people from Szczecin and Poznań are today indeed using Berlin-airports, and the EU gives new impulses to cooperation" but emphasised that "there is no determinism about regional development, which lies in the hands of local actors" (Hensel 2001, 4). This is of course somewhat simplified, as Warsaw still has strong control over cross-border policies, but the statement illustrates the central government's point of view in this question. A member of staff from the Polish interior ministry pointed out that while they "support the Oder partnership, they would rather see this cooperation take place within the German-Polish governmental commission" – in which the government in Warsaw has a stronger role (Rada 2008).

Szczecin's Geo-economic Orientations After 1989

According to an academic Polish regional expert, commentators, especially in Germany, tend to write a lot about cross-border cooperation in the Szczecin region, but there are not that many activities going on (Dutkowski 2009). One reason for this – or effect of this –may be that since the collapse of communist rule, Szczecin has not been a national or European centre of development (Dorsch 2003). It has clearly continued to be lag in comparison with similar-sized western Polish cities like Poznań, Wrocław and Gdańsk, which have experienced a period of strong economic boom, particularly before the current economic crisis. The city usually performs badly in GDP comparisons and liveability surveys (web Bulletinen 2009a). This is also true for the surrounding territory. Out of the 16 Polish Voivodeships, for instance, in December 2003, West Pomerania only ranked ninth with regard to the influx of FDI (Maack et al. 2005). At the same time, on the German side of the border, the nearest counties (*kreis*) are among the poorest ones in the Federal Republic.

In the mid-1990s, Mieszkowska (1996, 111–112) described Szczecin as a provincial capital with 410,000 inhabitants, noting that due to long prevailing political constraints, Szczecin did not continue its pre-war development to the west. She argued that "roads on its western outskirts were highly vulnerable to rapidly increasing traffic from the newly opened frontier and the sudden jump in car ownership". The infrastructure needed to accommodate new cross-border communications was not in place and had not been planned for.

During the years 2000–2003 alone, 40,000 jobs were lost in the West Pomeranian Voivodeship, half of them in the industrial sector. But since 2003, growth has been taking place. The main drivers of the economy are the food and furniture industries. Other sectors which are growing include medicine and optical devices, as well as tourism. Until 2003, maritime potentials had no significant effects on growth and labour market. Cooperation in the border region has so far taken place in the areas of construction, engineering, metal and wood processing, textile industry and retailing. But shared value creation chains are few. Most cooperation consists of exports from and imports to Poland. A greater amount of FDI is concentrated on the food, chemical, and furniture industries, but demand for the products of these investments does not stem from the adjacent German borderland (Maack et al. 2005) (even though the latter could benefit from such investment and manufacturing as well).

Despite recognising its structural deficits, some experts from Germany draw attention to the potential on the German side in areas such as the maritime economy, agriculture, renewable energy, enterprise-oriented services and tourism. At the same time, they acknowledge that half of the Vorpommern economy relies on the agricultural sector. They suggest that activating the industrial potential of the more industrially developed northeastern part of Brandenburg could lead to tighter cooperation, for example, in the chemical, paper and wood industries, in agriculture and in renewable energy (Maack et al. 2005).

Szczecin's airport has increasingly developed an international character. Since 2004, there have been flights to the British Isles and Ireland, including London, Liverpool, Edinburgh and Dublin, which received a large influx of migrant workers from Poland after it joined the EU that same year (Airport Szczecin Goleniów 2004). In 2012, there were also flights to Oslo and a range of charter flights to Egypt and Turkey (Port Lotniczy Szczecin Goleniów 2012). The airport handled over 500,000 passengers annually.

Szczecin's shipyard had long been the city's most significant employer. However, in the spring of 2009, the EU ruled that national subsidies to it violated internal market regulations and had to be repaid or else new owners found (Pop 2009). Initially, in summer, 2009, it looked like investors from Qatar were interested, a country with which Poland also signed a contract to import liquefied natural gas (LNG) for 20 years. The deal included the building of an LNG terminal at Świnoujście, planned since 2006. The Qataris withdrew, but construction of the terminal went ahead. In the autumn of 2009, it looked like the city would take over the shipyard and incorporate it into a special economic zone (web Bulletinen 2009b; 2009c), but in December 2011, the shipyard was declared bankrupt, and in early 2012 was undergoing liquidation. In late 2010, it was announced that a new gas pipeline would be built to connect the LNG terminal with the domestic gas pipeline network, enabling the transmission of natural gas throughout the whole of Poland. The project was cofinanced with 50 million euros from the European Energy Programme for Recovery (EEPR) and was intended to contribute to increasing the energy security of the EU member states through the cross-border infrastructural investments aimed at integrating the EU energy market. This was not only the first such port in Poland but the first in its region. In addition, in 2010, it was announced that the Faculty of Maritime Engineering at the Maritime University of Szczecin and the Faculty of Drilling, Oil and Gas at the AGH University of Science and Technology in Krakow would jointly teach a new course, LNG Transportation and Operation of Terminals. The first class graduated in March 2012.

The port of Szczecin-Świnoujście was doing rather well before the current crisis (which hit this sector particularly badly), even though until recently, it was the only large Baltic port to lack a handling plant for containers (Kirchner 2009). In 2006, much of the container traffic consisted of feeder transport from and to Hamburg and Bremerhaven (Kirchner 2009; Klauenberg 2009). Given the Pan-European role of those two ports and of Rotterdam, Szczecin-Świnoujście can at best expect to play a role as a regional distribution centre and even that in competition with Rostock-Warnemünde. There are at the same time strong political forces in centrally ruled Poland that disapprove of Szczecin's port authority turning towards Germany (Kirchner 2009).

The challenges of the shipyard and the port notwithstanding this paper assume that Szczecin's relative backwardness is also related to the city's geographically peripheral location. Typically, most states and political entities try to orient their peripheries towards their centres (Lundén 1997, 2004). One illustrative example of this is that from Szczecin there are more transport options to various distant settlements within Poland, let alone in the West Pomeranian Voivodeship, than to much closer ones on the German side. This can hinder the agglomeration from a more even expansion. At the same time, Szczecin's port suffers because it is distant from Poland's centres of economic growth, which are in the southern and central parts of the country – and nowhere in its borderlands – and are well connected to West European agglomerations by road, particularly in Southern Poland (Klauenberg 2009). This relative locational disadvantage is also observed by Musekamp, who notes that "… Szczecin (German: Stettin) is one of the largest cities in Poland, but at the same time it is one of the most indistinct on most Poles' mental maps. One reason is its location: Szczecin is in the extreme north-western corner of the country, on the border with Germany … Being a highly centralised state after 1945, Poland was not focused on its periphery; in the eyes of many Szczecinians today, the focus has not changed since the fall of Communism" (Musekamp 2009, 305).

A group of German experts have analysed the border region. Their findings are largely in accordance with the priorities of the city of Szczecin master plan. They differ on two points, however. One recommendation from the German side is that "the economic, technologic and human potentials of the German hinterland of the city should be taken more into consideration, together with a stronger, more integrative view of a common region". A second is that "the excellent maritime potential of Szczecin needs to be developed into the basis of the city's economy". The German authors find that today Szczecin is primarily becoming orientated along the axis to Berlin and/or a future corridor along the German border towards the Czech and Slovak Republics, as well as northwards to Skåne/Öresund (Maack et al. 2005, 18). Recently, the port has been taken over by a subdivision of Deutsche Bahn, DB Port Szczecin, which has modernised the port using EU money, and recently opened a new container terminal on 12 March 2012 and markets itself as a leading cold storage port; it describes itself as "the shortest distance from all north continental and Baltic ports to 5 capital cities: Berlin, Prague, Vienna, Budapest and Bratislava, and also shortest land passage against any other north continental/Baltic Port to main Adriatic ports Trieste and Koper for the Baltic - Adriatic landbridge,…Furthermore, Szczecin is the main hinterland port for Berlin" (DB Port Szczecin 2012).

What they refer to is the so-called Central European Transport Corridor (CETC), which is meant to "provide efficient cargo handling from the north to the south of Europe. It can create an area of economic infrastructure and social links between cities and regions and stimulate economic growth, especially among the new EU Member States" (CETC 2008–2010). The corridor is part of the E65 route connecting the Baltic with the Adriatic Sea but also links Malmö in Sweden with Chaniá in Greece.

Quo Vadis, Szczecin?

With regard to the CETC, officials in Szczecin and Skåne support its development (web Bulletinen 2009d), as do local businesses and the current operator of the Szczecin Port, DB Port Szczecin, though its long-term feasibility does not yet seem clear. Competing projects are already under construction, such as the Gdańsk-Budapest-Rijeka corridor or the Hamburg-Venice corridor. If the idea is to divert part of this north-south traffic through relatively impoverished, peripheral areas in order to avoid congestion on other corridors, then the development of the CETC might happen, but some experts question whether such transits bring any significant benefits to the areas they cross. According to a regional expert (Dutkowski 2009), the area's future lies rather in health tourism, as well as research and development (rather than as a logistical centre or a transport hub, for instance). There is some cooperation between the universities of Szczecin and Rostock and to a lesser extent Greifswald, but the potentials here are far from exhausted.

When it comes to the port and especially the shipyard, Szczecin's traditional bread givers, as the collapse and liquidation of the shipyard shows, Szczecin may find it difficult to compete with the neighbouring cities in Poland and Germany. After Germany lost Szczecin in 1945, the GDR built up Rostock as the country's main port. In Poland, the Gdańsk agglomeration was developed as the main port city. Today, Hamburg and Lübeck are the dominant ports not only in Germany but beyond, in Central and East Europe. Plans call for them to supply a hinterland which extends all the way to inside Russia. There are also physical geographic barriers to increasing the capacities of Szczecin's port. The river Świna is not deep enough for huge vessels and needs to be dredged regularly. The LNG terminal may to some extent be seen as a reaction to the forthcoming Russian-German Nord Stream pipeline ending up in Lubmin just 50 km west of Świnoujście, but both are the subject of some debate, as they could hamper the development of tourism.

Tourism, particularly health tourism, is growing on both sides of the border and constitutes one of the most significant sectors. Unlike several other above-mentioned industries, there also seems to be little cross-border competition in this field. The Polish side generally targets less affluent visitors. At the same time, customers from Germany are present on both sides of the border (Abraham 2009). Apart from wellness services (such as spas, hairdressing, personal grooming services and similar activities), health tourism is dominated by medical tourism, treatment of the elderly and general healthcare: for example, dentists in Szczecin increasingly receive clients from Germany (Herrick 2007) but also from Scandinavia, due to geographic proximity. There is arguably an additional development potential. The demographic outlook in Europe, not least in Germany and Poland, is that the trend towards an ageing population will continue, while healthcare and the healthcare sector are increasingly liberalised. In some countries and cases, EU patients can go to other member states if they can get treatments faster or cheaper and be reimbursed by their national healthcare providers.

Until recently, services constituted a relatively under-researched field within economic geography. Hermelin and Rusten (2007) have long been exceptions and note that more and more services are deterritorialised due to the spread of ICT. Health- and elderly care are arguably to a certain extent different in this regard, as in most cases physical contact between providers and clients is unavoidable. Polish healthcare practitioners used to emigrate en masse, but this is changing. Wages have increased in Poland, while the main destination countries (UK, Ireland) are particularly badly hit in the current economic crisis. At the same time, some clients cannot be transported over long distances. When it comes to elderly care, studies and surveys also show the importance of geographic proximity for the clients when choosing their new places of residence.

Conclusion

We have seen that regardless of political considerations, Szczecin's links with the outside world depended on the dominant means of transportation at a given time period. When maritime transport prevailed, contacts with other port cities flourished. When rail and road connections were built, links towards continental Europe became at least as important. More recently, low-cost flights are connecting Szczecin to other European cities. Generally, the leapfrog approach so far seems somewhat more correct in having predicted increased contacts with other metropolitan regions rather than with the direct neighbourhood, especially the German side.

It is clear that the geo-economic orientations of a city like Szczecin, with its rich but difficult history, are a matter of not only political but even theoretical debates. In these debates, path dependencies and historical references are used by commentators and experts. This can be both positive and negative. On the one hand, some necessities for cooperation with Germany exist due to geographic environmental realities (e.g. management of the Oder River and Szczecin Lagoon, new Polish settlers in the region and so on). On the other hand, some commentators may become blind to some realities, such as that the two sides of the former historical region (which has of course not at all always been a coherent piece of territory) have developed within different frameworks and conditions for 67 years and continue to do so today. Nevertheless, increasing cross-border mobility and especially the new group of cross-border settlers – if they become permanent – might contribute to making this still very sharp cultural and mental boundary less rigid.

Discourses around Szczecin's orientations seem to be influenced by the respective commentators' geographic and national background, with a sometimes quite explicit desire in Germany for "reviving the historical Pomeranian region". Some actors, especially in north-western Poland, present a similar line of thinking, but there seems to be a higher level of cautiousness east of the Oder. For example, there are only 4 road border crossings in the immediate Szczecin region (2014), although there were many more before 1945. A conflict of interest is to some extent apparent between the central powers in Warsaw, responsible for cross-border links among other things, and local ones – Szczecin and the Polish Voivodeship along the border – justifying the thesis that states with a liberal economy but regulated foreign trade, surrounded by the same type of states, tend to disfavour locations at the boundary (Lundén 2004). Indeed, the central administration seems, to some extent, to find itself in a dilemma about letting Szczecin go its own way while keeping it loyal to Poland: bad policies can be counterproductive and may result in economic hardship and an even more marginalised border region or one strongly leaning towards Germany. On the other hand, investment by the private sector, such as DB Port Szczecin, and a response to market forces may lead to further integration.

Szczecin then tends to fit with rescaling theory's suggestion that cities are increasingly "living their own lives" through interurban partnerships, networks, etc. At the same time, the self-manoeuvring capacity of a city in Poland is more limited than in, for example, Germany, due to different legal administrative state structures (which can of course be a fortune or misfortune with, for example, Berlin suffering from huge financial difficulties). The role of such structural differences is perhaps less taken into account by rescaling theory.

Central place theory seems to have considerable relevance in explaining some patterns of spatial relations in our case. Until recently, within Central and East Europe, all planning was the preserve of the states' national government, with little regard to market mechanisms. For the Szczecin area, Berlin is becoming an important transport hub, not least for flying. As well, Poles from Szczecin are suburbanising into Germany, and the city's healthcare practitioners are successfully attracting German and Scandinavian clients.

But this insight should not lead one to reconstruct historical regions. Instead, it is necessary to maintain a focus on analysing concrete areas of potential cooperation. What Szczecin's future foci of orientation ought to be should predominantly depend on which sectors the metropolitan area aspires to concentrate on and on the type of infrastructure preferred to sustain those links. If the health tourism potential is to be further exploited, the links to proximate, large and affluent metropolitan regions, like Hamburg, Öresund and to some extent Berlin and Poznań, need to be advanced. If more resources are to be invested in R&D that requires less local collaboration, as R&D is generally less distance dependent, even if closer cooperation might be necessary between, for instance, environmental research institutes in the region. A lower dependence on geographic proximity is also characteristic of some ICT-based services such as international companies' call centres and back-office financial administration, of which there are more and more in Poland. Thus, while acknowledging the necessity of physical contact in some fields, we should not be blinded by the ways geography and history can be used to legitimise or support various politico-economic contacts. It is with this in mind that one should approach the (re)constructed spaces of a "Pomeranian region", "Baltic region", "Central Europe" and the like.

Further Research

In Western Europe, the phenomenon of agglomerations growing across international borders has been present for several decades. Perhaps the most important milestone was the elimination of border controls through the Schengen Agreement's gradual implementation in the early 1990s. Since then, cities like Strasbourg, Lille, Karlsruhe, Maastricht, etc. have taken on a more cross-border character. More recently, parallel trends can be observed in East-Central Europe, with Trieste,

Oradea and Bratislava being just a few examples. The latter's increasing ties with Vienna are in some aspects similar to Szczecin's with Berlin: Slovak scholars noted the intensified influence of Vienna airport on Bratislava's air traffic after the opening of the border to Austria already in 1997 (Horňák and Kováčiková 2003). With more possibilities for at least local traffic, cities like Uzhhorod, Brest and Hrodna can at least hypothetically experience a similar development.

Acknowledgements I am grateful to the Baltic Sea Foundation at Södertörn University, Sweden, for the financial support of our project. I would like to thank my colleagues, Prof. Thomas Lundén, Prof. Anders Mellbourn and Dr. Joachim von Wedel for their valuable comments. I am also thankful to the Herder-Institut in Marburg, Germany, for providing me with a scholarship to study the area in their mighty archives.

References

Airport Szczecin Goleniów (2004_ Business and services – statistics. Available at www.airport.com.pl/en/statystyka.php. Last accessed on 11 Nov 2009

Brenner N (1999) Globalisation as reterritorialisation: the re-scaling of urban governance in the European union. Urban Stud 36(3):431–451

Brenner N (2003) Metropolitan institutional reform and the rescaling of state space in contemporary Western Europe. Eur Urban Reg Stud 10(4):297–324

Brenner N (2006) New state spaces: urban governance and the rescaling of statehood. Oxford University Press, Oxford

Buchweitz R, Dobbert R, Wolfhard N (2007) Eisenbahndirektionen Stettin, Pasewalk und Greifswald 1851–1990. Bernd Neddermayer, Berlin

Castells M (1996) The information age: economy, society and culture, vol 1, The rise of the network society. Blackwell, Malden

CETC – Central European Transport Corridor 2008–2010. About the agreement – opinions. Available at www.cetc.pl/index.php?lang=en&mod=oporozumieniu&action=opinie. Accessed 17 July 2009

Ciok S (2009) Oddziaływanie Berlina na pogranicze zachodnie Polski. In: Jańczak J, Musiał-Karg M (eds) Pogranicze polsko-niemieckie po 2004 roku. Nowa jakość sąsiedztwa? Wydawnictwo Adam Marszałek, Toruń, pp 135–152

City of Szczecin (2002) Strategia Rozwoju Miasta Szczecina (City of Szczecin Development Strategy). Available at www.szczecin.pl/Strategia/_pl. Accessed 18 Aug 2009

De Geer S (1912) Storstäderna vid Östersjön'. YMER 32:41–87, + figures

Deutsche Städte (2009) Herzlich Willkommen in Schwedt – Geschichte in Stichworten. Available at http://www.deutsche-staedte.de/schwedt/index.html. Accessed 11 Dec 2009

Deutsche Bahn (2012) Fahrplan und Buchung http://www.bahn.com/p/view/index.shtml. Accessed 22 June 2012

Deutsches Polen-Institut (2009) Polen in Ostdeutschland. Polen Analysen, No. 49. Deutsches Polen-Institut, Darmstadt. Available at www.laender-analysen.de/polen/pdf/PolenAnalysen49.pdf. Accessed 20 May 2009

Dorsch P (2003) Nationale und transnationale Vernetzung polnischer Städte und Regionen – Auf dem Weg zu einer nachhaltigen Stadt- und Regionalentwicklung. Discussion paper Nr. SP IV 2003-106. Wissenschaftszentrum für Sozialforschung, Berlin

Europakorridoren AB (2009) What is the European Corridor? Available at http://www.europakorridoren.se/e_om.asp?id=1. Accessed 20 July 2009

Flint C, Taylor PJ (2007) Political geography. World-economy, nation-state and locality. Pearson, Harlow

Götz N, Hackmann J, Hecker-Stampehl J (2008) Kartan i huvudet – ett tyskt perspektiv på Östersjöregionens mentala kartering. YMER 128:301–320

Güßefeldt J (2005) Die Raumwirtschaftstheorien von Christaller und Lösch aus der Sicht von Wirtschaftsgeographie und "new economic geography", Göttinger Geographische Abhandlungen. Verlag Erich Goltze, Göttingen

Hensel K (2001) Die Chancen der EU-Erweiterung im Fokus. Uni on (Zeitung der Europa-Universität Viadrina Frankfurt - Oder) 30:3–4

Hermelin B, Rusten G (2007) The organizational and territorial changes of services in a globalized world. Geogr Ann 89B(Suppl 1):5–11

Herrick DM (2007) Medical tourism: global competition in health care. NCPA Policy Report, No. 304. National Center for Policy Analysis, Dallas

Horňák M, Kováčiková Z (2003) Transport node of Bratislava in relationship with Pan-European multimodal corridors. In: Kowalczyk A (ed) Theoretical and methodological aspects of geographical space at the turn of century. Warsaw University, Warsaw, pp 143–150

Jańczak J (ed) Conflict and cooperation in divided cities. Logos, Berlin

Kirchner D (ed) (2009) Der Hafen Stettin und seine Hafenbahn. Historischer Arbeitskreis Stettin, Lübeck

Klauenberg J (2009) Modellierung und Typisierung von Distributionszentren in Seehäfen. Eine Untersuchung an Beispielen aus der Nordrange und dem Ostseeraum. Dr Kovač, Hamburg

Kommunalgemeinschaft Europaregion Pomerania (2007) Members of Euroregion Pomerania. Available at www.pomerania.net/main.cfm?l=en&rubrik=5&th=32#10. Accessed 12 June 2009

Kommunalgemeinschaft Europaregion Pomerania (2008) Was ist die Euroregion Pomerania? Available at www.pomerania.net/main.cfm?l=de&rubrik=5&th=30. Accessed 12 June 2009

Krugman PR (1991) Geography and trade. MIT Press, Boston

Land Berlin (2003) Positionspapier zur zusammenarbeit des Landes Berlin mit Mittel- und Osteuropa – Grundsätze und Schwerepunkte – vom Senat 14. 10. 2003. Available at www.berlin.de/imperia/md/content/rbm-skzl/europareferat/95.pdf. Accessed 23 Nov 2009

Land Berlin (2005) Positionspapier "Interessen des Landes Berlin in der Oderregion". Beschlossen vom Senat von Berlin, 15. 11. 2005. Available at www.berlin.de/imperia/md/content/rbm-skzl/europareferat/moe/positionspapier_oderregion.pdf. Accessed 16 Jan 2010

Lorot P (1997) De la géopolitique á la géoéconomie. Géoéconomie 1(1):23–35

Lundén T (1997) Makten över marken. En politisk geografi. Studentlitteratur, Lund

Lundén T (2004) On the boundary. About humans at the end of territory. Södertörns högskola, Huddinge

Lundén T, Mellbourn A, von Wedel J, Balogh P (2009) Szczecin. A cross-border center of conflict and cooperation. In: Jańczak J (ed) Conflict and cooperation in divided cities. Logos, Berlin, pp 110–121

Maack K, Grundmann M, Kreft J, Lewandowska A, Voss E (2005) Wachstumspol Stettin und Auswirkungen auf die Entwicklung der deutsch-polnischen Grenzregion. Hans-Böckler-Stiftung, Düsseldorf

Meynier A (1937) La construction des autostrades en Allemagne. Annales de Géographie 46(264):657

Mieszkowska K (1996) City profile Szczecin. Cities 13(2):109–116

Murashova A (2010) Latvia asks: who are we? The Baltic Times, vol. 10, No. 696, p. 5

Musekamp J (2009) Szczecin's identity after 1989: a local turn. In: Gelazis N, Ruble AR, Czaplicka J (eds) Cities after the fall of communism. Reshaping cultural landscapes and European identity. The Johns Hopkins University Press, Baltimore, pp 305–334

Piskorski JM (ed) (1999) Pommern im Wandel der Zeiten. Zamek Książąt Pomorskich, Szczecin

Piskorski JM, Wachowiak B, Włodarczyk E (1994) Stettin. Kurze Stadtgeschichte. PTPN, Poznań

Pop V (2009) Poland sells shipyards after EU ruling. EUobserver, 29 May 2009. Available at http://euobserver.com/9/28209. Accessed 26 June 2009

Port Lotniczy Szczecin Goleniów (2012) Dla Podróżnych. Available at http://www.airport.com.pl/pl/show/start-page. Accessed 22 June 2012
DB Port Szczecin (2012) Company profile. Accessed at http://portszczecin.deutschebahn.com/dbportszczecin-en/start/company/profile.html. Viewed 22 June 2012
Rada U (2008) Ehe oder Partnerschaft. Die Tageszeitung, 19 March 2008. Available at www.taz.de/regional/berlin/aktuell/artikel/1/ehe-oder-partnerschaft/. Accessed 24 Nov 2009
Sassen S (1991) The global city. Princeton University Press, Princeton
Savitch HV, Kantor P (2002) Cities in the international marketplace. Princeton University Press, Princeton/Oxford
Statistisches Bundesamt Deutschland (2012) Regional Statistik. Available at www.regionalstatistik.de/genesis/online;jsessionid=2062179E86A704F2520A1D0D9D002A47. Accessed 22 June 2012
Stokłosa K (2003) Grenzstädte in Ostmitteleuropa: Guben und Gubin 1945 bis 1995. Berliner Wissenschafts-Verlag, Berlin
Swyngedouw E (1997) Neither global nor local: "Glocalisation" and the politics of scale. In: Cox KR (ed) Spaces of globalization. Guilford Press, New York, pp 137–166
van der Heiden N (2007) City's foreign policy – competitiveness through citizens' exclusion? Paper prepared for the EURA conference, Glasgow, 14 September
Wahl J (2009) Der Finowkanal – Bau, Geschichte, Gegenwart. Wasser- und Schifffahrtsamt, Eberswalde, Available at http://www.wsv.de/wsa-ebw/wir_ueber_uns/wasserstrassen/finowkanal/index.html. Accessed 11 Dec 2009
Web Bulletinen (2009a) Poznan är Polens bästa stad att bo i. 23 June 2009. Available at http://bulletinen.org/wordpress/2009/06/23/poznan-ar-polens-basta-stad-att-bo-i/. Accessed 26 June 2009
Web Bulletinen (2009b) Varven är fortfarande inte sålda, 9 Jan 2009. Available at http://bulletinen.org/wordpress/2009/09/01/varven-ar-fortfarande-inte-salda/. Accessed 22 Nov 2009
Web Bulletinen (2009c) Szczecin kan överta varvet, 11 Nov 2009. Available at http://bulletinen.org/wordpress/2009/11/02/szczecin-kan-overta-varvet/. Accessed 22 Nov 2009
Web Bulletinen (2009d) E65 i Polen ska utvecklas, 7 Oct 2009. Available at http://bulletinen.org/wordpress/2009/10/07/e65-i-polen-ska-utvecklas/. Accessed 22 Nov 2009

Interviews and Communications

Abraham K (2009) Employee at the tourist information centre, Świnoujście – personal interview on 13 October at the centre
Dutkowski M (2009) Professor of geography at the Dept of Regional and Urban Planning, University of Szczecin – personal interviews on 16 July and 12 October at the department.
Michalski K (2009) Public official at the City Office of Szczecin – e-mail on 27 July

Chapter 10
Transformation Processes in the Former Black Triangle

Rick Glöckner and Isolde Roch

Abstract This paper reviews economic and demographic changes in those parts of Germany, Poland and the Czech Republic in the former 'Black Triangle', tracking differences and similarities in the trajectory of change. The region covers 34,000 km^2 and contains 5.7 million people. In 1989, it was one of the most polluted in Europe, but by 2002, remediation was largely complete. The area is mountainous and home to large deposits of minerals, particularly brown coal. Approximately 80 % of the Czech Republic's energy was produced in the region, as was much of eastern Germany's. The areas in the three countries started with similar problems, but have followed different development trajectories. In 1989, the three countries changed from centrally planned one-party systems to pluralistic, free-market democracies. There was a shift away from heavy industry and labour-intensive agriculture. As part of the EU, German regions benefitted from EU and German Federal assistance. The Polish and Czech regions had only limited assistance from the EU and their national governments. However, the pace of change was more rapid in the German regions, leading to depopulation and high unemployment. The area has passed through four stages of growth: 1989–1993, shrinking to health; 1994–1999, restructuring; 2000–2005, stabilisation on a new level; and 2006 onwards, divergence and growth. The first phase saw the closure of many firms and privatisation of the rest, with a rise in unemployment and outmigration. In Germany, decline persisted in the second phase, but in Poland and the Czech Republic, the situation improved. In the third phase, new firms and new growth clusters appeared. In the latest phase, while heavy industry has continued to decline throughout the region, growth has increased in the Czech Republic and Poland, reflecting both national trends and the local situation. However, there are still gaps in infrastructure. Today, there is considerable potential for cross-border cooperation and development. Saxony, the Land in which the German area is located, has extensive experience in corporate fund raising and applying for and

R. Glöckner (✉)
University of Bamberg, Bamberg, Germany
e-mail: rick.gloeckner@uni-bamberg.de

I. Roch
Leibniz-Institut für ökologische Raumentwicklung, Dresden, Germany
e-mail: i.roch@ioer.de

© Springer Science+Business Media Dordrecht 2015

J. Buček, A. Ryder (eds.), *Governance in Transition*, Springer Geography,
DOI 10.1007/978-94-007-5503-1_10

using EU aid. Labour costs in Poland and the Czech Republic are below those in Germany and below the European average. However, barriers to cooperation remain in the form of legal frameworks, lack of institutions to facilitate cross-border management of projects, different degrees of centralisation in the three countries and lack of a stable forum to build consensus and develop programmes.

Keywords Black triangle • Industrial restructuring • Economic transition • Cross border relations • Local policy formation

The 'Green Triangle' research study was carried out by the Leibniz Institute of Ecological Urban and Regional Development (IOER), the University of Economics in Prague and the Lower Silesian Centre for Local Growth (Dolnośląskie Centrum Rozwoju Lokalnego – DCRL) in Wrocław. It traced development in the triangle in the area of the joint borders of Germany, the Czech Republic and Poland. This paper deals with the economic and population changes which have taken place in the three national parts of the research area and their similarities and differences. First, the research area is introduced, followed by an overview of the project's research work. After that follows a description concerning the area's redevelopment, divided into specific phases. At the end, a short conclusion highlights the potential and constraints of further common development in the research area.

Introduction: The Former 'Black Triangle'

Location

The area once known as the 'Black Triangle' is located in the tri-border region between Germany, Poland and the Czech Republic. In the past, the term was mainly used in a fuzzy way. In this project, the borders, defining the research area, were determined through mutual agreement between the partners involved with the project taking into account administrative conditions and definitions used in the previous publications (for example Abraham et al. 2003). As defined in the 'Green Triangle' project, the research area covers approximately 34,000 km^2 inhabited by 5.7 million people (Glöckner 2008). The research area consists of NUTS III districts located within the German Land of Saxony, the Polish Voivodeship of Lower Silesia (Dolnośląsk) and the Czech Oblasts of Northwest (Severozápad) and North East (Severovýchod) which together constitute Northern Bohemia. They are shown in Table 10.1 and Fig. 10.1.

Until April 30, 2004, Saxony was on the European Union's external border. After the accession of the Czech Republic and Poland to the European Union in 2004, the region moved from a marginal position, a more central one within Europe.

Table 10.1 Administrative districts belonging to the research area

Germany	Poland	Czech Republic
Chemnitz, Stadt	Powiat grodzki Jelenia Góra	Okres Ceska Lipa
Landkreis Erzgebirgskreis	Powiat boleslawiecki	Okres Chomutov
Landkreis Mittelsachsen	Powiat dzierzoniowski	Okres Decin
Landkreis Vogtlandkreis	Powiat jaworski	Okres Jablonec
Landkreis Zwickau	Powiat jeleniogórski	Okres Karlovy Vary
Dresden, Stadt	Powiat kamiennogórski	Okres Liberec
Landkreis Bautzen	Powiat klodzki	Okres Most
Landkreis Görlitz	Powiat lubanski	Okres Semily
Landkreis Meißen	Powiat lwówecki	Okres Sokolov
Landkreis Sächsische Schweiz – Osterzgebirge	Powiat strzelinski	Okres Teplice
	Powiat swidnicki	Okres Trutnov
	Powiat walbrzyski	Okres Usti nad Labem
	Powiat zabkowicki	
	Powiat zgorzelecki	
	Powiat zlotoryjski	

General Conspectus of Geography, History and Economy

The Green Triangle is characterised by a complex geography. Low mountain ranges dominate the borders between Saxony and Northern Bohemia (Ore Mountains) as well as between Northern Bohemia and Poland (Giant Mountains). The crest line of the Ore Mountains with Klínovec (1,244 m above sea level) as the highest forms a natural border. To the south, it shelves steeply into basins around the cities of Most and Sokolov. In the north, it is more even and merges into the Ore Mountain Foreland. The Giant Mountains, which form the biggest part of the Polish–Czech border in the research area, are much steeper on the Polish side. They adjoin the Silesian Lowlands in the East. The Giant Mountains reach heights of 1,602 m above sea level (Sněžka) on the Czech side (Abraham et al. 2003). The River Elbe and the landscapes it has created, like the Elbe Sandstone Mountains, and Upper Lusatia determine the landscape of eastern Saxony and the bordering parts of the Voivodeship *Dolnośląskie*. These are other important parts of the research region's complex orography.

Because of its natural resources, the research area became a distinctive economic region with a long industrial history. By in the Middle Ages, ore and silver mining started in the Ore and Giant Mountains, which led to early and dense settlement, especially in settlements where silver was found. Thereupon, industrial branches, based on mining, were established. Their spatial dimensions led to a high density of settlements reaching the ridges of the Ore Mountains. This has been the foundation for the important role the region later played as a centre of industrialisation in continental Europe.

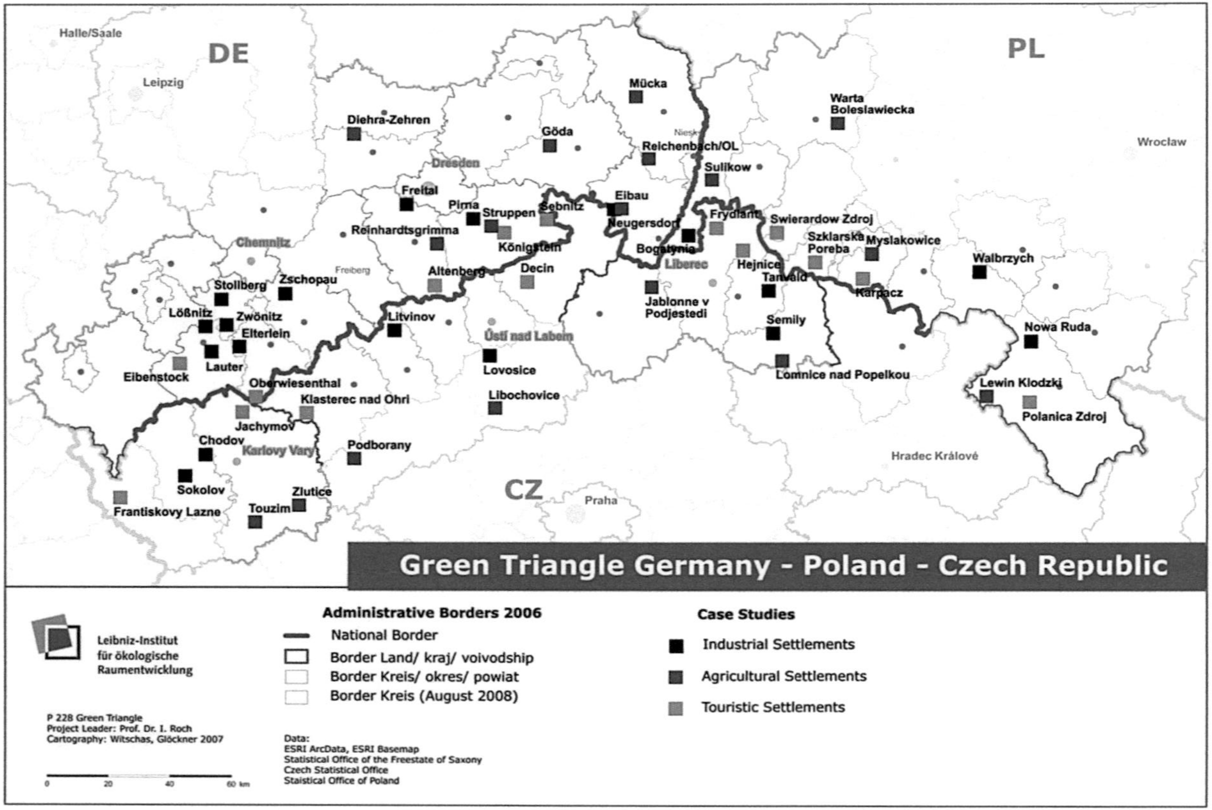

Fig. 10.1 The research area and the case studies (Source: design by Witschas and Glöckner, IOER, adopted by Glöckner)

Before World War II, the triangle was diversely structured. Besides agriculture and mining, light industry, machine building, the automobile industry, the chemical industry, electrical engineering and the pharmaceutical industry played important roles. Northern Bohemia and Saxony played important roles in the industrialisation in their countries. For example, 12 out of 32 Czech cities with more than 20,000 inhabitants were located between the Egerland and the Giant Mountains at the beginning of the twentieth century. Because of thermal springs along the mountain ranges, Northern Bohemia also became famous all over the world for its spas like Karlovy Vary and Mariánské Lázně in the nineteenth century. During and after World War II, the economy became dependent on heavy industry, especially coal mining, fuel and energy production (Prucha 1998). Existing connections between small and medium enterprises across the borders were cut because each country concentrated more on its individual state economies (Leick 2006). Coal-mining areas developed in the Northern Bohemian Basin in the Czech Republic, in Lusatia in Germany and around Turów in Poland, each of which functioned as a national centre for energy production and the chemical industry. Approximately 80 % of the Czech Republic's energy was generated in the basin around Most (Stutz 2005). The River Elbe was an important route of transport and an important factor for the establishment of the chemical, textile and iron industries in the Czech Republic (Prosek 1994). Therefore, Northern Bohemia became an over-industrialised region (Prucha 1998).

This high concentration of mining and industrial areas had major impacts on population development and the environment. As a result, the tri-border region gained its image as the 'Black Triangle'. At one point, 30 % of Central European sulphur dioxide emissions were produced in the research area. The air pollution led to one of the biggest forest declines in Europe. Between 1981 and 1987, approximately 15,000 ha in the Ore Mountains and approximately 11,000 ha in the Sudety Mountains fell victim to pollution (Espere Climate Encyclopedia 2003). Intensive coal mining in the area also caused huge changes in the relief. Open cast mining left basins and stockpiles behind. Although recultivation played a big role in planning and implementation policies during this period (Katzur 1979; Štýs 1996a, b), it was not possible to cope with the increasing need for coal and its growing importance as an energy source (Berkner and Thieme 2005), and the plans for recultivation and landscape restoration could only partly cover the actual needs. This caused serious damage to the landscape, leaving behind extensive brown field sites and industrial structures.

With the political changes in 1989, the region entered a new era determined by political and economical transformation processes, which not only changed the economic and political systems but also had major impacts on the living conditions of its inhabitants. With EU-accession, the tri-border region gained a new central position in Europe. At the same time, Poland and the Czech Republic became attractive locations for investments.

The IOER 'Green Triangle' Project

As mentioned in the introduction, the 'Green Triangle' project, under the leadership of the Leibniz Institute of Ecological Urban and Regional Development (IOER, project management – Isolde Roch), examines development in the tri-border region of Germany, Poland and the Czech Republic since the political changes of 1989, looking at principles, trends, and development goals which were used in the formulation and adoption of development concepts in the border regions at the beginning of the 1990s. With its project partners, the University of Economics in Prague and the DCRL, the project team examined the present state of development with reference to aspects of sustainability, ecology, economy and society. The project has followed existing projects and programmes, evaluated their success and constraints and examined future perspectives. Selected questions are:

- How did planning activities, concepts and goals influence transformation processes in different parts of the research area? What are the differences and similarities between countries?
- What processes of renewal and development were initiated? What constraints did they face and how were they implemented?
- How were the processes guided? What influence did political and financial instruments have on the development processes? Which instruments had the highest impacts?
- What standards for quality of life and environment were achieved? What are the differences and the similarities between the three parts of the research region?
- What are the future perspectives for further cross-border cooperation? What are their options in form and content?

For the evaluation, the research questions were divided into three thematic fields: ecology and environmental pollution, economic transformation processes and settlement development.

For each of these topics, comparable indicators were compiled, which showed development for different years (mainly 1989, 1995, 2000 and 2005). For some topics, data are missing or are incomplete for specific locations, areas or years, because the different national statistical offices did not provide comparable data sets.

Given the fact that not all settlements could be analysed in terms of their development, case studies were chosen in mutual agreement between the project partners based on the settlement's dominant economic function in 1989 (industry, agriculture, tourism; see Fig. 10.1). To display and estimate their development, interviews with mayors (or with other authorities, e.g. of the building departments) in connection with statistical data were used. This procedure made it possible to give more detailed information about the settlement development than just using statistical data, which mostly were only available for administrative districts bigger than municipalities. But, such statistical data have also been useful in completing a picture of change. The interviews followed a standard format to secure comparability in each country.

This paper concentrates on population development under the impact of the general transformation processes in the three parts of the former Black Triangle, using the qualitative and quantitative data gathered for the project as well as secondary literature. Specific developments are illustrated and compared to each other (Roch 2009a, b). For 1989 on, four phases of development are visible, which rely on synopses created for the fields of ecology/environment, economy, social aspects and instruments/process control. These phases of development relate to the following periods:

- 1989–1993: anticipating growth and shrinking to health
- 1994–1999: restructuring and shrinking
- 2000–2005: stabilisation on a new level
- 2006 onwards: divergence versus a win-win situation

For every subject area, the general transformation processes which have taken place in the partner countries are shown, followed by the actual impacts these processes had on the regional development, especially on the population and settlement development. Different approaches to the occurring problems are mentioned. After a comparison between the different processes and their outcomes, a short conclusion is drawn, and some development possibilities and constraints are presented, based on current results from the project 'Green Triangle' (Roch 2010).

Development Phases in the Former Black Triangle

After 1989, the three national economies in the research area faced a fundamental change from a socialist-centralised system based on heavy industry to a democratic society mainly based on services. They approached the challenges coming from different preconditions and dealt with the economic and political changes in different ways.

Between Anticipating Growth and Shrinking to Health (1989–1993)

Saxony

After the political change of 1989, a rapid structural change in the Saxon part of the Black Triangle set in. Because of the special situation of East Germany belonging to a now united German state, Saxony showed clear differences in development, compared to the other two parts in the research area in these early years. A new administration based on Western German models was installed quickly, following largely the model of Baden-Württemberg. In 1992, a new Saxon constitution was proclaimed, which contained important elements demanded in protests which took place before German unification. The administrative organisation of the Land of Saxony was based on existing federal laws (Hübler and Cassens 1993). The transformation of the economy was placed in the hands of the Treuhand Agency, which was responsible for the privatisation of former nationally owned facilities, following the principles of the free market. At the time of its inception, the Treuhand Agency owned over 95 % of the enterprise sector. With this method of economic transformation, the privatisation process was completed very rapidly. By 1994, the privatisation of small- and medium-scaled enterprises was nearly completed (Dornbusch and Wolf 1994).

In this first phase, a serious population decrease set in. The rate of decline lessened after a few years, but has continued throughout the period covered by our research, and is expected to continue in the future (Banse and Jeřábek 2009; Statistical Office of the Free State Saxony 2010). This has been a direct consequence of the economic breakdown after German unification. The mining industry, agriculture and heavy industry were affected the most. In these sectors, the decline of jobs was very high. For example, the number of people working in heavy industry (excluding construction) decreased in all administrative districts by 50–70 % from 1991 to 2004. The biggest share of these losses occurred from 1991 to 1992, when most enterprises were privatised. Unemployment rates hovered around 15–19 % in individual districts (Statistical Office of the Free State of Saxony 2010). There was no gradual transition but rather an economic collapse. Statistics show that new jobs did not appear in the expected or required dimensions although the building and the service sectors had an increasing share in the whole economy. The number of unemployed tripled from 1990 to 1992 to a peak of approximately 360,000 people in all of Saxony, which equals an unemployment rate of about 16 %. This was a clear maximum in the first phase (Bundesagentur für Arbeit 2010).

The loss of jobs and the resulting insecurity led to a rapid decrease of births, which reached an overall low in 1994. Another important consequence was a high level of emigration in these early years. In 1990, approximately 3 % of the Saxon population left the Land. In 1993, net migration was positive for the first time since 1989, but due to the decrease of the number of births, the population of Saxony continued to shrink. Because of these two negative developments, the Saxon research area lost approximately 6 % of its population (equalling some 228,000 people). The biggest annual loss occurred in 1990. From December 31, 1989 to December 31, 1990, the population decreased by 3 %, about 98,000 people (Statistical Office of the Free State of Saxony 2010).

Parallel to these processes, the rapid renewal and upgrading of the technical infrastructure led to perceptible advances. German Federal and Saxon Land financial aid programmes from 1990 onwards have been mainly responsible. They focused on funding the development of transportation infrastructure and the renovation of residential buildings. Another focus was aimed at dealing with the serious environmental problems and damage, especially air and water pollution, brownfield sites and the consequences of mining. Remediation took the form of the construction of sewage treatment plants, flue gas desulphurisation facilities and the changeover to modern heating systems, as well as the refurbishment of the Wismut plant and the post lignite mining areas. But the official aid programmes also addressed the negative effects of economic transformation on the labour force, e.g. the Saxon programme 'Labour and Qualification'. These aimed to overcome the rising number of unemployed workers and the rising share of underqualified people (SMU 1995).

The German administration had no concept of using EU-funds for the newly formed East German states, because the transition process was considered an inner-German issue. From 1992 to 1993, the East German states claimed their areas to be considered as Objective 1 regions. About 50 % of the structural funds going to Germany reached East Germany and 26 % of these funds reached Saxony. In this phase, the German part of the research region was not able to take part in European initiatives for regional development (Krämer 2002).

Czech Republic

Political, economic and social transformation in the Czech Republic were characterised by a 'dilemma of simultaneity' (Bertelsmann Stiftung 2003b). As in Saxony, the political system and the economy had to be transformed, but with the division of the Czechoslovak Federative Republic in 1992, the new Czech Republic additionally had to face the challenge of building a new nation state (Bertelsmann Stiftung 2003b). With the economical reform, adopted in 1990, the transformation was accomplished through a radical form of voucher privatisation, which proceeded rapidly. Some strategically important enterprises (like telecommunication and gas supply) stayed state owned. In 1991, the process started with the so-called small privatisation, which focused on small enterprises (Dauderstädt 2004). Within the Czech Socialist Republic, 97 % of enterprises were state owned, the highest level in the Central and Eastern European states. This was one reason for the voucher method. But an inadequate legal and institutional framework led to problems with this method of transformation so that in hindsight the voucher privatisation process has often been considered a failure (Bertelsmann Stiftung 2003b). The legal foundations for the whole process were laid from 1990 to 1992. A Ministry for Privatisation as well as a fund for the restitution of properties was established. Prices and the external trade were liberalised early in 1991. The banking system's transformation also started in 1991, but was not finished until 2001. In 1993, the Czech Republic joined the International Monetary Fund (IMF) so the country could benefit from IMF credits as well as from the World Bank to finance the transformation (Bertelsmann Stiftung 2003b).

In the research area, the share of people working in industry decreased from 60 to 44 %, while the share of people in services rose from 30 to 51 % in the first development phase (Glöckner, Slavik and Perduta 2009). Industrial production declined about 30 %. Because Northern Bohemia was so strongly characterised by mining and industrial branches, as well as by the oversupply of labour, unemployment increased more than anywhere else in the Czech Republic (Roch et al. 2000). But in comparison to the zones in the other two countries which were part of the former Black Triangle, unemployment stayed moderate and the population development mostly stayed stable at the NUTS III level (in Czech 'kraje'). Even in- and outmigration were mostly balanced at the NUTS III level (Czech Statistical Office 2010). Rental prices went up only moderately. New housing construction was slow to start, but the designation of business parks and wholesale facilities occurred relatively rapidly.

Attempts to deal with the negative consequences of the economic transformation were largely uncoordinated. This impeded the efficiency of institutions and mechanisms in increasing employment in problematic areas, like Northern Bohemia, and regional development faltered in the immediate post-transition period (Maier 1999).

Poland

In 1989, Poland installed the first noncommunist government in Central and Eastern Europe. In 1990, the administration adopted a National Programme for the Economic Stabilisation and Renunciation of the Centralised Economy. The subsequent transformation processes in Poland have often been described as shock treatment because free-market structures were founded in a very short time. On the one hand, this led to very fast changes and subsequent rapid economic growth. However, GDP initially decreased sharply in the first 2 years after the start of the transformation process, falling by -11.6 % in 1990. In 1990, a short hyperinflation of about 500 % took place. The unemployment rate rose to 16.7 % for whole Poland by 1993 (Dauderstädt 2004).

The privatisation of small enterprises started early in 1990 (Dauderstädt 2004) and was executed by using investment funds in a manner similar to the voucher privatisation programme in the Czech Republic. However, the transition to a private economy in Poland proceeded slowly, so that even in 2002, there were still noticeable deficits concerning the privatisation of big steel and mining companies (Bertelsmann Stiftung 2003a; Buras 2005).

In the research area, unemployment grew more rapidly and was higher than elsewhere in Poland because of the large share of workers in the mining and industrial sector. In 1993, 26 % of the employable population was without jobs. This led, amongst other outcomes, to an emigration of younger population groups, so that the age distribution worsened. But the overall population of the Voivodeship Dolnośląskie stayed stable, suggesting that internal migration, probably concentrated in larger central places, played a big role on the voivodeship level. From 1990 to 1998, the population grew, but only by about +0.6 % (Krätke and Borst 2004).

After 1989, the EU supported Poland with the PHARE programme (Poland and Hungary: Aid for Restructuring of the Economies), which was initially limited to Poland and Hungary, but from 1994, the EU decided to extend it to other accession candidates, like the Czech Republic. Its main functions were to promote administrative reform and support investments in infrastructure (European Community 1989).

Structuring and Shrinking (1994–1999)

Saxony

Negative economic development persisted during the second development phase. Only the building sector registered noticeable growth because of infrastructure needs, particularly since transportation and housing still needed considerable investment. In this phase, massive suburbanisation occurred, characterised by the construction of privately owned houses, shopping centres and business parks in the edges of the main conurbations areas (Schmidt et al. 1993). As a result, the mostly small and former agricultural settlements around bigger cities were the main winners of the internal migration, showing significantly increasing population (Glöckner 2008). In contrast, city centres lost noticeable amounts of inhabitants. For example, the inner city of Plauen had a share of 60 % unoccupied buildings. Thus, growth and shrinkage happened concurrently. Former areas of agricultural production outside the cities became dwelling and service zones.

Population loss and high unemployment, present during the first phase, continued. From 1994 to 1999, the population shrank by some 3 % which equals 105,500 people (Statistical Office of the Free State Saxony 2010). Unemployment remained high. In March 1996 the average unemployment rate in the research area was 18.8 % and in December 1999 was 19 %, with a temporary maximum of nearly 23 % at the start of 1998. These rates were around 2–3 % higher than the Saxony average. The lowest unemployment was mainly in bigger cities, like Dresden and Chemnitz, while the former highly industrialised districts showed the highest levels (Bundesagentur für Arbeit 2010). The age distribution worsened because the birth rates remained below the level of 1989, and emigration over the inner-German border to West Germany remained being high. Case studies showed that mainly the youngest age groups and groups in the 30–40-year-old age range, the employable population, were decreasing, so that the share of the oldest groups grew (Glöckner 2008). Population densities in the former industrialised and densely settled crest line of the Ore Mountains fell to those of rural areas during this time.

What changed for the better was that planning instruments and mechanisms came into force to control the physical transformation process and cushion its negative effects. The creation of a new planning administration building was essentially finished, and a base was created for the creation of a spatial planning policy. In 1994, the first Saxon regional development plan, including a landscape programme, was adopted. Its goals were taken up by the regional development plans which covered four regions: Upper Elbe Valley/Eastern Ore Mountains, Lower Silesia, Chemnitz/Ore Mountains and Western Ore Mountains/Vogtland. However, although shrinkage was visible in almost all aspects of the economy, municipalities often over-provided business and housing areas, still believing in growth. In this phase, most municipalities started to work out their land-use plans. But due to the change of administrative districts, many of them were not finished, and plans often stayed in a conceptual state.

With the administrative reform in the years 1994 and 1996, the administrative districts (NUTS III – Landkreis) were changed by enlarging their area and decreasing their number from 48 to 22 in Saxony. This was done to cope with a shrinking population, so that they would function more efficiently.

Throughout this phase, which is identical with the European Objective 1 funding period, Saxony was an Objective 1 region and could take part in the European Community's structural funding programmes. Approximately 7 billion DM were available for Saxony. The German 'Joint Agreement for the Improvement of Regional Economic Structures' mostly funded business investments and the protection of administrative jobs. This was criticised by the EU and Saxony because they considered it to be a too specified approach of funding. Therefore, the Saxony government uncoupled a part of the money from the Joint Agreement and used it mainly for the fields of education and the environment (Krämer 2002).

Czech Republic

This phase actually started with economic growth. Czech GDP first grew in 1994, and in 1995, it increased by 6.4 %. But in the years 1996–1997, the Czech Republic suffered a huge financial and economic crisis so that in 1998 GDP declined (Dauderstädt 2004).

From 1994 on, a second wave of voucher privatisations started, in addition to the privatisation of small enterprises which had started in 1991 At the end of this phase, in 1999, an enterprise restructuring agency had been founded (Dauderstädt 2004).

As a result of the recession, unemployment rose considerably after 1995. The most affected branches were agriculture, industry and mining, but they remained the backbone of the economy in the research area at this time. In 1999, the NUTS II area Severozápad, covering most of the Czech research area, had an unemployment rate of 13.5 %, the highest of all eight Czech NUTS II regions (Eurostat 2010). Thus, the disparities seen in the first phase grew. Northern Bohemia, a former mono-industrial region with a focus on energy production, became an area with strong structural deficits and a weak economic dynamic compared to the rest of the Czech Republic. Partly due to this, and complicating the situation, salaries started to differ from the national average, particularly in declining sectors, where they fell to relatively low levels (Glöckner, Slavik and Perduta 2009; Czech Statistical Office 2010).

At the same time, investments for rebuilding old industrial zones and building new ones and the improvement of infrastructure, dwellings and business facilities resulted in some economic activity, especially in the building sector. It was hoped that the development of new industrial zones would attract investors and employers into municipalities (Glöckner et al. 2009).

The population was characterised by sinking birth rates, which bottomed out in 1999. Changes in the age distribution, particularly a higher share of older age groups, were noticeable, but in comparison to Germany and Poland, this process was moderate. Cities and towns lost population due to a permanent stream of people moving out, but here also the process was weaker than on the German side. The case studies showed that in the main, former industrial and agricultural settlements were losing population, while in areas attractive to tourists, the population stayed stable or even showed positive growth (Czech Statistical Office 2010; Glöckner 2008). However, overall, within the Czech NUTS III regions, belonging to the research area, there was no significant population loss. From 1994 to 1999, the entire Czech research region grew by approximately 2,400 people, an increase of just under 0.2 % (Czech Statistical Office 2010).

In this phase, the Czech Ministry for Regional Development was founded (1996), and in 1998, the 'Principles of National Regional Policy' were adopted. This was a consequence of the lack of coordinated regional development in the first development phase. The principles aimed to formulate regional development policy with reference to European Union's regional policy. They also aimed to assist in preparation for EU-membership. The main principles were (Pallagst 2000) balanced regional development, the abatement of disparities and the promotion of economic and social development of the regions.

Consequently, plans for regions with structural deficits were developed. Municipalities started to work together, crossing regional and even national borders. This work was funded by European programmes like PHARE CBC. On German side, cross-border cooperation with the Czech Republic was financed by INTERREG II, based on jointly created concepts of development (Roch, Scott and Ziegler 1997). In 1994, ECU 25 million was provided for investments into the German-Czech borderland. They were used especially to improve the technical infrastructure, foster cross-border traffic and address environmental and socioeconomic issues (Roch et al. 1997).

Poland

This specific phase in Poland was characterised by annual GDP growth of around 5 %. The extreme inflation, which started in the first phase, was minimised, and after 1995, Poland became interesting to foreign investors. In 1994, national investment funds were established to handle the privatisation of bigger companies (the privatisation of small firms started in 1990 and mass privatisation came into force in 1993) (Dauderstädt 2004). The ongoing transition process took two forms: the first was closing large but technologically obsolete industrial enterprises and the second was modernising and reorganising others and helping them adjust to new economic conditions. However, local infrastructure showed no noticeable improvement (Buras 2005).

In the research area, the natural growth rate of the population and the net migration rate stayed negative, which led to a decrease of the population of about 4.4 % from 1995 to 1999 (Statistical Office of Poland 2010). Considering the fact that the population in the Voivodeship Dolnośląskie showed an average decline of about 2.4 % from 1995 to 1999, it is obvious that the economic and demographic developments in the research area are far below average, which is probably paralleled by changes in the labour market (Glöckner 2008). Although the overall economic situation in Poland improved, the absolute numbers of employees in all economic branches from 1995 to 2001 declined in the research area (numbers refer to NUTS III units – powiat). The biggest losses were in agriculture, with a decline of 25.3 to 55.5 %, depending on the powiat, and industry, with losses of 16.9 to 45.8 %. In comparison, the decrease in the service sector was relatively moderate falling between 0.85 and 20 %. Table 10.2 shows the differences between the three national regions (Glöckner 2008).

In 1994, Poland applied for EU-membership, and a contract for a partnership between Poland and the European Community was adopted. In 1999, a new Polish administrative structure came into force which reduced the number of regions (voivodeships), but added a new layer of administration between the local commune (gmina) and the voivodeship: the powiat (county). This was clearly related to the preparation for EU-membership because now Poland possessed an adequate number of NUTS II territorial units, which were an important requirement for EU funding (Buras 2005).

Also in 1994, a new planning law was adopted, which aimed to overcome old socialistic structures by introducing new forms of planning. The problem was that most local plans were old and out of date and resulted in rapid suburbanisation and unplanned splinter development, comparable to that taking place in Saxony. Until an amendment to the planning law in 2003, the old plans remained the basis for spatial development (Lorens 2005).

The PHARE programme as a means of co-financing INTERREG II projects supported cross-border cooperation in the Polish border regions. The main focus was the improvement of the infrastructure and environmental protection, comparable to Germany and the Czech Republic.

Stabilisation on a New Level (2000–2005)

Saxony

During this phase, economic activities in Saxony stabilised on an obviously lower level than in the year 1989. Step by step, clusters capable of competing in global markets developed, for example, in Southwest Saxony, where the automobile industry settled, and in the Elbe Valley around Dresden, where amongst others Germany's and Europe's centre of microelectronics is located (Schätzl 2008).

But processes of shrinkage were noticeable in most aspects of the economy and population. After a boom in the previous phases of the transition, the building industry also declined in tandem with the end of funding for construction of new privately owned flats and houses by the German Federal government. Connected to this, migration from inner cities to rural areas decreased so that population development in suburban areas which had previously been growing and profiting from new housing stagnated and even contracted (Glöckner 2008). The population, as well as the number of births, declined by approximately 4 % in this phase (Statistical Office of the Free State of Saxony 2010). The unemployment rate in the Saxon research area

Table 10.2 Relative change of employees in different economic sectors 1995–2001 (in percent)

Kreise/okres/powiaty	Agriculture	Industry	Services
Germany			
Chemnitz, Stadt	25,00	–34,84	1,33
Plauen, Stadt	–25,00	–30,86	6,06
Zwickau, Stadt	0,00	–10,03	14,50
Landkreis Annaberg	7,69	3,29	4,12
Landkreis Chemnitzer Land	–17,65	–0,56	8,63
Landkreis Freiberg	–28,57	–0,62	–0,56
Vogtlandkreis	–40,43	–6,76	3,07
Mittlerer Erzgebirgskreis	–32,00	10,86	–9,27
Landkreis Mittweida	–39,02	–8,04	4,14
Landkreis Stollberg	8,33	–7,28	5,99
Landkreis Aue-Schwarzenberg	8,33	–10,47	–0,32
Landkreis Zwickauer Land	–41,38	–7,54	5,44
Dresden, Stadt	–23,81	–16,32	11,90
Görlitz, Stadt	–16,67	–42,44	4,81
Hoyerswerda, Stadt	0,00	–54,63	17,19
Landkreis Bautzen	–20,83	–10,47	9,30
Landkreis Meißen	–25,00	–17,60	4,01
Niederschlesischer Oberlausitzkreis	–13,04	–22,58	13,99
Landkreis Riesa-Großenhain	–41,38	–8,09	5,84
Landkreis Löbau-Zittau	–13,04	–14,56	13,15
Landkreis Sächsische Schweiz	–3,70	–17,28	11,66
Weißeritzkreis	–24,32	–6,20	16,46

	Agriculture	Industry	Services
Poland			
City with powiat status Jelenia Góra	–48,79	–21,21	–8,84
Powiat zlotoryjski	–40,73	–29,33	–10,64
Powiat zgorzelecki	–25,71	–17,52	–0,85
Powiat zabkowicki	–43,46	–31,52	–18,10
Powiat walbrzyski	–42,70	–33,74	–14,70
Powiat swidnicki	–55,13	–38,19	–13,07
Powiat strzelinski	–36,29	–16,96	–14,41
Powiat lwówecki	–55,47	–34,60	–19,95
Powiat lubanski	–38,17	–45,78	–17,80
Powiat klodzki	–44,74	–41,34	–17,50
Powiat kamiennogórski	–33,00	–27,00	–15,61
Powiat jeleniogórski	–28,68	–32,24	–18,81
Powiat jaworski	–25,29	–28,78	–19,39
Powiat dzierzoniowski	–47,38	–38,06	–11,56
Powiat boleslawiecki	–30,30	–21,75	–6,02

	Agriculture	Industry	Services
Czech Republic			
Česká Lípa	–31,56	12,05	–8,27
Chomutov	–22,72	–15,48	–0,51
Děčín	–40,66	–2,89	n.a.
Jablonec	–43,79	–5,23	–8,74
Karlovy Vary	–40,04	–5,20	n.a.
Liberec	–42,00	2,44	5,80
Most	–4,95	–37,95	–0,96
Semily	–4,62	2,75	n.a.
Sokolov	–48,33	–15,56	n.a.
Teplice	–73,48	–28,94	0,35
Trutnov	–34,09	9,15	–8,64
Ústí nad Labem	–67,13	–39,92	n.a.

rose from an average rate of about 19 % to about 20 % from 2000 to 2005. Thus, the study region had an unemployment rate slightly higher than the Saxon average (Bundesagentur für Arbeit 2010).

The shrinkage due to the transition was finally recognised by the planning authorities in 2004. In 2004, a renewed regional development plan for Saxony came into force, introducing the principle of sustainability and recognising the changed socioeconomic conditions. Corresponding to the development of the population and the economy, the focus was set on the principle of concentration. In addition to formal instruments (federal planning, land-use planning, regional planning and land-use plans), informal instruments began to play a more important role (e.g. concepts of regional development, regional conferences). In numerous municipalities, land-use plans became binding because during this period, land-use plans were adapted and approved by the responsible authorities.

The reassessment and the partial demolition of existing urban quarters became the main focuses of funding. The programme 'Stadtumbau Ost' was set up to improve the quality of life. Therefore, the number of empty dwellings was reduced, and the balance of supply and demand in housing was re-established (Roch et al. 2010).

Czech Republic

In the Czech Republic, the third phase was characterised by a stabilisation of economic, ecological and social conditions. The economy boomed after the Czech Republic joined the European Community in 2004 as investments into the country grew. Big enterprises, like the telecommunication and gas companies, were privatised in 2002, and external trade as well as tourism grew. Retrenchment affected mainly agriculture industry (Dauderstädt 2004).

Although there were positive developments in questions of ecology and quality of life, GJDP growth in the administrative districts in the research area was still below the average Czech level (Czech Statistical Office 2010). Unemployment in the areas remained one of the highest in comparison to other NUTS II areas in the Czech Republic. The region Severozápad (Northwest) showed values around 11 % for much of the period, after peaking at 14 % in 2000, followed by a rise to 13.5 % in 2005 (Eurostat 2010). In the research area, the share of permanently unemployed people (two or more years without a job) remained one of the highest in the whole country. The population showed only slight shrinkage, falling just 0.6 % (equalling a loss of 8,884 people) from 2000 to 2005. The number of births grew by 6.77 %, and the net migration rate increased from 0.024 to 0.192 % (Czech Statistical Office 2010). However, within the area, the overall infrastructure and the state of border crossings remained poor.

The population in cities remained stable because of employment-related migration, and salaries slowly rose due to the slow stabilisation of the economy. The economic change also resulted in a slow improvement in the demographic situation in the research area. New dwellings were built and industrial estates rebuilt. Until 2004, instruments to help prepare for EU-membership, like ISPA (aid for transport), PHARE (regional economic assistance) and SAPARD (aid for agriculture), operated, but after 2004, the Czech Republic was eligible for structural funds from EU regional development programmes. Thus, new development happened in conjunction with funding from the European Union (e.g. Consolidated Funds of the EU for the Development of Housing) and national support from the Ministry for Regional Development (Glöckner, Slavik and Perduta 2009).

Poland

After the positive economic development up to 2000, Poland entered a phase of stagnation, with rising unemployment rates. In the Lower Silesian Voivodeship (Dolnośląsk), the unemployment rate rose from an already high 21.3 % in 2000 to a maximum of 26 % in the years 2002–2003. After this peak, it slowly decreased, falling to 22.8 % in 2005 (Eurostat 2010). The population in the Polish research area declined by some 2 % (a loss of 26,883 people, Statistical Office of Poland 2010). Like in Saxony and the Czech Republic, this value is misleading: most of the decrease was amongst adolescents and younger members of the labour force. As a result, the share of the population of childbearing age is decreasing, which suggests that problems may persist over the longer term.

After a poor start at the beginning of the third phase, economic development in the Polish region of the study area surged after Poland joined the EU in 2004. In connection with EU-membership, an increasing number of enterprise start-ups were noticeable, as well as a reduction in unemployment. Population loss halted in industrial and tourist areas and the birth rate improved, especially in places where new enterprises settled. Private house building increased (Glöckner 2008; Statistical Office of Poland 2010).

With a new national Planning Act, adopted in 2003, the local plan was to come into force as a legally binding document should. But such local plans were not drawn up in many municipalities because their creation was too expensive and time consuming. As a result, local development was mainly controlled by administrative decisions, which caused problems, such as growing corruption due to the lack of adequate legal guidelines (Lorens 2005).

Upon joining the EU, Poland gained the highest funding status. From 2004 to 2006, a centrally – that is, nationally – implemented phasing-in started with a series of operational programmes which implemented funding contracts. In the following funding phase, from 2007 to 2013, the offices of the Voivodeship Marshal (the executive organ of the Voivodeship) became the responsible authorities for funding contracts (Reents 2007).

Divergence Versus a Win-Win Situation (2006 Onwards)

In the latest phase, which we call 'divergence versus a win-win situation', the process of socioeconomic transition slowed and even stabilised, although there were different deficits in the three national parts of the research area. For the most part, traditional branches continued to decline throughout the tri-border region. However, the region has been transformed into a green and environmentally friendly border region – a 'Green Triangle'. In terms of infrastructure, there are still greater deficits on Czech and Polish side, especially in small or peripherally located municipalities. However, despite local problems in the two countries, since 2004 their economies have begun to grow more rapidly than neighbouring areas in Saxony. Comparing GDP growth rates in the three national regions between 2003 and 2006 on NUTS II level, it is clear that the Severozápad (Northwest) and Dolnośląsk (Lower Silesia) are growing much strongly than the districts in Saxony (see Fig. 10.2, Eurostat 2010), particularly after 2004, the year of accession to the EU.

Within Saxony, GDP growth has mirrored the overall German growth rate. However, given that the part of the research area within Saxony still suffers from structural deficits and that its overall economic situation is below the German average, it is becoming apparent that the catching-up process is stagnating. This situation is not expected to radically change in the future (Pohl 2000). The processes of economic stabilisation taking place in Poland and the Czech Republic have helped to foster the normalisation of population development – that is, increasing birth rates and in-migration or a halt to population loss – while in Saxony, the population is expected to decline about 10 % until 2020 due to a low population of childbearing age (see Fig. 10.3) (Banse and Jeřábek 2009).

In Saxony, in the face of continued population decline, spatial planning has focused on a 'lighthouse policy' which is a kind of growth-pole policy, aiming to strengthen chosen centres from which regional development should radiate. This main principle of Saxony's spatial planning, appearing in the Saxony regional development plan, was adopted in 2003. Another adjustment to cope with population loss was a reduction in the number of administrative districts in 2008. In the latest

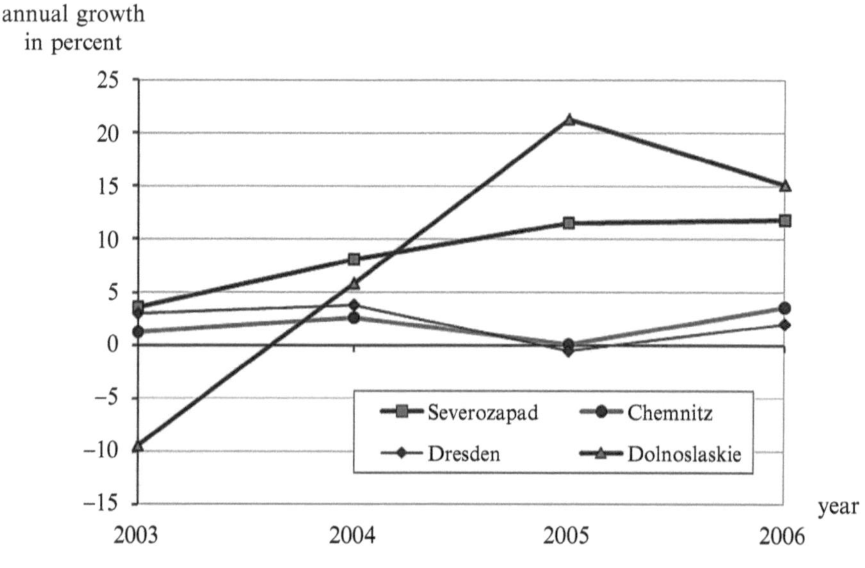

Fig. 10.2 Annual growth of the regional GDP (2003–2006, NUTS II level)

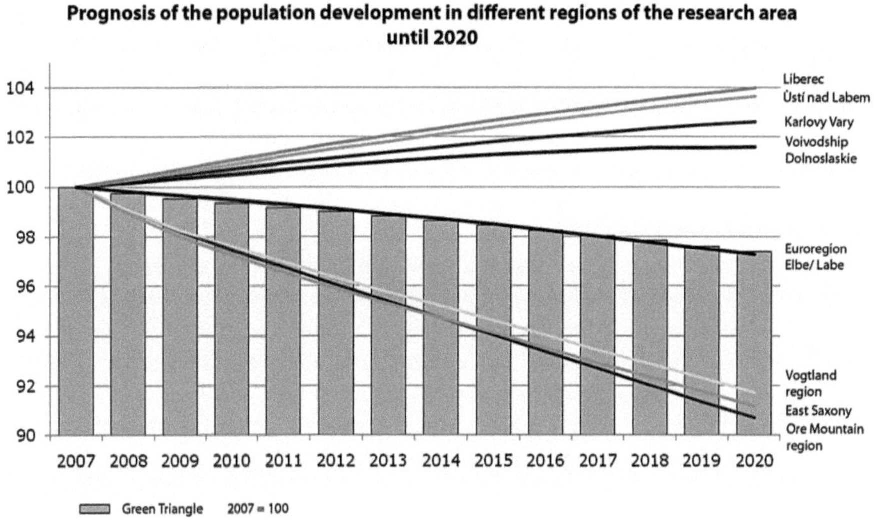

Fig. 10.3 Prognosis of the population development in the research area until 2020 (Source: Statistical Office of the Free State of Saxony, Statistical Office of Poland, Czech Statistical Office, model 1, date: 31.12.2005, design: Banse, IOER, adopted)

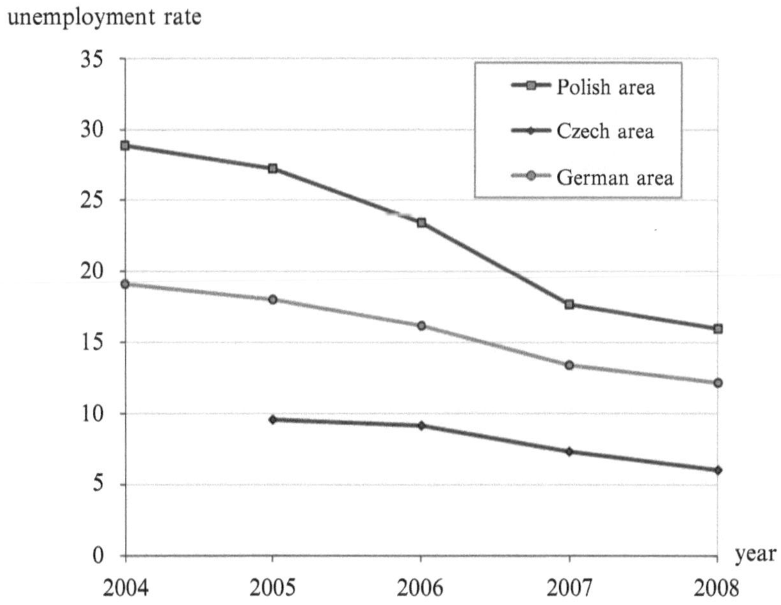

Fig. 10.4 Unemployment rates in the former Black Triangle (2004–2008)

phase, unemployment has decreased noticeably throughout the research area (see Fig. 10.4). It fell from 28.8 to 16.6 % in the Polish research area between 2004 and 2008. By contrast, in 2008, the overall unemployment rate in the Lower Silesian Voivodeship (Dolnośląsk) was just 10 %, which clearly shows that the Polish districts close to the border are disadvantaged. This is similar to the situation in the Czech research area which is also characterised by decreasing unemployment rates (approx. 10 % in 2004, 6 % in 2008), which are significantly higher than in the Czech Republic itself (4.7 % in 2008). In Saxony the unemployment situation is the same. While in the research area the unemployment rate fell to a value of about 12 %, the German average was at approx. 9 % in 2008 (Eurostat, Statistical Office of Poland, Czech Statistical Office, Bundesagentur für Arbeit 2010). These changes are shown in Fig. 9.4.

The authorities responsible for spatial planning in Poland and the Czech Republic are dealing with infrastructure deficits, as well as problems facing traditional branches of economy and the labour market. But given the growing dynamism in their economies and the stabilisation of population, the development perspectives seem to be more positive for these two parts than they are for Saxony.

The whole tri-border region is an EU Objective 1 region during the period from 2007 to 2013. The area has gained the highest level of EU funding, especially for infrastructure improvement, labour market restructuring, education and cross-border cooperation.

Comparative Summary

Although the part of the research area within Saxony appeared to be advantaged in the first two phases of the transition process because it was part of Germany after 1990 and had access to EU financial aid, its future perspectives are not as positive as those for the Czech Republic and Poland. The transition strategies in eastern Germany were very different from those in Poland and the Czech Republic. They caused the collapse of the economy in the German research area in a very short time. This collapse was not followed by a new and globally competitive economic base. Furthermore, Poland and the Czech Republic became advantaged as investment locations with their accession to the European Union (Born, Fichtner and Krätke 2006). Along with a declining population, Saxony shows noticeable concentration processes concerning the settlement structure, which induces labour supply shortages in peripheral areas. Economies in the research area in the other two countries started to grow after accession to the EU. Their infrastructural situation is steadily improving, which has had positive effects on the population and the economy. Whether this positive development will persist depends mainly on how well the area-specific potentials of the tri-border region are used to create a competitive European region based on a stable, self-containing economy (Roch 2009a).

The main similarities in the development of the three parts are:

- A change from a socialistic-centralised to a democratic society based on services
- A huge loss of jobs in traditional branches like mining, agriculture and industry
- Strong structural deficits in comparison to the average in the individual countries

- Obsolescence, outmigration and declining birth rates (especially in the early years and in very different values) in bigger dimensions than in the individual national heartlands
- At first, increasing unemployment rates, then decreasing unemployment throughout the last phase
- Objective 1 region status (in the case of Poland and the Czech Republic) since accession to the EU in 2004

The different starting points and different approaches to the socioeconomic transition caused the biggest differences between the individual regions. These are:

- Very different chronological sequences in the transition process
- Positive economic growth starting with the end of the third phase because of the accession of Poland and the Czech Republic to the EU
- A better perspective in population development in Poland and the Czech Republic
- A better overall infrastructure situation in Saxony because of financial aid from the German Federal government and the EU
- But spatial concentration and shrinkage in Saxony placing strains on the spatial planning system and hindering attempts to promote equal regional development

Although a new economic situation evolved in all three countries, the three parts of the study region will all need additional external financial aid from abroad because their economies are not yet self-sustaining and cannot remain stable on their own (Glöckner et al. 2009). If financial assistance is not available, disparities are expected to grow in the tri-border region. But it is also likely that the motivation for the three regions to work together on new common initiatives also will grow. So, in this point of view, the transition is not yet complete in the tri-border region. Saxony is in danger of no longer having access to subsidies; such aid is slowly being phased out (as, e.g. has occurred in the Leipzig district since the start of the latest funding period). Less aid from the EU may lead to lower investment.

Conclusion: Potential for and Constraints on Further Development

Potential

Although the economies of the three national regions show diverging development patterns, there are several relevant toeholds for cooperation to reach a common development in the observed region. Perhaps, the biggest potential for growth is the area's 'new' location in the middle of Europe. 'Toeholds' for the border region can be the metropolises of Berlin, Prague and Wrocław. This central location is and will be developed by the improvement of the accessibility of the region, e.g. by building the freeway A17/D 8 Dresden–Prague or the extension of the freeway A4 over the Polish border to connect Dresden and Wrocław. The accessibility of the tri-border region and its centres of national and international relevance is essential to foster its locational advantage, which could be converted into an economic advantage (Roch 2009a).

Another important potential is the existing human capital in connection still with competitive labour costs in Poland and the Czech Republic, which German and foreign investors already have discovered and used. Cross-border cooperation between companies located in the tri-border region is not yet on a very high level, but contains a high development potential (Jurzcek 2006).

There are a number of other potential strengths (Roch 2009a). These include using Saxony's long experiences with EU aids in corporate fundraising, the improvement of cross-border infrastructure and the potential to develop new ways of fostering structural change or the reuse of mining areas. They also involve cooperation in terms of tourism, particularly health spa tourism. In addition, a high potential quality of life around prospering centres, like Liberec, *Jelenia Góra* or also Wrocław and Prague, can provide the basis for further economic and social development in the region. Finally, the three parts of the region can engage in economic cooperation based on the individual strengths of each part of the three-nation region. In Saxony, it contains good infrastructure, a high level of education and extensive expertise in research and innovation. The Czech Republic and Poland are attractive for investment because of lower wages and social costs, growth prospects of local markets. Therefore, administrative and legal constraints have to be lowered or removed so that an economical cooperation will be easier to achieve and create.

Constraints on Development

Constraints on potential developments mentioned above are mainly organisational and legal ones, although there are existing examples of successful long-term cross-border cooperation, like the Euroregions. These constraints are mainly the following (Roch 2009a):

- Partnership between different countries must be developed step by step based on mutual interests and attendance, basing on dedicated protagonists and a manageable setting of the respective area.
- Difficulties in active cross-border cooperation often reflect a lack of time on the part of members of responsible committees, cities and regions.
- Difficulties in the legal frameworks for activities related to fundraising and project formulation and approval.

In the end, the success of cross-border cooperation for development in the research area depends on the recommendations and the motivation of the relevant groups of actors as well as the political will in the individual regions and districts to seize opportunities and promote development.

References

Abraham J et al (2003) Gemeinsamer Bericht zur Luftqualität im Schwarzen Dreieck 2002. Sächsisches Landesamt für Umwelt und Geologie, Dresden

Bundesagentur für Arbeit (2010) Labour market statistics. Available at http://www.pub.arbeitsagentur.de/hst/services/statistik/detail/q.html

Banse J, Jeřábek M (2009) Sozioökonomische Entwicklungsverläufe unter Berücksichtigung der Bevölkerungsentwicklung und des Arbeitsmarktes im sächsisch-böhmischen Teil des Grünen Dreiländerecks und der Euroregion Elbe/Labe. In: Roch I, Jeřábek M (eds) Grenzübergreifende ökologische Regionalentwicklung. In: Acta Universitatis Purkynianae of Jan Evangelista Purkyně University Ústí nad Labem, Ústí nad Labem, vol 4, Supplement, pp 37–51

Berkner A, Thieme T (eds) (2005) Braunkohlenplanung, Bergbaufolgelandschaften, Wasserhaushaltssanierung. ARL-Arbeitsmaterial, Nr. 323. Hannover

Bertelsmann Stiftung (2003a) Ländergutachten Polen. Bertelsmann Transformationsindex 2003. Available at http://bti2003.bertelsmann-transformation-index.de/169.0.html. Accessed 2 Feb 2010

Bertelsmann Stiftung (2003b) Ländergutachten Tschechien. Bertelsmann Transformationsindex 2003. Available at http://bti2003.bertelsmann-transformation-index.de/ 172.0.html. Accessed 2 Feb 2010

Born KM, Fichtner T, Krätke S (eds) (2006) Chancen der EU-Osterweiterung für Ostdeutschland, vol 321, ARL – Arbeitsmaterial. ARL, Hannover

Buras P (ed) (2005) Polens Weg von der Wende bis zum EU-Beitritt. Hohenheim-Verlag, Stuttgart und Leipzig

Czech Statistical Office (2010) Labour market statistics. Available at http://vdb.czso.cz/vdbvo/en/uvod.jsp

Dauderstädt M (2004) Transformation und Integration der Wirtschaft der postkommunistischen Beitrittsländer. In: Bundeszentrale für Politische Bildung (ed) Aus Politik und Zeitgeschichte – Beilage zur Wochenzeitung Das Parlament, Bonn, 2 February 2004, pp 15–24. Available at www1.bpb.de/files/PTPLAI.pdf. Accessed 11 Feb 2010

Dornbusch R, Wolf HC (1994) East German economic reconstruction. In: Blanchard OJ, Froot KA, Sachs JD (eds) The transition in Eastern Europe, 1st edn, Country studies. The University of Chicago Press, Chicago/London, pp 155–190

Espere Climate Encyclopedia (2003) Climate in cities – areas endangered by air pollution. Available at www.atmosphere.mpg.de/enid/3ou.html. Accessed 9 Dec 2009

European Community (1989) Council regulation (EEC) No. 3906/89 of 18 December 1989 on economic aid to the Republic of Hungary and the Polish People's Republic. Official Journal of the European Communities (OJEC), 23 December 1989, No. L 375, pp 11–12

Eurostat (2010) Labour market statistics. Available at http://epp.eurostat.ec.europa.eu/portal/page/portal/statistics/search_database

Glöckner R (2008) Entwicklungsverläufe ausgewählter Siedlungstypen im Vergleich - Ein Beitrag zur Evaluation umweltschonender Sanierungs- und Entwicklungsprozesse des deutsch-polnisch-tschechischen Dreiländerecks. Diploma thesis, propounded at Dresden University of Technology, Institute of Geography, Chair of Economical and Social Geography of Eastern and Southeast Europe. Dresden

Glöckner R, Slavik J, Perduta J (2009) Wandlungsprozesse in den nationalen Teilräumen aus Sicht der Prozesssteuerung. Presentation on the occasion of the symposium "Perspectives for the German-Polish-Czech Tri-border region – Potentials and Strategies", Prague, May 2009

Hübler K-H, Cassens H-J (eds) (1993) Naturschutz in den neuen Bundesländern. Chancen für die Landschaftsentwicklung. Bewertung der Naturschutzpolitik, Stand der Gesetzgebung in den neuen Bundesländern und in Berlin. Eberhard Blottner Verlag, Taunusstein

Jurczek P (2006) Wirtschaftliche Struktur und Entwicklung im sächsisch-böhmischen Grenzgebiet – Unternehmerische Einschätzungen, Handlungsbedarfe und Aktivitäten. In: Born

KM, Fichtner T, Krätke SS (eds) Chancen der EU-Osterweiterung für Ostdeutschland. Arbeitsmaterial der Akademie für Raumforschung und Landesplanung, Band 321, Hannover, pp 34–50

Katzur J (1979) Die Wiedernutzbarmachung der Bergbauflächen als Aufgabe des Landschaftsarchitekten. In: 50 Jahre Hochschulausbildung. Landschaftsarchitektur „Die Aufgaben der Landschaftsarchitektur in der entwickelten sozialistischen Gesellschaft". Wissenschaftliche Tagung an der TU Dresden – 5–6 July 1979. Schriftenreihe der Sektion Architektur, TU Dresden, Heft 14

Krämer R (2002) Aktiv in Europa – Die ostdeutschen Länder in der Europäischen Union. Brandenburgische Landeszentrale für politische Bildung, Potsdam

Krätke S, Borst R (2004) EU-Osterweiterung als Chance - Perspektiven für Metropolräume und Grenzgebiete am Beispiel Berlin-Brandenburg. Beiträge zur europäischen Stadt- und Regionalforschung, Band 1. Lit Verlag, Münster

Leick B (2006) Die Rolle grenzüberschreitender Unternehmenskooperationen im Grenzraum Südwestsachsen- Nordböhmen im Zuge der EU-Osterweiterung. Dissertation propounded at Freiberg University of Technology, Faculty of Economics. Freiberg. Available at https://fridolin.tu-freiberg.de/archiv/pdf/WirtschaftswissenschaftenLeickBirgit573501.pdf. Accessed 11 Jan 2010

Lorens P (2005) Stadtentwicklung in Polen – Aktuelle Trend und Herausforderungen. In: Altock U, Güntner S, Hunning S, Peters D (eds) Zwischen Anpassung und Neuerfindung - Raumplanung und Stadtentwicklung in den Staaten der EU-Osterweiterung. Reihe Planungsrundschau, Band 11. Berlin, pp 35–48

Maier K (1999) Wirksamkeit der bisherigen und zukünftigen Instrumente zum Abbau der regionalen Disparitäten in der Tschechischen Republik. In: Andrle A (ed) Gleichwertige Lebensbedingungen in Mitteleuropa - ein tragfähiges Konzept für die Raumordnung? Arbeitsmaterial der Akademie für Raumforschung und Landesplanung, Band 253, Hannover, pp 135–143

Pallagst KM (2000) Raumordnung in der Tschechischen Republik. Verlag Arno Spitz, Berlin

Pohl R (2000) Die unvollendete Transformation – Ostdeutschlands Wirtschaft zehn Jahre nach Einführung der D-Mark. Wirtschaft im Wandel 8:222–239

Prosek P (1994) Umweltbelastungen im Nordböhmischen Braunkohlenrevier. Geographische Rundschau 46(6):352–359

Prucha V (1998) Wirtschaftsstruktur und Wirtschaftspolitik in den tschechischen Grenzregionen. In: Schultz H (ed) Bevölkerungstransfer und Systemwandel – Ostmitteleuropäische Grenzen nach dem Zweiten Weltkrieg, 4th edn, Frankfurter Studien zur Grenzregion. Verlag Arno Spitz, Berlin, pp 305–320

Reents M (2007) EU-Strukturförderung 2007–2013 – Städtische Dimension der Förderung in Ziel 1 und Ziel 3. Presentation for "Begegnung polnischer und deutscher Urbanisten", 14–17 June 2007, Lebus. Available at www.srl.de/dateien/dokumente/de/eu-strukturfoerderung_2007-2013._staedtische_dimension_der_foerderung_in_ziel_1_und_ziel_3.pdf. Accessed 8 July 2009

Roch I (2009a) Entwicklungsphasen im sächsisch-böhmisch-schlesischen Dreiländereck. In: Roch I, Jeřábek M (eds) Grenzübergreifende ökologische Regionalentwicklung. Geoscape, Supplement, Vol. 4. Ústí nad Labem, pp 9–36

Roch I (2009b) Gemeinsamkeiten und Besonderheiten regionaler Entwicklung. In: Roch I, Jeřábek M (eds) Grenzübergreifende ökologische Regionalentwicklung. Geoscape, Supplement, Vol. 4. Ústí nad Labem, pp 162–182. Available at http://geo.ujep.cz/gu_pdf/geoscape_4_2009_supp.pdf

Roch I (2010) Möglichkeiten weiterer Annäherungen im deutsch-polnisch-tschechischen Dreiländereck zum gemeinsamen Vorteil. In: Binas E et al (eds) Wozu Region? Chancen und Probleme im Transformationsprozess strukturschwacher Regionen, 5th edn, Görlitzer Beiträge zu regionalen Transformationsprozessen. Peter Lang – International Academic Publisher, Frankfurt am Main, pp 181–194

Roch I, Scott J, Ziegler A (1997) Umweltgerechte Entwicklung von Grenzregionen durch kooperatives Handeln, IÖR-Schriften 024. Leibniz-Institut für Ökologische Raumentwicklung, Dresden

Roch I, Mundil R, Walz U et al (2000) Entwicklungskonzept für den Bezirk Karlovy Vary – Koncepce rozvoje Karlovarského Kraje. Leibniz-Institut für Ökologische Raumentwicklung, Dresden, 95 p

Roch I, Banse J, Effenberger K-H (2010) Die Lebensbedingungen Wohnen und Freiraumqualität – Vergleichende Untersuchungen von Stadtgebieten in Bonn und Dresden. In: Rosenfeld MTW, Weiss D (eds) Gleichwertigkeit der Lebensverhältnisse zwischen Politik und Marktmechanismus. Empirische Befunde aus den Ländern Sachsen, Sachsen-Anhalt und Thüringen. Arbeitsmaterial der Akademie für Raumforschung und Landesplanung, Band 351, Hannover, pp 33–70

Schätzl L (2008) Wirtschaftsentwicklung in Bonn und Dresden. In: Roch I, Banse J, Leimbrock H (eds) Freiraum- und Wohnqualitäten. Potenziale für den städtischen Umbau. Shaker Verlag, Aachen

Schmidt R, Wirth P, Banse J (1993) Entwicklungstendenzen im Umland großer Städte in den neuen Bundesländern am Anfang der 90er Jahre. IÖR-Schriften 01. Institut für ökologische Raumentwicklung, Dresden

SMU - Staatsministerium für Umwelt und Landesentwicklung (1995) Landesentwicklungsbericht 1994. Freistaat Sachsen, Dresden

Statistical Office of Poland (2010) Available at www.stat.gov.pl/bdren_n/app/strona.indeks

Statistical Office of the Free State of Saxony (2010) Available at www.statistik.sachsen.de/genonline/online/logon

Stutz B (2005) Europe's Black Triangle turns green. On earth – environment, politics, people. Spring 2005. Available at www.nrdc.org/onearth/05spr/triangle1.asp. Accessed 8 Dec 2009

Štýs S (1996a) Zelené Plíce Černého Severu. Bíly Slon, Praha

Štýs S (1996b) Proměny měsíční krajiny (Metamorphosis of a lunar landscape). Ekokonsult Pons, Most

Chapter 11
The Influence of Regional Identities on Spatial Development: A Challenge for Regional Governance Processes in Cross-Border Regions

Stefan Obkircher

Abstract How do regional identities influence spatial development, and what kinds of barriers do neighbouring identities create to coordinate regional programmes across borders? This paper tries to answer this question, looking at efforts to establish regional planning and development programmes along the Swiss-Austrian border. The cross-border region Alpine Rhine Valley is a meeting point for the cantons of Graubünden and St. Gallen, Switzerland, the Principality of Liechtenstein and the federal state of Vorarlberg, Austria. This work is part of an ongoing research project, based on a series of interviews of local residents in an attempt to determine their regional identities and how they relate to the broader cross-border region. The spatial structure of this area is polycentric: there is no clear central place, but a network of small and medium-sized regional centres has been established. The region shares a common cultural vernacular, but despite long-standing cross-border linkages, it has yet to develop a genuine supranational identity. Until recently, cross-border cooperation was initiated by higher levels of government, rather than municipal levels, and consisted of top-down rather than bottom-up regionalisation. More recently, municipal networks have been created, embracing governments on both sides of the border, leading to greater contacts and greater discussion. Regional identities on both sides of the border remain strong, but the growing number of cross-border planning initiatives and projects organised at the municipal rather than the regional or national level (bottom-up) seems to be slowly fostering the creation of a broader cross-border identity: the concept of the area as one region has come significantly closer, and there is a growing perception of the need to manage and develop this space jointly. However, in many ways, the region remains too large to identify with. Large regions are too far away from everyday issues and therefore barely tangible for the people! Scale makes the difference. Smaller regions are living regions. One conclusion could be that cooperation within regional development issues should focus more on small-scaled regions. People identify more readily with smaller areas and can embrace cross-regional cooperation more readily.

S. Obkircher (✉)
Department of Geography, University of Innsbruck, Innrain 52, A-6020 Innsbruck, Austria
e-mail: stefan.obkircher@uibk.ac.at

© Springer Science+Business Media Dordrecht 2015
J. Buček, A. Ryder (eds.), *Governance in Transition*, Springer Geography,
DOI 10.1007/978-94-007-5503-1_11

Keywords Regional identity • Cross-border cooperation • Cross border identity formation • Regional Governance

Introduction

Looking closely at the spatial development of the Austria-Switzerland cross-border region, one can identify an almost "borderless space". The municipalities and small towns have become a single agglomeration. Leaving or entering a new municipality is mostly only recognisable by means of place name signs or passing through the frontier barriers (Vision Rheintal 2010).

Cross-border regions are relevant to problems of European integration and an important spatial category in Europe. In such regions, the interactive effects of administrative, functional, mental and historic barriers have to be considered. In addition, the actors' underlying comprehension of space – as territory, as an interactive system and as a social construct – not only affects their routines but also characterises their actions in cross-border regions (increasing mobility; satisfaction of daily requirements, spread over the whole region, etc.). Borders are no longer merely separating lines. Although the functional importance of national borders in Western Europe is decreasing, their mental and constructed meaning is still a great factor for the success of cross-border regional development.

To face these challenges, several programmes have been initiated in the study area (European Interreg IV programme, Swiss Federal Agglomeration programme or Vision Rheintal). These processes are supported by specialist teams and completed by public participation. The aim of these approaches is to break up the existing hierarchical, top-down structures and introduce a more bottom-up strategy, including a less institutionalised coordination of regional development – closely connected to the paradigm shift from spatial planning to spatial development. A first analysis of these cross-border governance processes at the regional level shows that financial support can be a catalyst for cooperation but also that it needs to be complemented by a number of other factors if sustainable long-term effects are to be achieved. These factors include a certain degree of institutionalisation, social interaction, as well as cooperative attitudes and a common perception of relevant regional development issues. The influence of EU policy and national policy on cross-border development is of great importance, but the local actors follow their own agendas with priorities that are often quite different from those of the superordinate levels (see also Leibenath et al. 2008). Because of that, this paper focuses on how regional identities affect these spatial development processes.

Scope of Research

The research project presented here analyses the structural and institutional conditions needed for constructive cooperation and examines how social factors influence these spatial development processes. Those factors, or soft framework conditions, are subsumed by the term regional identity – describing the intensity of local bonds, spatial orientations and identifications that play a major role for the understanding of the actors' activities and actions and furthermore its influence on governance processes.

At the beginning of the article, I want to start – in addition to a brief presentation of the region – with a discussion about the spatial development processes in the study area in general: How does spatial planning in the subregions work and what is the effect of the border situation? Furthermore, the research interest in this context is the structural conditions of regional identities and how they are related to regional interests – the lived dimension of space. That means, which units in the everyday life of the inhabitants are relevant and how do they experience "their region"? The experienced region is not the same as the "top-down" constructed units as they are defined by stakeholders and politicians, as a study in the Zurich agglomeration has shown (Reutlinger et al. 2010).

The Case Study

The work described in this chapter is based on an ongoing study. To answer the research questions, the field work included the following components:

- An analysis of functional interactions in the region.
- An analysis of regional and national programmes, particularly projects with their main focus on the spatial planning and development of the Alpine Rhine Valley. This was based on participant observation, document analysis, the analysis of the existing planning instruments and qualitative interviews with experts.
- Mental mapping, photo surveys, workshops, and interviews with the inhabitants to get to know "their" region. One special focus was on intergenerational issues and differences between immigrants and natives.

The study is ongoing. Up to now (2012), the qualitative study has been based on 29 guided interviews with local residents from those parts of the Alpine Rhine Valley where the researched cross-border programmes are located. The interviewed people were between 25 and 91 years old. Data were also available from a previous survey on the topic "regional identities of juveniles in the Alpine Rhine Valley" (conducted 2008) and from a street survey on the topic "the image of the Alpine Rhine Valley" (conducted 2009; 329 questionnaires). The field work for this research took place between August 2009 and April 2010. The interviews include stakeholders and politicians involved in the cross-border programmes; this survey is still in progress and to be extended.

The Alpine Rhine Valley Cross-Border Region

The Alpine Rhine Valley cross-border region is a meeting point for the Swiss cantons of Graubünden and St. Gallen (Switzerland – CH, also SG), the Principality of Liechtenstein (LI) and the federal state of Vorarlberg (Austria – A, also Vlbg). This area has a polycentric settlement system. While a clear centre has always been missing, a network of small- and medium-sized regional centres has been established. The research area is adjacent to dynamic economic areas such as Zurich, Stuttgart and Munich, but also gains some of its character from nearby rural and touristic regions such as the Appenzell Canton or the Bregenzerwald region. Despite early industrialisation (nineteenth century), the valley's settlement structure had a rural character well into the twentieth century (Broggi 2006). During the last few decades, however, changes have occurred in the economic, social and settlement structure. Rural settlement structures have been transformed by urban and suburban expansion.

This transformation can be identified on the basis of the population and employee development in the study area. As can be seen in Fig. 11.1, between 1980 and 2006, there was a population growth of 23 % and a growth of employees of approximately 33 % (Meier 2010). According to forecasts, the population in the region will increase by another 160,000, or 36 %, between 2000 and 2020. Growth on both sides of the border accelerated after 1945, but has been greater on the Austrian side of the border than the Swiss side.

These transformation processes, however, do not proceed in a homogeneous area but in a highly heterogeneous region, which is historically broken up into small districts ("small chambered") and crossed by various borders (Fig. 11.2). Thus, the region can be viewed as a living entity or a mosaic of a variety of economic, geographic, cultural and social interconnections (Obkircher 2008). Additionally, inter- and intra-regional migration leave their marks within the regional social structure. At present, the region faces new challenges within the framework of globalisation and regionalisation. Thus, the Alpine Rhine Valley has to compete with other regions, mainly with the neighbouring large metropolitan areas. Currently, the region is challenged by problems of governing and coordinating regional development in a cross-border context. To successfully implement common initiatives, knowledge about existing intra-regional orientations, perceptions and identities is a necessary prerequisite.

As noted, the settlement structures of the region are merging through a variety of interactions, and it is becoming one living space. This is one of the most important

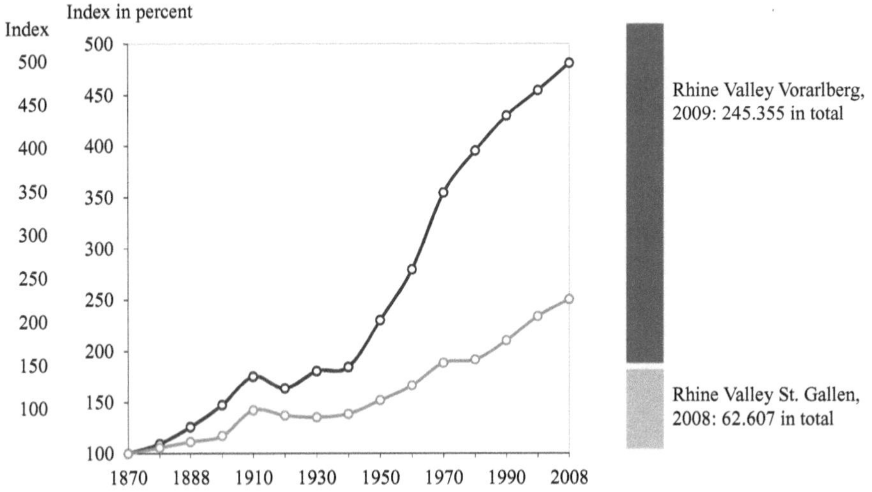

Fig. 11.1 Population development in the Alpine Rhine Valley cross-border region (www.Vorarlberg.at)

Fig. 11.2 The different administrative layers in the Alpine Rhine Valley

push factors in establishing cooperative structures; it is the pressure that "you have to". But although the Alpine Rhine Valley is a geographical unit, in many ways the Alpine Rhine Valley does not yet exist as a sociopolitical entity. As a border region, it is characterised by tension between political boundaries and associated areas of sovereignty on the one hand and closely functional linkages and mutual dependencies on the other hand. Despite the limitations at different hierarchical levels, one can find strong links between the regions. Neither economic linkages and mutual dependencies nor the action space of the inhabitants stops at border checkpoints. Functional networks in the border region have an impact on the regional self-image. The Alpine Rhine Valley is today in many parts characterised by urban sprawl, by the lack of clear spatial boundaries between provinces and by the lack of large continuous natural areas. The closeness of the intertwined municipalities of the Alpine Rhine Valley is, in everyday life, expressed by the significant number of cross-border commuters (Fritsche and Studer 2007). For example, there are 8,000 commuters from Vorarlberg to Liechtenstein and 7,500 from Vorarlberg to Switzerland (Fig. 11.3). In addition, the whole region is undergoing many changes related to the European integration policy.

In light of these developments, land-use planning in the region faces significant challenges. This is because land use and land-use change in the region are very dynamic. This change is apparent in the expansion of the suburban or "Zwischenstadt" settlement pattern and in the resolution or superposition of suburbs onto traditionally rural-influenced settlement patterns. What is the meaning of Zwischenstadt? Zwischenstadt is a rural city as well as an urbanised landscape. Zwischenstadt can be defined as "diffused city" or "dispersed city". It is neither town nor country, but

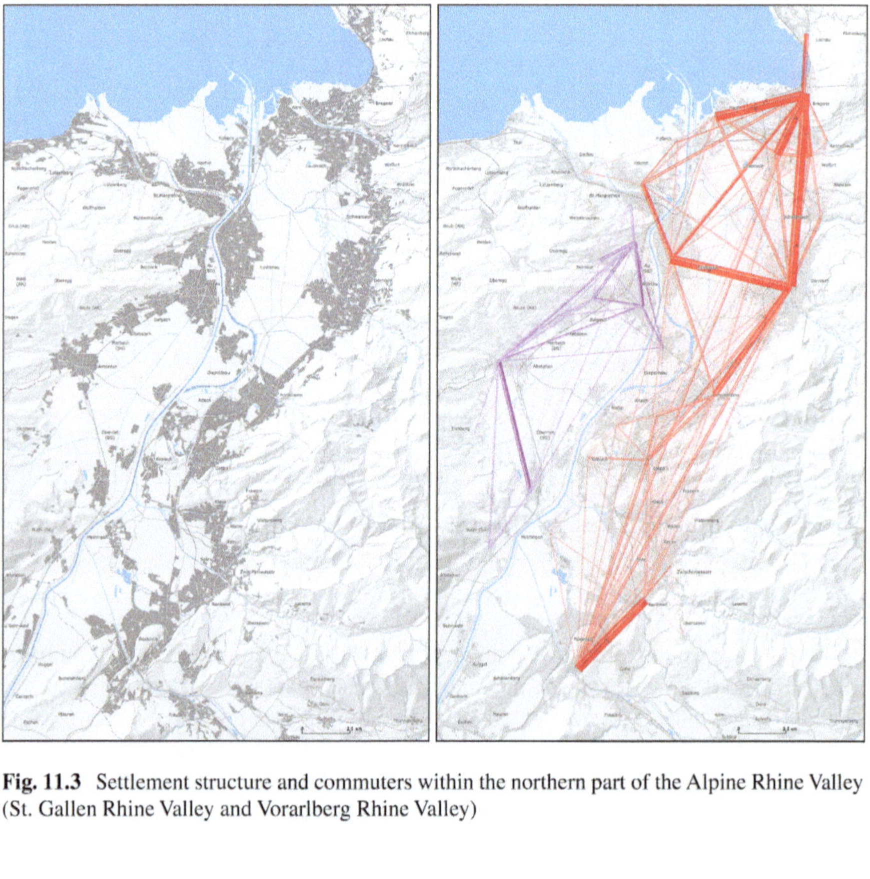

Fig. 11.3 Settlement structure and commuters within the northern part of the Alpine Rhine Valley (St. Gallen Rhine Valley and Vorarlberg Rhine Valley)

possesses characteristics of both (Sieverts et al. 2005). Local area plans in the region also clarify that today about a quarter of the Alpine Rhine Valley is already designated as a settlement area (Broggi 2006). The consequences of further urban sprawl are incalculable and associated with land-use conflicts and a need for high investments in operating and capital costs for public infrastructure (Meier 2010).

Within the European context, this is not a new problem for current spatial development processes in agglomeration areas, but before anything else, the Alpine Rhine Valley cross-border region must learn how to deal with this situation. In the past, inhabitants have been living in the country side. Now they are living in an agglomeration, next to each other – facing problems of their neighbours that are becoming their own problems.

Spatial Development in the Adjacent Subregions

Regional planning level in the canton of St. Gallen is much more regulated than in the federal state of Vorarlberg. The different role and importance attached to spatial planning in the two subregions highlight distinctions in how we can understand spatial planning and planning cultures. Thus, a significant challenge to international cooperation in spatial planning is to first develop a common understanding of the role and tasks expected of land-use planning, as well as its practical constraints, and second to find a constructive approach to deal with the diversity of planning systems and services. Cross-border cooperation must take into account the diversity of stakeholders and planning instruments and be sensitive to different understandings of planning, planning cultures and concepts. This is a major challenge in terms of communication (Meier 2010).

Although one common planning law for the whole region and one common administrative structure dealing with cross-border cooperation seem highly unlikely under the current conditions, the analysed "planning systems" are open to more intense interaction. Therefore, it would be desirable to establish one common vision/guiding principles of regional planning and a common approach to the spatial development of the Alpine Rhine Valley.

Borders and Border Regions: A Spatial Planning Trip from the European Union to the Alpine Rhine Valley

"The Alpine Rhine Valley is a border area that separates and unites. The mobility of people and economic networking on the one hand, and the territoriality of development issues and the (planning) systems on the other hand, create a formative ambivalence" (Meier 2010, 23). Borders often divide regions and areas which are functionally intertwined. For the people in these regions, the borders can act as barriers as well as offer them opportunities, especially through taking advantage of different framework conditions and imbalances. Borders are reflected in land use and development and are visible, for example, on traffic flows and patterns of settlement development (Meier 2010). This can be observed at all sorts of boundaries, but is particularly evident at national borders. Regions along national borders have received increased attention in Europe in recent decades. A central theme is regional cross-border cooperation, which in turn is closely linked to processes of European unification.

At the European level, it is argued that "Territorial Cohesion can only be achieved through an intensive and continuous dialogue between all actors dealing with territorial development. This process of cooperation can be called territorial governance. The private sector (especially locally and regionally based entrepreneurship), the scientific community, the public sector (especially local and regional authorities), non-governmental organisations, and different sectors need to act together in order to make better use of crucial investments in the European regions and contribute to tackling climate change" (EU Ministers responsible for spatial development 2007, 2).

This meaning of the concept of cross-border regional governance has grown in recent years, not least because it is promoted and required by the European Union. Of particular importance is that the objective of territorial cohesion is determined by the EU treaty of Lisbon: European Territorial Governance is used to interlink (spatial) actors (European Union 2007). For spatial planning issues in the cross-border region "Alpine Rhine Valley", development in the direction of territorial and regional governance is causing a paradigm shift, consisting in particular in the participation of nongovernmental organisations, economic actors and civil society. This paradigm shift is accompanied by the development of a new culture of planning, which aims to change the relationship between planners, stakeholders and inhabitants. It is characterised by the increasing replacement of static, sovereign, technocratic and restorative approaches of spatial planning by dynamic, cooperative, flexible and competitive approaches (Broggi 2007). At the same time, new approaches focus more on the process of spatial developments itself.

Challenges for Cross-Border Governance Processes in the Alpine Rhine Valley

Regional governance is often characterised as weakly institutionalised, consisting more of network-like forms of cooperation among regional actors. It regards regional development as a community task (Jessop 1998). The term governance is not to be confused with Government: "Government" refers to the state institutions and the formal rules and procedures for policy control and management. The value of "governance" approaches can be seen in the light of processes. Therefore, it is important to have a closer look at the controlling and coordinating structures of these processes. In other words, how are spatial development processes in the

Alpine Rhine Valley organised, and how can the nature of the interaction of the regional actors be described?

To an increasing degree, cross-border regions have to meet the challenges of governing and coordinating spatial development processes beyond the national frontiers. Regional governance is a basic concept for acting in a cross-border context (Diller 2005; Gualini 2003). Moreover, since the 1980s, the region as an operational level for spatial development has continually – and mainly in the field of regional development strategies – gained importance (Diller 2005; Pütz 2004; Benz and Fürst 2004). The development of the so-called new regionalism embodies the renewed significance of the regional level (Keating 1998). However, despite the recognisable "revaluation of the regional", regions remain part of a multilevel system. Knowledge about the connections and interplay of the various levels is vital for the implementation of sustainable regional development. Regional governance aims at adjusting decisions at various spatial levels in a way that enables sustainable development. In addition to decentralisation, a structural change of the forms of action, from a horizontal to a vertical structure, is necessary (Diller 2005; Pütz 2004).

Regional governance as a framework or instrument for sustainable regional development processes aims to improve state and communal activities in terms of "services for the region" and share responsibility for developing the region with private industries and civic organisations. In this context, the term "good governance" is relevant (Pütz 2007). Governance processes can cause increasing disharmony with old institutional structures and require at least a minimum of institutionalisation and even new forms of institutionalisation (Fürst 2007; Diller 2005). At the level of spatial development, governance takes on a collective and institutionalised form of regulation within which the actors participate through a formal or informal network (Pütz 2004). In this context, Diller poses the question of how far regional governance replaces or amends traditional political structures (2005). Successful governance processes may be indirectly observed through the establishment of appropriate structures and networks as well as the constellations of actors. Regional governance can only be considered durable if the inhabitants are integrated and a collective spatial reference is established.

Cross-border governance processes, such as territorial cohesion, cooperation and integration, pose challenges. These are included in the concept of "Europeanisation". In the study area, several programmes have been initiated to deal with these challenges. The two programmes analysed for this study are presented in detail below.

After the adoption of the agglomeration policy *Agglomeration Programme Rheintal* in Switzerland in 2001, cooperation among Swiss and Austrian municipalities was launched. The aim was to enable the sustainable development of urban areas and to coordinate cross-departmental themes within those agglomeration areas (according to ARE – Bundesamt für Raumentwicklung 2010). Development must be implemented by the canton and the municipalities of the agglomeration together, and the creation of a cross-border assembly (for cross-border agglomerations such as the Alpine Rhine Valley) is a compelling requirement. The Swiss federal government explicitly welcomed the inclusion of foreign partners in border regions. The agglomeration programme is a prerequisite for possible co-financing of spatial planning projects from Swiss federal infrastructure funds. The project is coordinated by an internal regional manager. It is based on the following guiding slogan: "One living space (Swiss and Austrian Rhine Valley), cross-regional cooperation, river Rhine as a connecting element, networked polycentric region, regional awareness and identification".

In 2004, the Austrian federal state of Vorarlberg launched the project *Vision Rheintal* (Vision Rheintal 2010), a regional development process which is shared among the federal state and its municipalities (Austrian municipalities). The aim is to break down existing hierarchical structures. In a two-year participative process, a mission statement for spatial development and regional cooperation was developed. At the end of this phase, the state and the municipalities strengthened their joint responsibility for the region by signing the *Rhine Valley Contract*.

In this context, the regional assembly is an important instrument for dealing with regional issues. Once or twice a year, the members of the state government, the members of the state parliament and the mayors of 29 municipalities meet to discuss the latest results of implementation and to advise on the next steps. The project is coordinated by an internal and an external regional manager. The aim of this compact or project is to create a networked polycentric region, promote cooperation within the region, promote cross-border cooperation and create an interconnected living space, foster and enhance a regional awareness and regional identity.

Regional Identity in the Adjacent Subregions

In the context of regional development processes in the Alpine Rhine Valley region, especially those with cross-border characteristics, the importance of regional identities is stressed. On the one hand, the target is to strengthen common identities, and on the other hand, a regional identity is supposed to be an essential catalyst for undertaking joint actions within the (border) region.

In Alpine Rhine Valley cross-border region, the varying spatial orientations, perceptions and thus regional identities are of great importance in everyday life and economic life. They strongly affect the chances of implementing sustainable regional development. However, little information exists as to how regional identity and orientation patterns in the Alpine Rhine Valley vary and about the factors by which they are influenced or altered. As well, there is less information about how regional identities are constituted and how they influence regional development processes in the cross-border region. Until now, these issues have not been considered in political and planning decisions within the Alpine Rhine Valley. Nevertheless, a stronger consideration of these topics is a prerequisite for improving the acceptance and chances of implementing realisable development concepts not only at the local and regional levels but also in cross-border contexts.

What is Regional Identity?

This work is based on a three-dimensional understanding of space as a territory, as an interactive system and as a social construct (Reutlinger et al. 2010). But what does this mean? Space as a social construct is understood as a relational space, which is established through a (common) regional identity. This view treats the region as a product of interactions and perceptions (Löw 2001; Läpple 1991), where the physical environment and regulative aspects, like national laws, have to be considered. This constructed space manifests itself in a common culture and common attitudes and approaches. The image of the region is the communicated symbol of it.

What is regional identity? In times of social heterogeneity and a diversification of lifestyles, the region has the potential to be an adequate scale for action and an appropriate anchor for social relations (Reutlinger et al. 2010). Regional awareness and regional identity are discussed in social geography from different angles. Among other things, this includes what Blotevogel et al. (1989) have described as "die Gesamtheit raumbezogener Einstellungen und Identifikationen, fokussiert auf eine mittlere Maßstabsebene" (the totality of spatial attitudes and identifications, focused on a medium-scale level). In this concept, regional identity has an affective dimension. Regional identity depends on the intensity of bonds, spatial distances and social differentiation. This understanding of regional identity is, however, a priori reduced to one-dimensional, spatial conditions. Sociocultural issues, such as communication and other social activities, can hardly be taken into account. Furthermore, this attempt to locate regional identities in a territorial space risks the problematic consequence of a generalisation of the social world within a territorial cutout. One example is the commonly held view that the inhabitants of the Alpine Rhine Valley are reticent and unapproachable. Therefore, the geographer Gerhard Hard suggests that we approach regional identity not through the abstraction of space but on substantive characteristics (1987). In other words, the challenge is not to extrapolate a regional identity from the region but rather to extrapolate from the regional identity to the region – to allow the regional identity or identities to define the region. "Der Raum, auf den sich Regionalbewusstsein bezieht, ist nur dessen Symbol, aber nicht dessen Inhalt" (the space which is the subject of regional awareness is merely a symbol of the actual components which make that space) (Bahrenberg 1987, 150). Using this framework, the "region" becomes a socially constructed region, and regional identity is readable as part of social processes. "Regional identity is, in a way, an interpretation of the process through which a region becomes institutionalised, a process consisting of the production of territorial boundaries, symbolism and institutions" (Paasi 2003, 478).

However, the importance of the physical-material environment in establishing regional identities must not be entirely ignored. For example, the intensity and quality of identification with a region is related to interaction within the physical area of people's own spaces of action. This is based on localisation which does not per se equate with regional identity. The objectively existing, "real" region is affectively connected with pictures and symbols: dialect, emotion, sound, etc. "Mit jeder Ortsbezeichnung werden nicht nur Lage- und Sachinformationen, sondern in der Regel auch Wertzuschreibungen, Emotionen und Präferenzurteile assoziiert" (each name of a place is not only associated with its position and factual information but usually also with the value given to it, emotion and preference ratings) (Weichhart 1992, 43). These physical and material structures serve as a kind of carrier of social messages and thus constitute a fundamental feature in the construction of regional identities. Conversely, if political-administrative or functional boundaries limit the experiences, actions and social relations of people, the physical structure of places and regions cannot act as a carrier. This becomes obvious in border regions, where the boundaries of a regional identity, however, it is constructed, overlap with national borders. Often, an identifying relationship with the neighbouring region hardly exists (Obkircher 2010).

Regional identity can give people a perception of stability and security. The concept of *creating regional identity* is relevant here. Similar to the concept of "alltägliches Geographie-Machen" (Werlen 1997), where ongoing human interactions constantly (re)create our geographical space, it is necessary to promote the symbolisation "hidden" behind the geographical cutouts and to connect it with a more active understanding of identity. Shared everyday experiences and actions can support the development of common regional identities. Social relations within places of action help to create a regional identity around these places and regions (Nohl 2006).

Stakeholders and Their Perception of the Region

To illustrate the characteristics of regional identities in the context of the Alpine Rhine Valley, I want to outline two explorative examples: first, the perception of responsible stakeholders and politicians, and second, the regional identity and regional awareness of young people in the region.

The *mental map of the stakeholders from the federal state of Vorarlberg* has a clear pole, Lake Constance (see Fig. 10.3; illustrated as a star). In addition, the river Rhine is more or less perceived as a constraint on intra-regional interaction and as a borderline. Accordingly, the stakeholders in Vorarlberg refer almost exclusively to their own state, from their point of view in the valley, while the neighbouring country is largely unknown (Sauter et al. 2002).

The *mental map of the stakeholders from the canton of St. Gallen* shows, however, a weaker perception of the border compared to the Vorarlberg point of view. Although the St. Gallen Rhine Valley region is most familiar, the neighbouring subregion Vorarlberg Rhine Valley is known as well. The city of St. Gallen (which is outside of the Alpine Rhine Valley) and Lake Constance form two important poles; the connecting axis (see Fig. 11.4; the city and the lake are illustrated as a star) is an important element of their mental map (representing Sauter et al. 2002).

"Die in unseren Köpfen fragmentierte Landkarte des Alpenrheintals erschwert die Kooperation über Grenzen hinweg: weiße Flecken auf unserer mentalen

Fig. 11.4 Mental maps of stakeholders and politicians (*left*=Swiss, *right*=Austrian)

Landkarte lassen uns gemeinsame Chancen nur undeutlich erkennen" (the fragmented mental map of the Alpine Rhine Valley complicates cooperation across boundaries: white spots on our mental map allow us to only vaguely recognise common opportunities and potentials) (Sauter et al. 2002, 11). It is of great importance to keep in mind such differences in the perception of the region and borders if cross-border cooperation, especially in the field of spatial planning, is to be realised in the near future.

This leaves one question unacknowledged: To what extent are the regional identities and mental maps of the inhabitants similar or different? This is all the more relevant for future cooperative projects. On the basis of these investigations, we can determine if the Alpine Rhine Valley and its subregions are viewed as a unified territory within people's perception and awareness or if the region is just a planning division (top-down constructed).

Perceptions of the People Living in the Alpine Rhine Valley

Below, in Figs. 11.5 and 11.6, I show how people perceive "their region", using some examples from the photo survey. Of course, their perception is not about the objects shown on the pictures but about the symbolism, the social meaning and the content of these pictures. This information was supplemented through additional interviews. Social life in the analysed border region corresponds neither to the image of the more socially distant neighbourhood conditions of a city, perhaps described as "anonymous co-existence", nor to the image of close interpersonal relations – "they know each other" – in the countryside (see also Fig. 11.5; showing the Zwischenstadt). Important personal relationships are less and less localised within the immediate surroundings, and the network of personal relationships is spatially dispersed (see also Reutlinger et al. 2010).

This can be seen in looking at the mental maps of juveniles. The *mental map of Swiss juveniles* is different to the *mental map of Austrian juveniles*. Swiss juveniles

Fig. 11.5 Symbolisation of the Alpine Rhine Valley (Compiled by author)

Fig. 11.6 Mental Maps of juveniles (*left*=Swiss, *right*=Austrian; results from the survey 2007)

perceive the state-border as weaker and describe favourite places and highlights in the neighbouring subregion (see Fig. 11.6; the cities of Dornbirn and Hohenems have an important function – illustrated as a star). Austrian juveniles have a strong connection to the Vorarlberger Rhine Valley, but their Swiss neighbours are hardly known. The official language in the Swiss Alpine Rhine Valley and the Austrian Rhine Valley is German, but the Swiss dialect is different from the Austrian dialect. Although the juveniles are aware of this, it does not affect the formation of a common regional identity. However, one important inhibiting issue is the lack of information about the other side of the border.

Nevertheless, the valley is to some extent still seen and perceived differently, depending on the origin of those interviewed. For instance, political boundaries and conditions such as access to recreational areas and historical roots play important roles. These alignments and connections are not static but subject to change, especially due to political and infrastructural developments.

Effects on the Spatial Development Processes

There is evidence of significant functional interaction in this cross-border region. Settlements are merging, and a polycentric settlement structure is visible. There is also evidence that some regional identities go beyond the borders of municipalities and (in Switzerland) cantons and even extend across the river Rhine (the border between Austria and Switzerland; the EU-border). But cooperation within this region still does not seem to be successful and long-lasting. The failure to achieve successful cooperation raises several questions. Are alternative approaches to regional development needed, and if so, why? Have regional governance processes already been established in the region? Is there any evidence of bottom-up processes? Finally, in what way do regional identities influence the governance of regional developments in cross-border regions? In other words: What is the importance of regional identities for the success of regional development processes?

Spatial planning depends on information, which "..... Wirklichkeiten beinhalten, um angemessene zukünftige Möglichkeiten zu erschliessen und kreativ zu verwirklichen" (examines realities in order to appreciate appropriate future opportunities and creatively develop them) (Breckner 2006, 440). These realities – those lived and perceived, in everyday life – are in turn crucial when attempting to draw conclusions on the development and construction of regional identities in the Alpine Rhine Valley region and thus to understand and assess the impact of identities on regional development. "Soft factors" such as personal emotional ties to a region and existing social relations, as shown in the border region of the Alpine Rhine Valley, may constitute an important basis in this context.

First Conclusions

The results of the research in the Alpine Rhine Valley are based on those surveys of the cross-border spatial development processes analysed so far.

The Alpine Rhine Valley is an ideal study area. As has been demonstrated, many small-scale political units meet and overlap in this region. Pilot studies have revealed the complexity of the spatial references (Obkircher 2008; Meier 2010). Within the Alpine Rhine Valley, there is a diverse network of cooperation between (space relevant) actors. Nevertheless, there is a great need for improvement because those networks and existing attempts at cooperation cannot yet resolve conflicts nor "hit the hot topics" due to their largely informal and under-institutionalised structures. On the other hand, neither formalised, cross-border institutions nor institutionalised projects in the Alpine Rhine Valley are able to develop common regional strategies. They are too often held back by their own administrative structures, by competitive factors and, especially, by different regional perceptions and awarenesses (in other words regional identities).

One side product of analysing regional identities is that one can visualise the symbols, or in other words the *unique selling points* of a region, which in this case is a "spongy Zwischenstadt". The Alpine Rhine Valley is neither town nor countryside, but still special in the perception of the inhabitants. The defining characteristics of the Alpine Rhine Valley listed by respondents were similar, regardless of age, location on either side of the border or immigrant background.

The range of institutional structures and cross-border projects in the Alpine Rhine Valley is large. They cover different administrative levels (local, regional and national) as well as diverse geographic units. Although only a few institutional structures are directly connected to spatial planning, many deal indirectly with space-related issues and therefore complement the range of cross-border cooperation in the field of the spatial development of the Alpine Rhine Valley (Meier 2010). In general, the perceived motives for cooperation are funding or a relative level of disadvantage. However, the long-term reliability of these as motivating factors in regional cooperation is questionable. In this context, all projects concerning the Alpine Rhine itself are particularly important for cross-border cooperation in the region. The motivation is unique: the River is a "vital line" and of great importance to a common identity. Thus, it is easier to start programmes regarding the reduction of hazard risk or water supply.

The cross-border concept of regional governance may be understood as consisting of ways to reach an inward consensus and an outwardly competitive orientation. This is important, because regional differences in the Alpine Rhine Valley are usually found in institutional and administrative structures and less in topographic or economic disparities.

Current regional governance processes are especially influenced by their respective administrative systems (St. Gallen and Vorarlberg). Legislation in both countries allows only a limited amount of self-organisation (e.g. bottom-up organisation and participation) within regional development processes. Public administrations and institutions influence the implementation of cross-border spatial planning. They can serve as a basis for understanding the "imprinting pattern" of the region.

This can be seen in the projects which were examined in the region. Austrian project partners are from the region itself, while Swiss project partners are external. Therefore, different priorities can be seen as a consequence of different regional affiliations. These differences obstruct everyday work. To guarantee a successful regional partnership, at least one project partner per county should come from the region. This will ensure that goals better reflect local priorities and ensure that any actions are more robust.

Individual and exceptional relationships between stakeholders define cooperation and networks in the Alpine Rhine Valley. This implies the need for appropriate financial and political support for these "region-makers". Region-makers can be understood as being people who are to a great extent committed to the needs of the region and understand this task due to an often exceptional cross-border regional identity without regard to either side of the Rhine.

Of course, it should be noted that not every project can be self-organised or self-governed. There are limits. The reason is a limited knowledge of the factors which affect the cross-border region Alpine Rhine Valley and the fundamental desire to protect one's own subregion. Programmes such as "Vision Rheintal" or the "Agglomeration Programme Rheintal" can reduce these barriers and build up a common regional awareness and possibly a common regional identity. According to our findings, they already do so.

The broad participation of stakeholders and inhabitants is a central pillar of these new forms of governance for regional development. The ability to act depends on the diverse actors of regional development in the Alpine Rhine Valley and in particular on the quality of participatory approaches. There are three requirements. The first is that there must be an open design process: it is not just about approval procedures and master plans. Collaborative tools such as regional conferences can be helpful for developing new, creative solutions.

The second is that the region must be established or given more powers as a spatial unit with the ability to act on its own, without recourse to a higher level of government. However, regions must not become institutionalised units and just another level of government. It is important to preserve institutional flexibility so as to adapt to ongoing spatial development processes; otherwise, the regional level becomes just another administrative unit, creating just another kind of segregation.

The third is that the perceived, mental space of inhabitants must be taken into account. In other words, the regional identities of the inhabitants must be recognised. If you can identify yourself with your region, you will be more willing to take responsibility for it. This is a crucial factor in regional governance which determines the ability of the region to act as a region. Regional identities are an essential catalyst for activating (border) regions.

To underline this hypothesis, I want to refer to the following insight. Spatial planning projects in the St. Gallen Alpine Rhine Valley have a strong quality of codetermination. People can decide on the project contents, but in Switzerland, participation during the creation and design phase of projects is lacking. That means participation is viewed as part of the decision process (e.g. information meeting, referendum). The planning process is not based on a two-way discussion or consultation. By contrast, in Vorarlberg, participation is more like a process of creating choice (e.g. the Wisdom Council). The Vision Rheintal process, for example, is considered a best practice example – "citizens plan their region". This promotes trust and responsibility for the region and helps create a core spirit which helps create and strengthen regional identities.

Finally, according to the initial research results, *regional identities relate to regional governance processes* in different ways.

Firstly, prior information about "the mood or atmosphere within the region" helps to adjust the methods of participation and can greatly simplify efforts to promote engagement. For example, at the beginning of a policy-making process, planners can identify historical conflicts between municipalities which still influence the recent regional identity of the inhabitants and which may affect the success of an ongoing project.

Furthermore, the research results suggest that a region's capability for action is limited by the size of the region itself. Successful regions relate less to territorial or functional areas, but rather to a socially constructed cognitive space. The whole Alpine Rhine Valley is too large to identify with. It is too far away from "everyday issues" and therefore hardly tangible for the people!

A "region" on a map is not necessarily a region in the minds of residents or in a cultural, economic or even political sense. Size makes a difference. Large regions can be too diverse, but smaller ones can become a living region. One conclusion could be that cooperation within regional development issues should focus more on small-scaled regions. Within those regions, the inhabitants have the identification required to make a true region. This increases the chance of implementing cooperative projects, particularly when it comes to sensitive issues, such as merging public services such as schools or nursing services.

Looking at the results to date, it appears that within the study area, the concept of the "region" has come significantly close to embracing the Rhine Valley as one living space, and there is a growing perception of the need to manage and develop this space jointly. Furthermore, local residents' identification with the broader region has increased as border crossings and regional cooperation have grown. Various interlinked cross-border projects have been initiated or are currently in the starting blocks. However, there will certainly be challenges in the future, including the question of the nature and strength of regional identity.

To conclude, I want to thank Josiane Meier for a professional exchange and for the opportunity to review and share her research results. In her research, she is analysing spatial planning structures in the cross-border region Alpine Rhine Valley. Her findings have provided valuable additions and new perspectives for my own research. In addition, it should be noted that the results presented in this chapter are only a preliminary insight. Further evaluation is needed.

Acknowledgements This study has been prepared with financial support by the Regional Research Institute for Social Science, Bregenz – Vorarlberg.

References

ARE – Bundesamt für Raumentwicklung (2010) Agglomerationsprogramm. Available at http://www.are.admin.ch. Accessed 1 Aug 2010

Bahrenberg G (1987) Unsinn und Sinn des Regionalismus in der Geographie. Geogr Z 75:149–160

Benz A, Fürst D (2004) Region – regional governance – Regionalentwicklung. In: Adamaschek B, Pröhl M (eds) Regionen erfolgreich steuern. Regional Governance – von der kommunalen zur regionalen Strategie. Bertelsmann-Stiftung, Gütersloh, pp 11–66

Blotevogel HH, Heinritz G, Popp H (1989) „Regionalbewusstsein" Zum Stand der Diskussion um einen Stein des Anstoßes. Geogr Z 77(2):65–88

Breckner I (2006) Die Produktion suburbaner Räume und die Rolle räumlicher Planung. In: Selle K (ed) Praxis Stadt- und Regionalentwicklung. Analysen, Erfahrungen, Folgerungen. Rohn, Dortmund, pp 438–449

Broggi M (2006) Die räumliche Entwicklung des Alpenrheintals – Analysen und Thesen in einer Übersicht. In: Broggi M (ed) Alpenrheintal – eine Region im Umbau: Analysen und Perspektiven der räumlichen Entwicklung. Verlag der Liechtensteinischen Akad. Ges, Schaan, pp 13–27

Broggi M (2007) Raumplanung und –forschung und öffentliche Kommunikation. In: Sauter M (ed) Nachhaltige Raumentwicklung. Michael Imhof Verlag, Petersberg, pp 18–28

Diller Ch (2005) Regional Governance by and with Government: Die Rolle staatlicher Rahmensetzungen und Akteure in drei Prozessen der Regionsbildung. Habilitation Dissertation (TU-Berlin)

EU Ministers Responsible for Spatial Development (2007) Territorial Agenda of the European Union. Towards a more competitive and sustainable Europe of diverse regions. Agreed on the occasion of the Informal Ministerial Meeting on Urban Development and Territorial Cohesion in Leipzig on 24/25 May 2007. Available at www.eu-territorial-agenda.eu/ReferenceDocuments/Territorial-Agenda-of-the-European-Union-Agreed-on-25-May-2007.pdf. Accessed in 4 Aug 2010

European Union (2007) Treaty of Lisbon amending the treaty on European Union and the treaty establishing the European Community, signed at Lisbon, 13 December 2007. Off J Eur Union 50(C306):1–272

Fritsche A, Studer L (2007) Grenzüberwindung im Dreiländereck. Wahrnehmungshorizonte und Mobilitätsverhalten der Bevölkerung in der Bodenseeregion. Roderer Verlag, Regensburg

Fürst D (2007) Regional Governance. In: Benz A (ed) Handbuch Governance. Theoretische Grundlagen und empirische Anwendungsfelder. VS Verlag, Wiesbaden, pp 353–365

Gualini E (2003) Cross-border Governance: inventing regions in a trans-national multi-level polity. DISP 152(1):43–54

Hard G (1987) „Bewusstseinsräume"– Interpretationen zu geographischen Versuchen, regionales Bewusstsein zu erforschen. Geogr Z 75:127–148

Jessop B (1998) The rise of governance and the risks of failure: the case of economic development. Int Soc Sci J 155:29–45

Keating M (1998) The new regionalism in Western Europe. Territorial restructuring and political change. Edward Elgar, Cheltenham

Läpple D (1991) Gesellschaftszentriertes Raumkonzept. Zur Überwindung von physikalisch-mathematischen Raumauffassungen in der Gesellschaftsanalyse. In: Wentz M (ed) Stadt-Räume. Campus, Frankfurt am Main/New York, pp 35–46

Leibenath M, Korcelli-Olejniczak E, Knippschild R (2008) Bridging the gaps? – cross-border governance between top-down policies and local needs and limitations. In: Leibenath M, Korcelli-Olejniczak E, Knipschild R (eds) Cross-border governance and sustainable spatial development. Springer, Berlin/Heidelberg, pp 187–193

Löw M (2001) Raumsoziologie. Suhrkamp Verlag, Frankfurt am Main

Meier J (2010) Regionale grenzüberschreitende Zusammenarbeit in der Raumplanung. Das Beispiel Alpenrheintal. Berlin (in print)

Nohl W (2006) Heimat als symbolischer Aneignungsprozess. Konzeptionelle Überlegungen und empirische Untersuchungen. Naturschutz und Landschaftsplanung 38:140–145

Obkircher S (2008) Wahrnehmungsbilder des Alpenrheintals – Räumliche Orientierung als Indikator für regionale Identität. Montfort – Vierteljahreszeitschrift für Geschichte und Gegenwart Vorarlbergs 60:98–123

Obkircher S (2010) Heimat. In: Reutlinger Ch, Fritsche C, Lingg E (eds) Raumwissenschaftliche Basics: Eine Einführung für die Soziale Arbeit (in print)

Paasi A (2003) Region and place: regional identity in question. Prog Hum Geogr 27:475–485

Pütz M (2004) Regional Governance. Theoretisch-konzeptionelle Grundlagen und eine Analyse nachhaltiger Siedlungsentwicklung in der Metropolregion München. oekom, München

Pütz M (2007) Modewort oder neues Paradigma: regional governance – definitionsmacht eines Begriffs. Raum 68:22–25

Reutlinger Ch, Schöffel J, Lingg E, Obkircher S (2010) S5-Stadt – Kontur einer alltäglich gelebten Agglomeration. ETH CASE (E-Reader) (in print)

Sauter J, Schlegel H, Zech S, Helfer F, Aliesch P (2002) Freizeit und Erholung im Alpenrheintal – ein Projekt zur Förderung der grenzüberschreitenden Zusammenarbeit in der Raumentwicklung im Alpenrheintal (A, CH, FL). Chur, Hohenems, Schaan, Buchs. Available at www.vorarlberg.gv.at/pdf/vorstudiebericht.pdf. Accessed 4 Aug 2010

Sieverts T, Koch M, Stein U, Steinbusch M (2005) Zwischenstadt – Inzwischen Stadt? Entdecken, begreifen, verändern. Verlag Müller+Busmann, Wuppertal

Vision Rheintal (2010) 29 Gemeinden. Ein Lebensraum. Available at http://www.vision-rheintal.at. Accessed 1 Aug 2010

Weichhart P (1992) Heimatbindung und Weltverantwortung. Widersprüchliche oder komplementäre Motivkonstellationen menschlichen Handelns? Geographie Heute 100:30–44

Werlen B (1997) Sozialgeographie alltäglicher Regionalisierungen. Band 2: Globalisierung, Region und Regionalisierung. Franz Steiner Verlag, Stuttgart

Part III
Administrative Systems, Services Delivery and Local Finance

Chapter 12
The Empowerment of Local Democracy and Decentralisation of Service Delivery in Local Government Reform: Evidence from Portugal

Carlos Nunes Silva

Abstract Local government reforms in Europe over the last few decades show a gap between central government promises of decentralisation and practice. The political rhetoric of central government favours increased local government autonomy and supports the growth of local democracy. This contrasts with a de facto centralising trend, whereby central government defines the priorities and targets of local government. In many countries, local funding powers remain limited, and central government remains the main source of local government revenue. This chapter examines recent changes in the structure, functions, finance, and regulation of central-local relations, in Portugal, in the period from 2005 to 2009, up to the onset of the 2009 financial crisis. In 2000, a newly elected government promised and implemented significant reforms to local government. The government promised increased local autonomy and the empowerment of local democracy, local government expectations. At the same time, the government also created new inter-municipal institutions to be responsible for planning and other specialist functions, particularly the administration of regional transportation systems. The government also strengthened regional identities by harmonising judicial and administrative regions. The evidence suggests the existence of a mixed trend. Service delivery has been somewhat decentralised, mediated by central government. However, issues of local democracy and the participatory rights of local citizens were relatively neglected. Decentralisation took place alongside a centralising trend, associated with institutional reorganisation and with changes in the legal framework of key components of the spatial planning system. Direct control by central government of local government was replaced by indirect control through regulations. Local governments remain restricted in their ability to challenge legislative decisions, and decisions affecting local competences and budgets are often taken by state

C.N. Silva (✉)
Institute of Geography and Spatial Planning, Universidade de Lisboa, Edifício da Faculdade de Letras, Alameda da Universidade, 1600-214 Lisbon, Portugal
e-mail: cs@campus.ul.pt

© Springer Science+Business Media Dordrecht 2015
J. Buček, A. Ryder (eds.), *Governance in Transition*, Springer Geography,
DOI 10.1007/978-94-007-5503-1_12

budgetary institutions rather than by the national legislature. The evidence also suggests that the response to the fiscal crisis inhibited the full development of local autonomy.

Keywords Local government empowerment • Devolution • Local government reform

Introduction

Local government reforms in Europe in recent decades suggest the existence of a gap between central government political discourse on decentralisation and its practice (Lowndes 2002; Tranvik and Fimreite 2006; Wollmann 2004). The political rhetoric of central government in favour of increased local government autonomy in support of an empowerment of local democracy has been followed in Europe by a trend towards the definition, by central government, of priorities and targets for local government (Fleurke and Willemse 2006). This contradicts the notion of local self-government and local political autonomy, a situation in which local councillors are supposed to decide policies, targets, and priorities based on the preferences of local citizens.

Building on this background, the chapter examines recent changes in the structure, functions, finance, and regulation of central-local relations, in Portugal, in the period from 2005 to 2009, corresponding to the XVII Constitutional Government, supported by an overall majority of the Socialist Party in the national parliament. In doing so, it addresses the following research question: has the reform or 'modernisation' of local government, proposed by central government, in 2005, and implemented in the following years, enhanced local government political autonomy and empowered local democracy or was it mainly a decentralisation of service delivery from the state to the municipalities?

In Portugal, the local government institutional framework is still based on the 1976 Constitution. The constitution was adopted after the overthrow of the dictatorship in 1974. Despite seven constitutional revisions, in 1982, 1989, 1992, 1997, 2001, 2004, and 2005 (CRP 2005), and calls for a further revision between 2009 and 2013, the level of control assigned to central government and the level of autonomy provided to local government are unchanged (Silva 2004, 2006, 2009a). In addition, the call-for regional tier of self-government has not yet been implemented. According to the XVII Government's programme (PCM 2005a, b, c), the creation of Administrative Regions was to be made through a national referendum, taking into consideration the five planning regions in mainland, corresponding to the 5 NUTS 2 regions. At the same time, there was to be a reorganisation of the deconcentrated departments of central government, according to the same regional map, as well as the reinforcement and stabilisation of the CCDR functions (CCDR, Comissão de Coordenação e Desenvolvimento Regional—Regional Planning Commissions, one for each NUTS 2, except in the two autonomous regions: Azores and Madeira), to enable the coordination of central government policies at the regional scale. For these reasons, continuity rather than rupture is the main feature that emerges when we examine what happened in the last three decades in the vertical structure of the State. The local electoral system, the internal structure of municipalities and parishes, the portfolio of municipal competences, and the local finance system have experienced changes, of one sort or another, but none represented a rupture with what was defined in the 1976 Constitution and in the common legislation that followed it. The same can be said about numerous policy areas under the direct responsibility of local government. However, the political discourse in this period has favoured more decentralisation, and successive local government reforms were intended to increase local autonomy and the share of local government in the public administration sector and to improve local democracy, in particular citizen engagement in local politics. This does not mean that local decision-making became less constrained by central government or that there has been a continuous expansion of local autonomy resulting in more advanced forms of self-government. On the contrary, evidence suggests a centralising trend in the context of new managerial practices, in which central government retains an important control role, by setting up the core values and priorities, for example, within EU funding schemes, or in central-local partnerships at the municipal level. This is similar to what has been found in other European countries (Goldsmith 2002; Fleurke and Willemse 2006; Denters and Rose 2005; Geddes 2006).

In addition, and besides these types of formal control, central government can exercise control by other means, perhaps more informal. These include the control of local government income and expenditure through the local finance system (Oulasvirta and Turala 2009), administrative regulations and standards (e.g. through the nature of school curriculum and activities, types of social services, etc.), and the level of access to central government decision-making. In other words, there has been a move from the prevalence of forms of direct to those based on indirect control through innumerable central government regulations, a move also found in other European countries (Goldsmith 2002; McAteer and Bennett 2005) and in other parts of the world (Chandler 2005). Notwithstanding this, the system continues to be based on the principle of general competences and on the right to self-government.

Consequently, central government retains a major influence in local government policies, not just in financial terms but also through numerous and detailed norms and regulations, despite the political rhetoric on decentralisation.

Central Government Policy in the Period 2005–2009: Goals and Proposals

Portuguese local government is based on the principle of general competence and as such can adapt its policies to changing circumstances. One of the public administration goals of the XVII Constitutional Government (2005–2009) was the reform or 'modernisation' of local government, which included the transfer of more competences to municipalities and parishes along with the corresponding financial resources, strengthening the traditional multifunction model of local government (PCM 2005a) especially Chapter III, section 'VI. Modernizar a Administração Territorial Autárquica', pp. 126–129: Modernisation of Local Government). The following constitutional government (2009–…), supported again by the Socialist Party, maintained most of the goals established in 2005, in the field of local government reform (PCM 2009: chapter V – Administração autárquica – Aprofundar a descentralização, uma administração autárquica para o Séc. XXI – Local government – Deepening decentralisation, a local government for the XXI Century, pp. 95–98). Decentralisation of new competences focused on basic infrastructure (sewage and waste), social equipment, the urban environment, and social policy (old people, health, drug addition, education, professional training, children, immigrants, and disabled people).

In some of these cases, particularly in the social policy field (e.g. primary education), the main goal appears to have been streamlining the local delivery of social services rather than enhancing local democracy and municipal political autonomy. As the evidence available shows, unitarian welfare states tend to be the sole provider of welfare services (Tranvik and Fimreite 2006), making local government responsible for the delivery of these services, while the central level continues to define targets and standards. In other words, when, as in the case of social services decentralised to municipalities, service standards are established by central government (e.g. some functions within primary education), municipalities cannot establish priorities according to their independent political will. The decentralisation of managerial competences in the field of social equipment and social services and in other policy sectors is important for empowering local government, but this does not necessarily mean strengthening truly local self-government, as defined in the Constitution.

The traditional strategy adopted by central government was to achieve certain policy goals (e.g., in the field of urban regeneration, housing rehabilitation, social services, etc.), through special programmes, targeted at sectors or areas with problems. This was also followed by the government elected in 2005 (e.g., within its policy for cities 'Política de Cidades – Polis XXI, 2008-13', adopted in April 2007, following previous programmes such as PROSIURB, POLIS, URBAN, etc.), a practice adopted in other countries. Wollmann (2004), for example, refers the case of France, where central-local relations make extensive use of this approach to policy-making through projects or programmes, definition of local targets, and contracts. Nonetheless, the possibility for local government to use specific central government programmes, such as those in the broad field of urban regeneration, presented new opportunities for municipal spatial planning and municipal housing policies. However, the objectives and standards were defined by central government, which, according to the criteria of Fleurke and Willemse (2006), is not the same as an increase in local self-government autonomy. Local councillors, as noted, were not in a position to decide priorities and targets based on the preferences of local citizens.

A second target for the reform or 'modernisation' of local government was the local finance system, in particular the size of the block grant and other fiscal revenues. This was expected to make municipalities financially stronger and more independent. In the end, the dependence of municipalities on revenues from loans and building construction taxes was to be reduced, without lowering the existing level of local financial resources (see PCM 2005a, Chapter III, Section VI – Modernizar a Administração Territorial Autárquica). In practice, due to the financial crisis which started in 2009, and despite the decentralist tone of political discourse, on both the income and the expenditure sides, the central government exerted additional control over local government, deciding the level of unconditional and conditional grants.

A third area of reform was local government internal organisation (institutional framework, debureaucratisation, e-government, professional training, etc.). This included relations between municipalities and other public and private entities, such as municipal and inter-municipal enterprises, public-private partnerships, and municipal concessions. Paradoxically, this complex web of organisations that makes up the new local governance landscape, and which is responsible for delivering public services at the municipal level, has expanded in the last few decades and represents, to some extent, a dilution of municipal autonomy. The adoption of policies, priorities, and targets is not done solely by its elected representatives, based on the preferences of local citizens, but by other organisations as well or through complex networks of central-local and public-private institutions.

A fourth objective of central government had to do with the gradual implementation of a new model of local governance, including a new local political leadership model, based on homogeneous political executives and on deliberative boards with more powers to control the activity of the executive and on effective intermunicipal forms of cooperation. (For a comparison with other local political leadership models, see Berg and Rao 2005.) It proposed, for municipalities, a parliamentary system with executives elected by majority representation instead of the present direct election of the executive and proportional representation (Silva 2009a). Since 2005, the president of the executive council cannot be re-elected more than twice (three consecutive terms or 12 years). This restriction applied only to Mayors and Parish Council chairpersons, and after four years, they could again run for the same posts. In 2006, new legislation was adopted introducing the principle of gender parity in electoral lists (Law 3/2006, 21.08.2006). A minimum quota of 33.3 % was set for each sex in the electoral list. This quota was not compulsory in small parishes with less than 750 inhabitants, or in small municipalities with less than 7,500 inhabitants.

Inter-municipal relations were another target in the reform of local government. The Gam-ComUrb model, implemented by the previous social-democrat governments, was unanimously considered inadequate by all stakeholders (Silva 2002, 2006). However, the proposed new forms of municipal associations for the metropolitan areas continued to lack political legitimacy as they are elected indirectly by the members of the municipal assemblies.

Central-Local Relations in Practice

Despite the overall political commitment towards decentralisation from central to local government, clearly articulated in the programme of the XVII Constitutional Government, in practice central-local relations experienced ups and downs in the 2005–2009 period. Considering the public statements made by the ANMP, it seems that for municipalities, the outcome of the reform fell short compared to what was expected. Nonetheless, in some policy areas, decisions taken by central government strengthened the autonomy of municipalities (e.g. in the creation of municipal police), while other decisions expanded municipal competences (e.g. in the education sector). However, in other cases decisions taken by central government implied a reduction of local government resources and its capacity to intervene or would have implied that if implemented (e.g. the proposed new 'Regime Geral dos bens do domínio público do Estado, das Regiões Autónomas e das Autarquias Locais'). In addition, in some policy areas, municipalities found themselves subject to new and different forms of control by the EU regulations, especially through monitoring, evaluation, and financial control on the use of EU funds received by the municipalities.

In the first case, reinforcement of municipal autonomy, the central government implemented its plan for the decentralisation of competences and the corresponding resources in a number of sectors. One example is the transfer of competences in the education sector, including the decentralisation of competences related to nonteaching staff and school buildings at the 2nd and 3rd education levels and additional activities to enrich the normal academic curriculum. However, at the end of 2008, the ANMP considered that the number of municipalities involved – 90 – and the activities/competences transferred was less than what they expected.

In other cases, the decentralisation of new competences was supported by research and planning as a first step for further decentralisation in the field of social policy. One example was the decision to prepare municipal social charters, 'Cartas Sociais Municipais', expected to be concluded in 2011 for each municipality, including an inventory of all social equipment needed for children, old citizens, disabled persons, and actions to be taken against social exclusion. Another example is the introduction of monitoring and evaluation procedures in these policy fields. For example, the Observatory of Decentralisation and Local Education Policies, a partnership between central government, municipalities, and university research centres, was created in 2009. This Observatory was created to monitor the transfer of competences from central government to the municipalities in the education sector (OPLE 2009). Another facet was the participation of municipalities in central government policy programmes, although without direct involvement in the decision-making process, such as the participation of municipalities in 'Contratos Locais de Segurança' (CLS). This included only the provision of local information, identification of local issues in the field of community safety, and related issues. A protocol between central government and the ANMP was signed in August 2008. One contract had already been signed with the municipality of Oporto. However, one year later, in July 2009, only 20 CLS had been signed in 7 out of 18 districts.

Local autonomy was also expanded through the revision of existing national policy programmes and projects. One example is the programme for the creation of municipal police forces. The requirement for a contract between central and local government was abolished. Under these contracts, central government covered the cost of police operations. In practice, due to lack of financial resources, only a small number of municipalities were able to create their own police organisation. By 2008, only 30 contracts had been signed between central government and municipalities for the creation of Municipal Police, and only 2 had received financial support from central government. The decision to abandon the requirement of a central-local government contract and an increase in the proportion of road fines retained by the municipality from 30 to 55 % (According to Decret-Law 197/2008, article 7) allowed much greater autonomy than the previous legal framework (Decret-Law 197/2008, 7.10.2008).

Notwithstanding these positive developments, central-local relations in this period were also marked by moves towards increased centralisation. At least in three cases municipalities and their national association reacted vigorously against the withdrawal of municipalities from key positions in the regional or local institutional structures belonging to central government. One example is the reorganisation of ICNB (Instituto da Conservação da Natureza e Biodiversidade/ Institute for Conservation and Biodiversity), implemented by the XVII Constitutional Government. Formerly, municipalities were represented on both the Board of Directors and the Consultative Board. In the new legal framework (Decret-Law 136/2007, 27.04.2007; Portaria 530/2007, 30.04.2007), municipalities were only to be members of the 'Strategic Council of the Protected Area', a purely consultative board. This reduced the already small influence municipalities had in the management of protected areas. In addition to this loss of ability to influence policy-making within protected areas, some new ICNB powers have the potential to override municipal competences in the field of spatial planning inside protected areas. This has caused central-local conflicts in the past; they are likely to continue and increase in the coming future.

A second illustration of this loss of local government influence is the reform of Tourism Regions, approved in 2008 (Decret-Law 67/2008, 10.04.2008). There was disagreement about the organisation of regional divisions in this sector, which includes five regions and six Poles of Tourism Development. The ANMP was in favour of just five tourism regions, coincident with the five planning regions in mainland Portugal. Municipalities were somewhat disappointed with the reduction in the number of municipal representatives on the boards of these new entities (e.g. the ANMP argued that at least 50 % of the Assembly members should be mayors).

A third example was the delay in forming 'Regional Councils', a structure within the CCDR (the Regional Planning Commissions), in which municipalities are represented. Decret-Law 134/2007, 27.04.2007, defines the organisation and competencies of the CCDR. Regional councils were slow to form. At the start of 2008, the regional councils had still not been created, and the ANMP made public the dissatisfaction of municipalities with that fact. By April 2008, only the Regional Council in Região Norte/Northern Region had been constituted. Among other aspects, the delay in the implementation of the regional councils prevented municipalities from participating as a group in decisions about the implementation of major investment projects in their region which had been approved by central government under a special regime (the so-called PIN – Projectos de Interesse Nacional) (Decret-Law 174/2008, 26.08.2008). They were also unable to participate in the coordination of public investments in their region, as well as in the preparation of regional development plans, sector plans, or regional spatial plans, developed and implemented by central government for the region or for some subregions. Also, due to this, municipalities were not formally involved in the preparation of PIDDAC - Plano de Investimento e de Despesas de Desenvolvimento da Administração Central, the annual central government investment plan. This plan defines which investments will be made by central government in each municipality.

There was also strong opposition from the national local government association (ANMP) towards decisions taken by central government in areas that fall outside the competences of municipalities such as reforms of the spatial organisation of the judicial system and the reorganisation of the network of local health centres (Decret-Law 28/2008, 22.02.2008; Decret-Law 81/2009, 2.04.2009), which were defined and implemented despite alternative proposals made by the ANMP. The spatial organisation of the judicial systems was to have five districts coincident with the limits of the five planning regions, expected to coincide in the future with the five administrative regions. These five districts had a total of 39 divisions, called 'Comarca', a clear reduction in the number of courts (Proposal of Law 124/2008, 12.03.2008).

Central-local relations were also affected during this period due to changes in the planning legislation. New legislation on the ecological and agricultural reserves (REN – Reserva Ecológica Nacional (National Ecological Reserve) (Decret-Law 166/2008, 22.08.2008); RAN – Reserva Agrícola Nacional (National Agriculture Reserve) (Decret-Law 73/2009, 31.03.2009)) gave more powers to technical departments within central government and did not strengthen the competences of the municipalities, as one would have expected from the political statements of central government.

In addition to these direct impacts on local governments' capacity to influence public decision-making within the spatial planning system, legal norms and the institutional reform of the ICNB will probably impact on the fluidity of the already highly complex local planning system (Silva 2000).

According to the Constitution, central government shall pass new legislation on key issues for local government only after consultation with the National Association of Municipalities (ANMP). On numerous occasions, the ANMP claimed to have had a very short period of time to comment new legislation proposals submitted by central government according to the Constitution. That was the case, for example, with the new law proposed by central government for the planning and management of harbour areas, which would, from the ANMP point of view, interfere with and reduce municipalities' spatial planning competences. Municipalities argued that the planning and management of harbour areas needs to be articulated and subordinated to municipal spatial master plans and other municipal spatial planning instruments.

Another example is the decision to adopt control development around the future Lisbon International Airport, in Alcochete, taken without consulting the national association of municipalities (ANMP), as is required by law (Law 54/98, 18.08.1998, article 4). On other occasions, this consultation process has been perceived and treated as mere formality. For these reasons, this is another area in which reform and improvements seem to be necessary. Associated with this is the request that municipalities should be heard before major public works are decided by central government within their territory.

Crossing this debate on centralisation-decentralisation are two key dimensions of local government institutional architecture: the local finance system and the inter-municipal cooperation model. The local finance system has always been a key area in central-local relations. In this period (2005–2009), despite the approval of a new local finance law, in 2007 (Law 2/2007, 15.01.2007, Local Finance Law), a number of problems continued to affect municipalities' financial capacity. In the new local finance system, municipalities became entitled to a transfer of up to 5 % of IRS (personal income tax), as part of the non-conditional transfer from central state budget to municipal budgets. Since then, delays in the monthly transfer have led to repeated protests by the ANMP, especially because in some cases this is an important or even a vital component of the municipal revenue. The total amount of unpaid transfers, in January 2010, was estimated at 33 million Euros, by the ANMP. In addition to this, central government decided not to transfer the 5 % of IRS to municipalities in the Autonomous Regions of Azores and Madeira, based on the argument that it is now the responsibility of their Regional Governments, as the IRS collected in those two autonomous regions became part of the income of the Autonomous Regions (Law 1/2007, 19.02.2007, Regional Finance Law). In addition, the need to balance the national budget, and the resultant cutbacks in spending due to the financial crisis from 2009 onwards, had an impact in the implementation of the new local finance system. In response, the ANMP proposed, in 2009, the creation of a Fund for Local Investment as an instrument to support local economies affected by the 2008–2009 financial and economic crisis (ANMP 2009). For these reasons, the reform of local finance system did not represent a radical departure from the previous situation. Continuity rather than rupture is the main characteristic. Nonetheless, while in theory the proposed changes seemed to represent progress for municipal autonomy, in practice, the consequence may have been more centralism, as the complaints made regularly by the national local government association suggest.

For inter-municipal cooperation, one of the objectives of the programme of XVII Constitutional Government was the abolition of the previous institutional framework, and the introduction of new forms of inter-municipal associations (Silva 2002, 2006) contains an overview of the previous inter-municipal model, also known as the Gam-ComUrb model. In 2008, the central government transferred competences to new inter-municipal associations. This was partly related to the implementation of the municipal component in the QREN. QREN is the acronym for Quadro de Referência Estratégico Nacional (Nacional Strategic Reference Framework). It is the framework for the European Union financial support to Portugal, in the period 2007–2013. About 70 % of all investments in the QREN regional programmes are municipal investments, to be coordinated by the new municipal associations, organised according to geographical divisions based on NUTS 2 and NUTS 3 regions (defined in Decret-Law 68/2008, 14.04.2008), based on NUTS 2 and NUTS 3 divisions. However, municipal associations continued to lack adequate resources to properly function and execute their responsibilities within the QREN, as well as in other areas, such as urban regeneration programmes, water and waste infrastructure, renewable energy projects, supra-municipal transport networks, and development plans.

Although this was an important step in the decentralisation process, the decision was taken two years later than that of fiscal decentralisation, and for that reason, municipalities will have a shorter period (two years less) to carry out investments (2009–2013, instead of 2007–2013), requiring a higher annual municipal investment level. As a result of this, municipalities will need easier access to credit, for these investments within the QREN, and municipal participation in the same investments may be lower (see also ANMP 2010). According to ANMP, these loans are neutral for the public deficit. At the end of 2008, no contract had been signed by municipal associations within the QREN. In part, this was partly due to delays in the constitution of the new municipal associations based on the NUTS 3 region. The ANMP recommended, on several occasions, that municipalities should join these associations, as soon as possible, as a precondition for the development of this process.

Central-local relations have also been affected by delays in the implementation of other supra-municipal institutional structures. This is the case, for example, of the AMT (Autoridade Metropolitana de Transportes/Metropolitan Transport Authorities), a process inherited from previous governments. From the central government's point of view, the creation of these transport authorities is seen as decentralisation, but for municipalities, it can imply further centralisation in the sense that it will be another higher tier with functional control of this sector and of municipal spatial planning, namely, in setting up planning goals for transport infrastructures and services. This is taking place in a sector which is already highly centralised (Silva 2009b).

In addition to these positive and negative developments in the framework of central-local relations, there are a number of other issues that if changed would improve the autonomy and capacity of municipalities. Municipalities should be able to directly address the Constitutional Court on issues related to local government when the legislation adopted by central government is considered unconstitutional. This is increasingly important, as central control over local government tends to move, as in other European countries, from direct formal control to indirect control through regulations. If municipalities are allowed to challenge legislation, that would reinforce the capacity of municipalities and of their national association to react against policy measures that reduce municipal autonomy and harm the interests of local citizens.

Municipalities also argue that revising or amending basic local government legislation (e.g. the legislation on competences and on local finance) through decisions taken within the annual state budget should not be allowed. This practice indirectly reduces local autonomy, a pattern also found in other European countries. The local government's legal framework should only be revised by an overall qualified majority in the national parliament. In addition, the supervision and the control of the legality of municipal actions should be the sole responsibility of an independent entity and not, as it is now, a competence shared by courts and central government departments.

Conclusion

The evidence examined in this chapter suggests that the local government reform was not as radical as anticipated by municipalities, considering the public statements made by the local government association, on many occasions. It seems that more radical progresses towards full local autonomy and a greater degree of self-government were expected by Portuguese municipalities. The gap between political discourse and the practice of central government in several areas within central-local relations is probably the result of the prevailing economic and financial conditions in this period, namely, budget constraints, as well as traditional power asymmetries in central-local relations.

Another conclusion, found in other European countries, suggests that central government increasingly exercises control through detailed administrative regulations for the implementation of specific policies and the approval of municipal projects by central government departments, especially within the EU financial framework, and through strict financial regulations. Additionally, there has been a shift from direct hierarchical and formal control to indirect forms associated with central-local partnerships, both public-public and public-private. In other words, there is an ongoing move from a relatively clear division of power between central and local government levels, as established in the Constitution and in the Local Government Act, to a more complex network of central-local cooperation. This has been described as a move from local government to local governance, in which municipalities are just one of several agencies responsible for delivering public services.

In sum, the evidence collected in this exploratory study suggests a mixed trend. The level of service delivery has decentralised, in part controlled by central government, which remains responsible for setting national priorities and targets for these programmes and projects, and there has been a centralising trend associated with institutional reorganisations and with changes in the legal framework of key components of the spatial planning system. Contrary to what we could anticipate from Socialist Party political discourse on local government 'modernisation', issues of local democracy and participatory rights of local citizens have not received the same priority within the overall local government reform. While growth and improvement of local service delivery is a vital dimension for a more balanced central-local relationship, it is equally important to enhance local democracy, promoting community participation in service provision and in goal setting as well.

References

ANMP (2009) Fundo para o Investimento Local (Fund for Local Investment). Associação Nacional de Municípios Portugueses, Coimbra

ANMP (2010) Memorando de Entendimento entre o Governo da República Portuguesa e a Associação Nacional de Municípios Portugueses para Promover a Execução dos Investimentos de Iniciativa Municipal no âmbito do Quadro de Referência Estratégico Nacional 2007–2013 (Memorandum of Agreement between Central Government and the National Association of Portuguese Municipalities regarding the implementation of municipal investments within QREN 2007–2013). Associação Nacional de Municípios Portugueses, Coimbra

Berg R, Rao N (eds) (2005) Transforming local political leadership. Palgrave, New York

Chandler J (2005) Comparative inter-governmental relations: models that need to travel. Local Gov Stud 31(3):269–284

CRP (2005) Constituição da República Portuguesa – VII Revisão Constitucional. Lisbon: Imprensa Nacional-Casa da Moeda

Denters B, Rose L (eds) (2005) Comparing local governance. Trends and developments. Palgrave, New York

Diário da República/Government Journal of Legal Acts (several years)

Fleurke F, Willemse R (2006) Measuring local autonomy: a decision-making approach. Local Gov Stud 32(1):71–87

Geddes M (2006) Partnership and the limits to local governance in England: institutionalist analysis and neoliberalism. Int J Urban Reg Res 30(1):76–97

Goldsmith M (2002) Central control over local government. A Western European comparison. Local Gov Stud 28(3):91–112

Lowndes V (2002) Between rhetoric and reality: does the 2001 white paper reverse the centralising trend in Britain? Local Gov Stud 30(3):303–330

McAteer M, Bennett M (2005) Devolution and local government: evidence from Scotland. Local Gov Stud 31(3):285–306

OPLE (2009) Protocolo que cria o Observatório das Políticas Locais de Educação. Ministério da Educação, Secretaria de Estado da Administração Local e Associação Nacional de Municípios (Agreement between Central Government and the National Association of Portuguese Municipalities for the creation of OPLE – Observatory of Local Education Policies), 15 April 2009

Oulasvirta L, Turala M (2009) Financial autonomy and consistency of central government policy towards local governments. Int Rev Adm Sci 75(2):311–332

PCM (2005a) Programa do XVII Governo Constitucional 2005–2009 (Programme XVII Constitutional Government 2005–2009). Presidência do Conselho de Ministros, Lisboa

PCM (2005b) Discurso do Primeiro-Ministro na apresentação do Programa do XVII Governo na Assembleia da República (Prime-Minister discourse in Parliament – presentation of the XVII Government Programme – Opening session), 21 March 2005. Presidência do Conselho de Ministros, Lisboa

PCM (2005c) Discurso do Primeiro-Ministro no encerramento da apresentação do Programa do Governo (Prime-Minister discourse in Parliament – presentation of the XVII Government Programme – Closing session), 22 March 2005. Presidência do Conselho de Ministros, Lisboa

PCM (2009) Programa do XVIII Governo Constitucional 2009–2013 (Programme XVIII Constitutional Government 2009–2013). Presidência do Conselho de Ministros, Lisboa

Silva CN (2000) O sistema de gestão do território em Portugal (Spatial planning system in Portugal). Cadernos Municipais – Revista de Acção Regional e Local 74(XIV):21–35

Silva CN (2002) Gestão de Áreas Metropolitanas: o "Modelo GAM-ComUrb" (The government of Metropolitan Areas: the GAM-ComUrb model]). Cadernos Municipais – Revista de Acção Regional e Local 81(XVI):42–63

Silva CN (2004) Portugal: local government system. Institut de Ciències Politiques i Socials, Universitat Autónoma de Barcelona & Diputació de Barcelona, Barcelona

Silva CN (2006) Decentralisation and regional governance in Portugal: self-government or inter-municipal co-operation? In: Salet W (ed) Synergy in urban networks? Sdu Publishers, The Hague, pp 204–226

Silva CN (2009a) Local political leadership in Portugal: exceptionalism or convergence towards a 'Mayoral Model'? Lex Localis – J Local Self-Gov 7(3):243–256

Silva CN (2009b) Portugal: el centralisme en la presa de decisions apaivaga els conflictes en matèria d'infraestructures de transport? (Portugal: centralism in decision making appease the conflict in terms of transport infrastructure?), vol 32. Idees – Revista de Temes Contemporanis, Barcelona, pp 184–217

Tranvik T, Fimreite AL (2006) Reform failure: the processes of devolution and centralisation in Norway. Local Gov Stud 32(1):89–107

Wollmann H (2004) Local government reforms in Great Britain, Sweden, Germany and France: between multi-function and single-purpose organisations. Local Gov Stud 30(4):639–665

Chapter 13
Municipal Bonds in Hungary: Constraints and Challenges

Gábor Kovács

Abstract After the collapse of communist rule, Hungarian local governments had to meet new challenges and expectations. Due to the decentralisation process, local governments' revenues decreased significantly in the last two decades, while the level and scope of services provided decreased less rapidly, if at all. Consequently, local governments, short of resources, had to develop their ability to raise funds needed to meet local spending needs. This study examines the general theoretical framework of municipal borrowing, comparing the economic and social advantages of bank loans versus municipal bonds. It also evaluates the role of the national government's central administration in regulating local authority indebtedness. It then presents and analyses the characteristics of Hungarian local government bond financing, assessing the appropriateness of bonds as a local government fundraising tool. The indebtedness of Hungarian local authorities has increased drastically since 2006, mainly through bond financing. A substantial proportion of these bonds were issued in foreign currencies because interest rates were lower. The subsequent economic downturn and devaluation of the Hungarian forint has left many local governments worse off because their indebtedness has increased in forint terms. Bond financing is often considered simply an alternative form of borrowing, and Hungarian local governments do not fully benefit from the flexible features offered by bonds. In today's Hungary, bond financing offers opportunities to broaden local government financial freedom and reduce the financial risk related to borrowing, but they also pose risks, particularly in the absence of local expertise. Regulation of local government borrowing is still mainly based on the coercive effect of a credit limit. This is a one-sided approach to the problem, and current Hungarian regulations fail to meet the expectations of the most important players in the European capital market, namely, institutional investors. New regulations are needed which can enhance the beneficial characteristics of bonds, while preserving the security they provide to local governments and prospective investors.

Keywords Local government finance • Municipal borrowing • Municipal debt management

---

G. Kovács (✉)
Kautz Gyula Faculty of Economics, Department of Economics, Széchenyi István University, Győr, Hungary
e-mail: kovacsg@sze.hu

© Springer Science+Business Media Dordrecht 2015

J. Buček, A. Ryder (eds.), *Governance in Transition*, Springer Geography,
DOI 10.1007/978-94-007-5503-1_13

Introduction

Following the 1990s, after the termination of the economic transition process, the Hungarian local government system had to meet new challenges and expectations. Due to the decentralisation process, local governments' revenues decreased significantly in the last two decades in real value, while the level and scope of services provided did not decrease at the same time. Therefore, local governments, which are short of resources, have had to perform what is probably one of their most important tasks: to develop their resource absorption and fundraising capacity, which would ensure the necessary financial tools to implement development.

One of the methods of using outside resources is borrowing, among which fund acquisition via local government bond issuance should be listed. The enhancement of the resource-deployment capacity of the local authority sector is a fundamental condition to ensure the necessary development resources at the local level. The extension of the capital absorption capacity at subnational level is also of high importance in relation to opportunities to obtain funding from the EU.

This chapter presents and analyses the main characteristics of Hungarian local government bond financing and assesses the usefulness in Hungary of bonds as a local government fundraising tool. The sample studied includes all the Hungarian bond issues between 2006 and 2008. This is deliberate: the aim is to examine local government behaviour before the onset of the current credit crisis in 2008–2009. As can be seen from the analysis below, since then the credit crisis began, the fiscal position of many Hungarian local governments has worsened, particularly because of a large number issued bonds in Swiss francs or euros rather than Hungarian forint. This is because the interest rate on francs and euros was significantly lower than that on forint, thus lowering borrowing costs over the term of the bond. However, devaluation since 2008 has actually increased indebtedness in forint terms. Hungarian local governments were not unique in this regard: within Hungary, in 2009, 60 % of all housing loans were in foreign currencies (European Covered Bond Council 2010). Moreover, in the United States, which has a long history of local government bond issues (municipal bonds), a growing number of local governments have overborrowed and been forced to default on their bonds (Walsh 2012).

This chapter attempts to summarise the lessons learned from local authority bond financing in Hungary and to develop some future perspectives regarding the viability and role of municipal bonds in the finance of Hungarian local governments. The timeliness of this research is primarily proved by the fact that compared to the mid-2000s, considerable changes have happened to the size of indebtedness of Hungarian local governments. Local authorities' external indebtedness has increased drastically since 2006, mainly in the form of bond financing.

The Role of Municipal Bonds in Local Government Finance

In the first, theoretical part of my paper, I introduce the most important types of municipal bonds, discuss the economic advantages and disadvantages of their use and summarise recent regulations of the Hungarian central government on the issue.

Why Issuing Bonds by Local Authorities Makes Sense

Borrowing – and issuing bonds – represents one possible and important way to finance local capital projects. The most important arguments for borrowing by local governments and against other forms of financing (e.g. against raising the level of local taxes or levying new local taxes) are as follows:

Long-term debt allows subnational governments to acquire or build capital improvements more quickly than they could on a pay-as-you-go basis. Borrowing over time is an effective way to overcome the problem of an inequitable burden of costs among tax payers. It allows more equitable payment schemes, since users can be made to pay for the capital cost of facilities as they are used over time. Most users will pay for the benefits either through local taxes or directly through user charges, and thereby an optimal allocation of resources can be achieved.

Benefits from accelerated local development can overshadow the cost of borrowing. When carrying out the investment quickly, operational costs (related to the given service) can be reduced (Swianiewicz 2004). Borrowing can also stabilise the required budget resources. The volume of capital spending in local government units fluctuates from one year to another. If capital projects are financed from current revenues, the demand for funding resources changes over time as well. In countries where a large proportion of local revenues is raised through local taxes, an irrational fluctuation of local taxes rates may result.

There are also *costs and risks* in the case of borrowing. Long-term debt limits a subnational government's future budget flexibility. Unwisely used, it can burden citizens with high taxes or service charges. Many countries permit long-term debt only for capital spending and not for meeting operating deficits (sometimes called the "Golden Rule").

Borrowing to meet short-term financing needs can provide opportunities for banks and subnational governments to develop working relationships and allow bankers to become familiar with the governments' financial affairs. Provided that the financing is repaid within the budget year and that carrying debt beyond the budget year is prohibited, there is no a priori reason to limit such financing to capital spending (Freire-Petersen 2004).

Borrow from a Bank or Issue Debt?

Even if borrowing appears to make more sense, the type of borrowing that is most appropriate to finance capital projects needs to be considered. The simplest way might be to borrow through a local or national bank. An alternative method is to issue debt in either the domestic or international capital markets.

Possible Advantages of Bond Financing

In the case of bond financing, sub-sovereigns can get all the funds they need upfront through the bond offering. Funds are not subject to partial payments based on a bank's monitoring of the project construction progress. In other words, the whole credit sum is available immediately, allowing local governments to quickly commit the capital.

Domestic bond markets can also provide an added source of financing which can tap into the wealth of a wide range of players, from individual investors to pension and mutual funds. The marketability of bonds induces the fact that theoretically more funds from potential investors are available for local governments, and several potential investors' savings can be mobilised by investing in bonds. Another consequence of marketability is that bonds generally have a longer repayment period than bank loans, and they may indeed provide a cheaper source of capital, especially when the offering is backed up by a robust dedicated revenue source. Bonds are lower in price than bank financing, due to lower interest, because of their marketability. In addition, a bond issue allows a higher volume of fundraising compared to bank loans which also creates more favourable conditions for investment.

The more flexible cash flow of the bond is also a crucial argument for bonds as opposed to bank loans. In general, in the case of bond financing, the whole sum of the capital payment is due only at the end of term. Therefore, if a fiscal problem should arise, repayment of a bond issue can be financed by a new bond issue, and so the obligation of repayment can be "rolled on".

The distant nature of the relationship between bondholders and issuers can enhance both the efficiency and transparency of government operations. At the same time, there is an opportunity – taking advantage of strength of potential local linkages – to raise and use funds for expansion at a more favourable rate through the public issue of bonds.

The Drawbacks of Bond Financing

Advocates of bond issuance, however, do not reckon with the fact that sub-sovereigns need to be familiar with *risk management* (e.g. interest rate and foreign exchange rate risk). The conditions related to staffing, experience and material resources required for well-founded local fiscal management are still missing in many places.

Auditing and other transaction costs (procedural fees for supervision, trading costs, the fee for taking it to the stock exchange, etc.) increase costs and make this instrument a great deal more complicated. In addition, issuing bonds increases the administrative costs of borrowing. Local governments must provide more detailed information for the investors than they would in the case of bank loans. For example, it is necessary to prepare a prospectus, publish a report every year to the Supervisory Commission and publish any extraordinary information where important changes have occurred in the project or in the local government.

Generic Categories of Municipal Bonds

In their broadest definition, municipal bond markets refer to borrowings by sub-sovereign public entities, directly or through public corporations, to fund governments (El Daher 1997). Municipal bonds can be classified and analysed primarily based on the securities pledged to repay the debt (Petersen-Valadez 2004). Concerning the pledges for the repayment of the bonds, three generic categories of long-term bonds can be distinguished: general obligation bonds, project revenue bonds and dedicated revenue bonds (Freire 1999).

In the case of *general obligation bond*, repayment is guaranteed by the "full faith and credit" of the issuing government. Issuers generally back general obligation bonds with all their revenue-raising powers. This means that the full taxing authority of the issuer is pledged to pay back the bonds. In addition, in the case of general obligation bonds, the local government itself acts as the issuer, and the local government's negotiable assets are counted against liabilities.

Project revenue bonds are secured only by the expected stream of revenue from the project being financed. These are secured by user fees or dedicated taxes rather than the general taxing power of local governments. The issuer may be a sub-sovereign or a public authority, such as a water authority, which is independent of the government. The expected future income coming from the project is the primary asset for repayment. On the part of the issuer, expected repayments from the bond are based on the surplus income arising from the completion of the project. Security can be provided by liens on real property owned by the public corporation, and repayment can also be made by mortgaging that property.

In the case of dedicated revenue bonds, repayments are guaranteed by a particular revenue stream which is unrelated to the project being financed. For example, a bond may be backed by the pledge of funds from intergovernmental transfers, which the sub-sovereign is due to receive, or by specific tax revenues such as liquor, sales or gas taxes.

In practice, the picture is more cloudy. Some bonds do not fit precisely into any one of the above categories. This is because in some cases, for example, dedicated project revenue bonds (and dedicated revenue bonds) are also guaranteed by the local government which owns the public corporation.

Municipal Bonds in Hungary: A Historical Overview

The transition to a market economy created growing difficulties in financing the central government budget. This led to the transfer of a greater share of public service provision to the local level. The *Act on Local Authorities* assigns responsibility to local governments for providing an extraordinarily wide range of public services, even when compared internationally. Since transfers from the central budget are less and less able to cover the investment needs of local governments, there is an even stronger demand for external funds that can be obtained through borrowing.

After an early and short-lived "bond-boom" in the middle 1990s from 2002, indebtedness started to increase. One of the reasons was the favourable macroeconomic environment (low inflation rate, moderate interest rates). The increase was also due to a decrease in revenues derived from the privatisation and sale of locally owned assets and a growing need for investment which could not be met by central government. Since accession to the EU accession in 2004, indebtedness has increased further, although it has not yet exceeded 2.5 % of GDP. The characteristics and features of a new "bond-boom" which started in 2006 are discussed later in the empirical part of this study in more detail. The extent of bond issues and local government debt is shown in Fig. 13.1.

Regarding the borrowing of local governments, the proportion of long-term loans has been increasing since the 1990s relatively to short-term funds. This is shown in Fig. 13.2. The volume of municipal bonds issued during the last 10 years

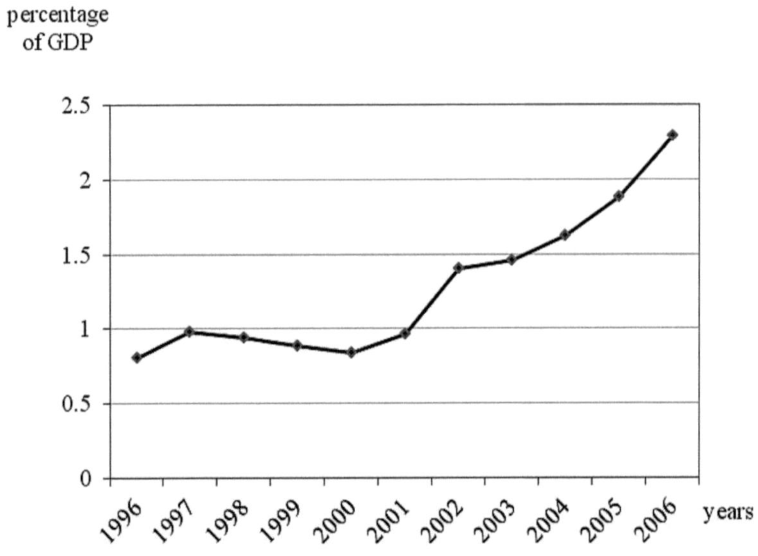

Fig. 13.1 Borrowing of local governments in Hungary in the percentage of GDP (1995–2006) (Source: Hungarian National Bank)

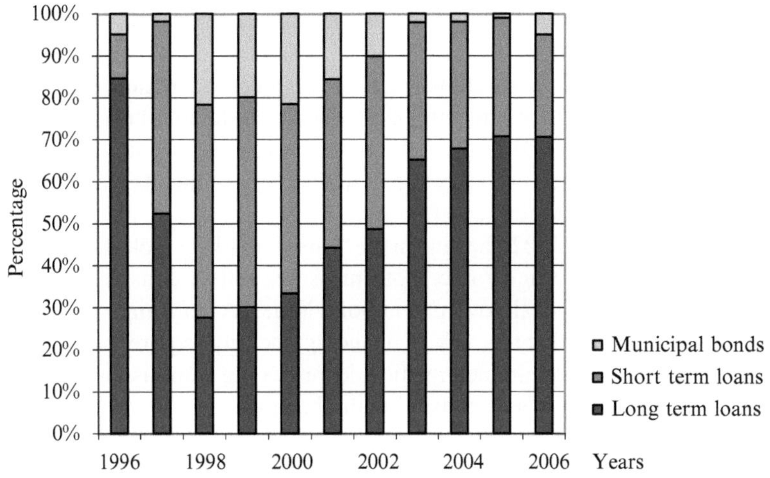

Fig. 13.2 The structure of borrowing (1996–2006) (Source: Hungarian National Bank)

is characterised by high volatility. After a slight drop at the beginning of the 2000–2010 decade, a considerable increase could be observed.

Examining the features of municipal bonds issued by local governments in Hungary, it can be stated that almost all bonds were placed privately, and the buyers were commercial banks. Municipal bonds could therefore be considered "bank loans in disguise". One possible explanation is that in Hungary, capital market financing is a less accepted and widespread solution than borrowing from a bank. However, despite administrative overheads, the total cost of issuing bonds – mainly due to a reduced interest rate – is less than that of obtaining a bank loan. At the same time, the banks have provided a guaranteed market for municipal bonds.

Legal Regulations Regarding the Issue of Municipal Bonds

The *Act on Local Self-Government* (1990/LXV) allowed the free borrowing – hereby also the issuance of bonds – of municipalities without the permission of central government. Legal restrictions regarding the size of borrowing and the nature of the securities which must be pledged to repay the debt were formulated in 1996:

- Municipalities are not permitted to meet their debt service obligations from personal income tax revenues, the normative state contribution, central subsidies or by selling core assets.
- Total debt (including bank loans, municipal bonds, lease, third-party obligations, commitments) cannot exceed the Corrected Current Own Revenues, which is 70 % of the positive difference between current own revenue and short-term liabilities.

In 1996, a *Bankruptcy Law* for municipalities (Municipal Debt Adjustment Act, Law XXV) was prepared and came into force. The law defines a debt adjustment process whose objective is to allow local governments to regain their financial health while at the same time protecting the rights of creditors. The Municipal Debt Adjustment Law defines and restricts the risk of investing in municipal bonds by imposing a definite financial and moral cost on local governments which default on debt or other payments (Makay 2004).

On the 1st of January 2002, the law on the "capital market" took effect and a new decree on "bonds" came into force. Both measures reflect considerable changes within the regulation of local government bond issues. A very important argument for bonds in Hungary is that this way of financing does not require the local authority to announce public procurement, which decreases the administrative burden on the one hand and the issuer's responsibility on the other.

Current Loan Financing of Hungarian Local Governments

After pioneering attempts at the beginning of the twenty-first century, the number and the total value of municipal bond issues have increased significantly, particularly since 2006 in Hungary. Accordingly, the period from 2006 can be considered a new period in the history of the Hungarian municipal bond issues, which is why I have decided to describe the characteristics of municipal bond financing for the period between 2006 and 2008.

Since the bonds issued during the studied period were almost always private issues and their buyers were commercial banks, a global analysis of local government loan financing is crucial. As the bond issues of the period increased the indebtedness of the local government sector to an unprecedented extent, I analysed the liquidity position of the sector and the characteristics of the application of external resources.

I have also described and evaluated the most important characteristics of Hungarian bond issues. As the bond issues of the period incurred significant fiscal risks in the municipal sector, it is crucial to evaluate the central regulation of loan financing and to reveal the connected deficiencies.

The Liquidity Position of Local Governments

I have described and characterised the increasing rate of local government debt by quantitative index numbers. As well as accounting for the constitution of local government debt, I have attempted to explore the primary reasons for local government indebtedness. It is important to examine the form of the financial instruments used to borrow and the extent of bond financing in local government borrowing. In addition I describe how indebtedness may influence individual institutions and actors in the local government sector differently and what kind of fiscal risks it involves.

Breakdown of the Net Financial Property of Local Governments

Figure 13.3 clearly indicates how the borrowing has accelerated since the end of 2006. Compared to the total amount recorded at the beginning of 2002, the liability portfolio of local governments multiplied approximately sixfold by the end of 2008. At the same time, it is noticeable that the net financial wealth did not decrease in proportion to the increase in borrowing. This is because increases in the value of financial assets paralleled increases in the value of borrowing – that is, the bulk of the money raised from borrowing were deposited with banks, while the remainder financed project costs (Homolya and Szigel 2009). Thus, as well as borrowing from banks through bond issues, local governments were present on the credit market as resource providers. This role made them "VIP" customers with banks (Gál 2009).

The volume of the extra deposits realised in connection with loan financing may amount to as much as HUF 200 billion (Vigvári 2009). By investigating the financial asset line of local governments' asset and liability statements, we can establish that the dynamism of the increase in money instruments was mainly due to an increase in the long-term liability portfolio. The bulk of the money from the loans was deposited with banks or placed in central government securities. Long-term bonds also allowed local governments to buy out some of their short-term debt (Gál 2009).

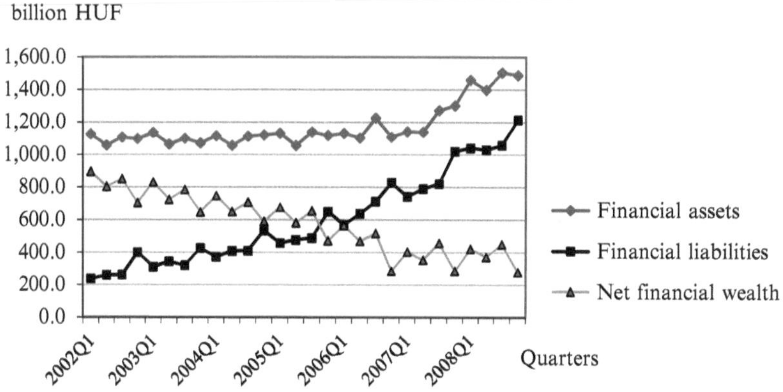

Fig. 13.3 Conformation of net financial wealth of local governments (Source: Own construction based on Hungarian National Bank's data)

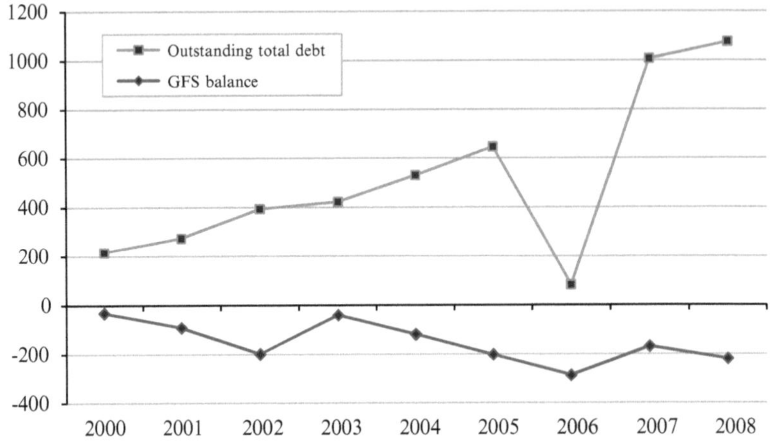

Fig. 13.4 The budgetary balance (GFS) of municipalities and the formation of outstanding total debt between 2002 and 2008 (in billion HUF) (Source: Own construction based on Gál (2009). Note: 2008 balance excludes privatisation revenues. Graph includes only data through the third quarter of 2008)

Reasons for Debt

The wave of debt, which started at the end of 2006 can be traced back to a variety of factors. The government's bill restraining local government borrowing and a fear of additional restrictions were primary causes (Bill No. T/4320 for the modification of Act No. LXV of 1990, 9 November 2007).

The upward long-term trend in borrowing can be explained mainly by the permanent discrepancy between mandatory tasks of local governments and the size of grants from the central government. Nonetheless, as is shown in Fig. 13.4, between 2000 and 2008, there was no strong correlation between the size of local governments' balance of payments and outstanding total debt (Vigvári 2009). As the research of Vigvári (2007) confirmed, the increase in debt was also not due to the increasing volume of investment grants received from the European Union.

Instruments of Indebtedness

When examining the individual components of local government financial liabilities, we can establish that the long-term (investment) loan portfolio did not change significantly between 2006 and 2008. As is shown in Fig. 13.5, short-term loans were extremely volatile and, over time, increased by only a small extent. Most of the debt increase was through the issue of long-term bonds.

However, although the increase in the volume of the short-term debt portfolio was marginal in absolute terms, it was significant in relative terms. This may be

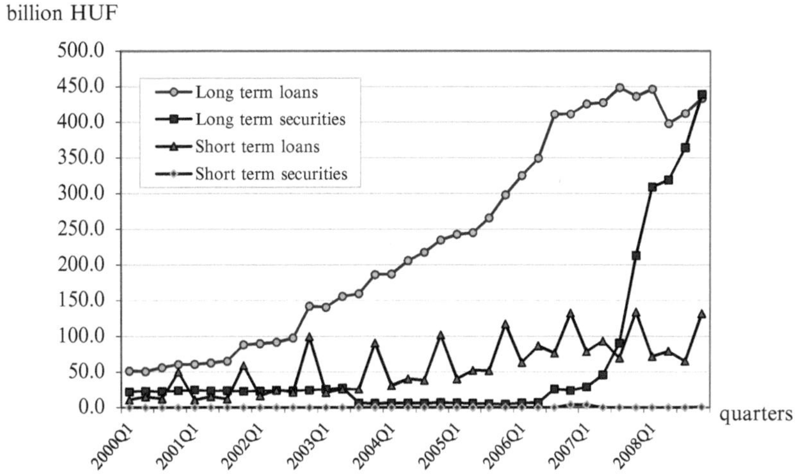

Fig. 13.5 The formation of local government loan financing between 2002 and 2008 (Source: Own construction based on Hungarian National Bank's data)

because municipalities were trying to exceed the credit limit. Concurrently, the long-term securities' portfolio increased drastically due to the wave of municipal bond issues from the end of 2006.

Between 2006 and 2008, another typical characteristic of local governmental borrowing was the rocketing increase of loans denominated in foreign currencies. The rate of financing in foreign currency increased by over 60 % between 2005 and 2008. Among foreign currency liabilities of local governments towards commercial banks, foreign currency credit portfolio increased by 50% and it exceeded HUF 90 billion at the end of 2008. The bond portfolio denominated in foreign currencies had shown a significant increase which was even higher. While the volume of the foreign currency bonds was negligible prior to 2005, it increased up to 15-fold to HUF 350 billion at the end of 2008 compared to the portfolio volume at the end of 2006.

Municipalities Involved in Debt

Not all local governments are indebted to the same extent. Homolya and Szigel (2009) found that indebtedness was primarily noticeable within a relatively small group of just 74. However, the number of local governments whose liabilities exceeded their own revenues doubled from 2006 to 2007. Over and above this fact, the volume of the liabilities of these 74 bodies accounted for half of the debt of the entire sector.

The extent of indebtedness differs between kinds of local government institution and incurs different risks, as Vigvári (2009) highlights. Accordingly, some local governments have excellent financial abilities. On the other hand, large city councils with considerable resources often run into debt very quickly. As well as being responsible for a large share of bond issuances, majority of hidden debts and conditional liabilities can be traced to them, too.

County governments are also among the most important national bond issuers. The worsened financial position of these actors and their relatively high budget deficits characterise the recent period. The fiscal risks of this segment are increased by that fact that they often became over-indebted due to their poor fiscal abilities and do not have sufficient revenue to meet their future liabilities.

The extent of the debts of Budapest's city-wide government is truly alarming. What is interesting is that, in comparison with that of the city-wide government of Budapest, the financial situation of the majority of the districts can be considered stable, the rate of their debt having not increased drastically in recent years.

Local Government Bond Issue Today

As we have seen, the drastic increase in the volume of local government debt started in 2006, primarily due to the issue of local government bonds. While up to HUF eight billion worth of bonds were issued in 2006, bond issue value in 2007 nearly reached the HUF 200 billion mark, and in 2008 it exceeded this amount (if calculated at 176 HUF/CHF and 266 HUF/EUR).[1] Consequently, the portfolio of bonds issued in HUF reached the HUF 50 billion mark, while the value of those denominated in foreign currencies increased above HUF 350 billion (Fig. 13.6).

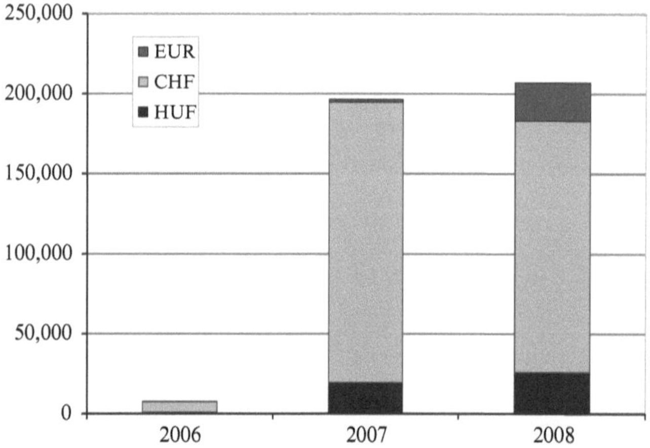

Fig. 13.6 Bonds issued by local government sector between 2006 and 2008 (in million HUF). (Source: Own construction based on Financial Supervisory Authorities data. Note: calculated at 176 HUF/CHF and 266 HUF/EUR)

[1] In June 2012, the Swiss franc was worth 240 Hungarian forint; the euro was worth 288.

Table 13.1 The total nominal value of the issued local government bonds per denomination (2006–2008)

Denomination	2006	2007	2008
HUF (in thousands)	900,000	19,497,966	25,969,000
CHF	38,519,371	996,123,591	893,403,144
EUR	0	6,000,000	89,297,722

Source: Own construction based on Financial Supervisory Authorities data

According to the report made by the Public Expenditure Survey Committee, 20 % of municipalities issued bonds in 2007 (among local governments which were audited by the Committee), while only one municipality issued bonds between 2004 and 2006 (Állami Számvevőszék 2008). The amount of bonds issued by individual governments ranged from HUF 200 million to HUF 5,000 million. Local governments planned to allocate 63 % of the income from bond issue to building up reserves, 15 % for operations and 22 % for debt services. Issued bonds had variable rates of interest and grace periods from 1 to 6 years. However, as is shown in Table 13.1, the issuance of bonds denominated in euros and Swiss francs increased dramatically in a relatively short period of time. This was because interest rates on euro- and Swiss franc-denominated instruments were significantly lower than those for bonds issued in forints.

In almost every case, the repayment of the principal of municipal bonds starts after passing of the *grace period*. According to Gál (2009), the selection from among competing banks organising the issue and listing of bonds is often decided by the length of the grace period. When examining bonds issued between 2006 and 2008 in foreign currencies, it is evident that the share of bonds issues in Swiss franc (CHF) and euros (EUR) soared and squeezed HUF-based bonds out (shown in Fig. 13.7). Comparing the two foreign currencies, the Swiss franc was more popular at the beginning of the period, while the euro overtook it at the end of the period. But bonds in Swiss franc were clearly dominant overall.

When we examine the currency of bonds on the basis of the number of issues, only 2 euro-based bonds (0.7 %), 37 HUF-based bonds (13.5 %) and 235 CHF-based bonds (85.8 %) were issued between 2006 and 2008.

An investigation of the length of the term of the issued bonds (Fig. 13.8) also shows an interesting result. The shortest term was 4 years, while the longest was 25 years for those bonds issued between 2006 and 2008. It is clear that the terms of bonds increased compared to earlier issues, and this increase was continuous between 2006 and 2008. While the average term of bonds was 16.73 years in 2006, it was 17.2 years in 2007 and 19.3 in 2008. During the researched period, there appear to have been no bond issues with terms shorter than 10 years. Most had 20-year terms. Nearly half the issues had 20-year terms in 2006 while the rate increased to roughly 60 % in 2007 and 2008. The appearance of extremely long-term

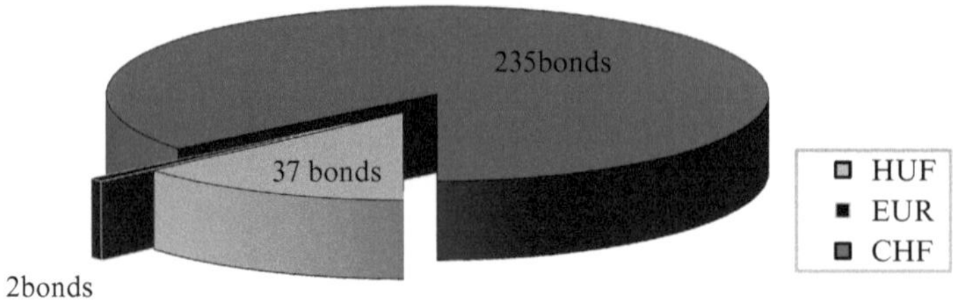

Fig. 13.7 Currencies of local government bonds between 2006 and 2008 (*Source*: Own construction based on KELER (KELER provides quasi-wholesale services and infrastructure to the players and intermediators of the Hungarian capital market))

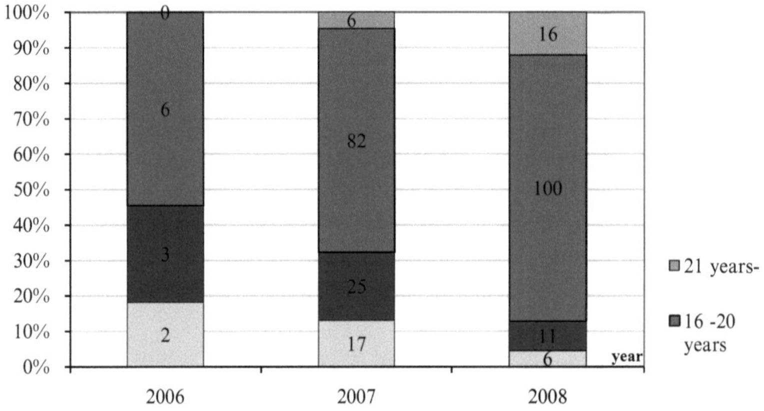

Fig. 13.8 Distribution of the duration of the local government bonds (numbers = number of issues) (Source: Own construction based on KELER)

issues is an important characteristic. Six bonds with a term longer than 20 years were issued in 2007, while 16 were issued in 2008.

Current Problems of Regulation

Problems and deficiencies of regulation are linked both to the drastic debts of local governments and to the significant increase in the fiscal risks within the sector. The fiscal risks to the subnational system appear in several forms: liquidity risk, lack of funds and in extreme cases insolvency (Vigvári 2009). The regulation system used by the central government could not function properly within the changed macroeconomic and money market environment. Problems in the budgeting process particularly related to obtaining revenue to meet local expenditure requirements and to forward the objectives of economic policy revealed gaps in bond financing regulation and also affected loan financing.

The Abnormal Function of Credit During the Credit Squeeze

As was mentioned above, limits on local government borrowing were approved in the middle 1990s, at roughly the same time as the Act on debt service of local governments. Since then, there have been no significant modifications to the regulations. The regulation of local government borrowing still relies mainly on the *coercive effect of the credit limit*. This is an unjustified, one-sided approach of the problem. On the basis of experience, it would be better to apply more sophisticated ways and means of controlling borrowing, either by counting local government borrowing as part of overall public sector borrowing (i.e. having a national limit on total public sector borrowing by all levels of government) or by defining specific limits for different settlement categories.

Vigvári (2009) underlines the fact that the public finance information system of Hungary only shows the redeemed guarantees. This situation results in that the traditional index numbers indicating the indebtedness of local governments can reflect a better but false image of the real solvency situation. Moreover, the regulation – though unintentionally – left a loophole: the debt limit doesn't apply to liquid loans. In practice this situation lets local governments roll debts forward and allows them to finance current budget deficits through borrowing.

Fast-spreading *financial innovations* have also played an important role in altering the credit balance of local authorities (Vigvári 2009). For example, guarantees provided within public-private partnerships (PPP), financing by bills of exchange and factoring all fall within the range of financial innovations. In the case of PPP, application banks provide a loan to the private operator providing the local service. The source of repayment is the service fee or rental paid by the local government to the private operator. According to the effective public acts, such borrowing should be treated as corporate borrowing, even though the ultimate guarantor of the loan is the local government (Gál 2009). These kinds of financial instruments and the resulting debt are not listed among those elements of debt which should be taken into consideration when determining if a local government has reached its credit limit.

At the same time, Vigvári (2009) points out that local government borrowing has generally not been restricted by this rule but by the willingness of banks to purchase bonds. According to the relevant reports of the Public Expenditure Survey Committee for recent years, debt limits did not serve as an effective barrier for towns with county rights and for the twenty districts of the capital city, Budapest. According to the Public Expenditure Survey Committee's report for 2007, 96 % of local governments taking on debt remained within the provisions of the credit limit specified by law (Public Expenditure Survey Committee's report in Állami Számvevőszék 2007).

Enforcement of the law is another issue. For example, in 2004, four municipalities successfully issued bonds despite having exceeded their credit limit (Balás-Hegedüs 2004).

Local Government Insolvency Law

Another important instrument for regulating local government borrowing is the Act on Debt Settlement Procedures for Local Governments, in other words the local government bankruptcy act. Overall, the act was well received as practical experiences demonstrated that the act was able to function effectively, highlighting its protective role. The act helped to increase financial discipline. Between its introduction and 2004, there were 60 cases that were settled out of court, and courts applied debt settlement procedures in 11 cases (Jókay-Szepesi-Szmetana 2004; Kopányi et al. 2004). On the basis of yearly reports made by the Public Expenditure Survey Committee between 2005 and 2008, there were one to three solvency restitution and debt settlement procedures in 3 of those years (Vigvári 2009). Consequently the local government bankruptcy act provides an excellent basis for assessing credit risks from bonds, but it cannot solve problems caused by the opaque financial reporting system, nor compensate for the lack of other, alternate institutional solutions to risk management.

Local Government Bonds as Secure Investments

The issuance of bonds eases the administrative burden on municipalities and at the same time forces issuers to provide financial information needed to satisfy prospective investors. However, the process of issuing bonds remains focussed primarily on meeting formal requirements. As noted, local government bonds are superior to simple borrowing due to their security aspect. The marketability of bonds allows them to tap into the savings of numerous market participants. This can reduce the general costs of borrowing, and more flexible, tailor-made terms of repayment are possible. Consequently it is important to have regulations which can enhance the beneficial characteristics of bonds while preserving their security aspect.

Present Hungarian regulations do not meet the requirements of *regionalisation*. Nor do they meet the expectations of the most important players in the European money market, namely, institutional investors. Furthermore, they do not support the fundraising efforts of local organisations. The capital market act does not allow regions or multipurpose micro-regional associations to raise money by issuing bonds, and it significantly restricts such possibilities for municipal service corporations.

Another characteristic of national regulation is that the Public Procurement Act does not require obligatory public procurement procedures during a bond issue, unlike regulations regarding normal borrowing. Consequently national regulators prefer the issuing of bonds over borrowing from banks because the administrative costs and their responsibilities are reduced. Although this reduces the cost of local government borrowing, it complicates the assessment of local government indebtedness by outside players. Conditions cannot be renegotiated or modified, and the only way to *eliminate* liability is for the issuer to repurchase the bonds (Gál 2009).

Conclusions

The drastic increase in the volume of local government debt started in 2006, primarily due to the issuance of local government bonds. Among Hungarian local governments, bond financing has followed a similar pattern in terms of bond issuance, bond purchases, investment by local government and the borrowing periods. Hungarian local governments are not yet able to fully benefit from the flexible features offered by bonds. Bond financing is often considered simply an alternative form of borrowing. Local governments remain unaware of the potential advantages of bonds as security. In today's Hungary, bond financing offers unrecognised opportunities to broaden local government financial freedom and reduce the financial risk related to borrowing.

The regulation of local government borrowing is still mainly based on the coercive effect of a credit limit. This is an unjustified, one-sided approach to the problem. Current Hungarian regulations do not meet the requirements of regionalisation – that is, of devolved regional policy making and policy implementation – and fail to meet the expectations of the most important players in the European capital market, namely, institutional investors. They fail to support the fundraising efforts of local organisations. New regulations are needed which can enhance the beneficial characteristics of bonds while preserving the security they provide to local governments and prospective investors.

Acknowledgement Dr Gábor Kovács's research activity was supported by TÁMOP 4.1.1/A-10/1/KONV-2010-0005 Programme (Universitas-Győr Alapítvány) and Eötvös Hungarian State Fellowship (Magyar Állami Eötvös Ösztöndíj).

References

Állami Számvevőszék (State Audit Office of Hungary) (2007) Jelentés az Állami Számvevőszék 2007 – évi tevékenységéről. 0803 J/5493, April 2008. Available at http://www.asz.hu

Állami Számvevőszék (State Audit Office of Hungary) (2008) Jelentés a helyi önkormányzatok gazdálkodási rendszerének 2007. évi átfogó és egyéb szabályszerűségi ellenőrzéséről. Report No. 0822, July 2008. Available at http://www.asz.hu

Balás G, Hegedüs J (2004) Local government borrowing in Hungary. In: Swianiewicz P (ed) Local government borrowing: risks and rewards. Open Society Institute, Budapest, pp 79–126

El Daher S (1997) Municipal bond markets. World Bank Infrastructure Notes, No. FM-8a. The World Bank, Washington, DC

European Covered Bond Council (2010) EMF Hypostat 2009. European Mortgage Federation, Brussels. Available at: http://www.ecbc.eu/uploads/attachements/27/65/Hungary.pdf. Viewed 26 June 2012. Later editions available at: http://www.hypo.org/Content/Default.asp

Freire M (1999) Credit rating and bond issuing at the subnational level. The World Bank, Washington

Freire M, Petersen J (2004) Subnational credit markets in developing countries. Oxford University Press, New York

Gál E (2009) Theoretical and practical problems of assessment of local governments' liabilities (Az önkormányzatok adós- és követelésminősítésének elméleti és gyakorlati problémái). PhD dissertation manuscript

Homolya D, Szigel G (2009) Local governments' borrowing – risks and banks' behaviour (Önkormányzati hitelezés – kockázatok és banki viselkedés). MNB-Szemle, September 2009, pp 20–119

Jókay C, Szepesi G, Szmetana G (2004) Municipal bankruptcy framework and debt management experiences 1996–2000. In: Kopányi M, Wetzel D, EL Daher S (eds) Intergovernmental finance in Hungary. A decade of experience 1990–2000. LGI, Budapest, pp 593–616

Kopányi M, El Daher S, Wetzel D, Noel M, Papp A (2004) Modernizing subnational government system. In: Kopányi M, Wetzel D, EL Daher S (eds) Intergovernmental finance in Hungary. A decade of experience 1990–2000. LGI, Budapest, pp 15–75

Makay M (2004) The prospects for municipal bonds. In: Kopányi M, Wetzel D, El Daher S (eds) Intergovernmental finance in Hungary. A decade of experience 1990–2000. LGI, Budapest, pp 571–592

Petersen J, Valadez M (2004) Borrowing instruments and restrictions on their use. In: Freire M, Petersen J (eds) Subnational capital markets in developing countries. Oxford University Press, New York, pp 45–71

Swianiewicz P (2004) The theory of local borrowing and the West-European experience. In: Swianiewicz P (ed) Local government borrowing. Risks and rewards. Open Society Institute, Budapest, pp 1–23

Vigvári A (2007) Is fiscal policy a wonder weapon? A few ideas related to fiscal rules for local governments (Fiskális politika csodafegyver? Néhány gondolat az önkormányzatokra vonatkozó költségvetési szabályok kapcsán). Pénzügyi szemle, 3–4. szám, 522–529

Vigvári A (2009) Financial risks in local government system (Pénzügyi kockázatok az önkormányzati rendszerben). ÁSZ KUT–study. Available at http://www.asz.hu/ASZ/jeltar.nsf. Accessed 12 Dec 2009

Walsh MW (2012) Crushed by borrowing. New York Times, 26 June 2012, section B, p 1

Chapter 14
Municipal Property During the Post-socialist Transition in Slovakia

Ján Buček

Abstract When local governments gain significant new powers, their main concern is the maintenance and expansion of their assets. This has been particularly true among the transition countries of Central and Eastern Europe. Looking at Bratislava, the capital of Slovakia, this paper examines changes which occurred after the collapse of communist rule in 1990. Slovakia underwent four main phases of transition. Each had its impact on local government assets and asset formation. The first was a phase of restitution and de-étatisation, when central government restored properties to local governments when those governments were themselves re-established, often transferring control of formerly state-owned enterprises. This was followed by a phase of asset consolidation. Firms were rationalised, broken up, sold off, or closed down. Assets were inventoried, and some sold to support local spending. A third phase was the establishment of a new, decentralised system of government. Local governments received a dedicated income based on income tax and gained the freedom to impose local taxes and fees. At this time, schools and other public buildings, and surplus facilities, such as army bases, passed into local government hands. The final period, which has recently ended, was again one of consolidation. Transfers of property and other assets took place against a backdrop of rapid economic change. Land buildings often became unexpectedly valuable, forcing local governments to choose between retaining them to generate long-term income or selling them to reduce debt, repair and improve infrastructure, and expand the city's asset base. Local governments had to become familiar with borrowing practices and learn to manage local debt. Change required new institutions to oversee the management and privatisation process at both the national and local levels and forced local governments to become familiar with the operation of property and debt markets. Local governments had to decide which services should be privatised, which should be run as public-private partnerships, which should remain under municipal control, and which should be run jointly with other local governments. The gradual process of change allowed most Slovak governments to successfully cope with increased powers. However, over the longer term, local governments need

J. Buček (✉)
Department of Human Geography and Demography,
Comenius University, Bratislava, Slovakia
e-mail: bucek@fns.uniba.sk

© Springer Science+Business Media Dordrecht 2015

J. Buček, A. Ryder (eds.), *Governance in Transition*, Springer Geography,
DOI 10.1007/978-94-007-5503-1_14

to further develop expertise in strategic planning and in asset management to ensure that they have rainy-day reserves and to protect and expand existing asset-derived revenue streams.

Keywords Municipal asset management • Asset management strategies • Municipal asset management oversight • Public private partnerships

Introduction: A Review of Issues in Local Government Asset Management

This chapter describes and evaluates the use development of municipal assets during the transition period in Slovakia. Slovakia, one of the post-socialist Central-East European countries, underwent radical changes in the field of local government (e.g. Buček 2006). The development of municipal assets is one of the less covered issues (among rare exceptions, there is work by Papcúnová and Balážová 2006). This chapter reviews the relevant legal framework. This is followed by an evaluation of changes from a period of uncertain and insufficient local government assets in the early nineties to the present period of larger and more diversified assets. The value and structure of municipal assets in Slovakia is briefly discussed. Using Bratislava, the capital of the Slovak Republic, as an example, one can see practical features of municipal asset formation, structure, and local government approaches. Developments in Bratislava resemble processes in other cities. The chapter focuses especially on issues of real-estate property, financial investments and reserves, and municipal companies. It also attempts to identify the changing importance of different kinds of municipal assets to city administration and city development. Particular attention is paid to the latest period, including a discussion of the impact of the financial crisis which started in 2008 on municipal assets and asset management.

The data used is taken from the final accounts of the Slovak Republic, and official documents from the Bratislava city government, including final accounts and annual reports. A document published by a team of authors led by the Mayor of Bratislava from 1990 to 1998, P. Kresánek (Kresánek et al. 1998), has been a particularly useful source of information. As well as including an effort to provide a long-term outline of municipal assets in Slovakia, the chapter attempts to provide greater details for the period from 2000 onwards. Data on municipal assets from the early 1990s is unreliable. Data from before 2009 are shown in Slovak crowns (due to the fluctuating exchange rate, they are not converted to euro), while more recent data are shown in euros, the national currency since 2009 (for orientation, the final exchange rate was €1 to 30.126 Slovak crowns). Within the text, the terms 'municipal assets' and 'municipal property' are used as being synonymous. They include all forms of local government ownership. Real (estate) property is used to describe what is usually the most extensive part of municipal assets – land and buildings. This chapter also refers to other asset-related items such as inventory, receivables, and financial accounts. Bratislava has a citywide government but also 17 district governments. In the case of Bratislava, the terms 'citywide government' and 'city government' are used to indicate that we are not dealing with a lower level of local government in Bratislava.

Local governments need to have administrative freedom and control over their assets if they are to successfully function over the long term. There is a long tradition of local government self-administration and ownership: it is perhaps the most common and historically developed case of public ownership. Various local government tasks, public administration, and public service provision require a diverse range and a large number of buildings and facilities and a range of equipment. The size of assets and their need and use is closely related to local government powers. At the same time, municipal assets influence long-term development and the overall perception of local public-sector roles. In recent decades, there has been an extensive retreat from the direct provision of services by local government, often influenced by the adoption of new public management approaches, in favour of contract-based private delivery and outsourcing, and public-private partnerships (e.g. Diefenbach 2009; Bovaird 2006). This is inevitably accompanied by a reduction in ownership of assets. For example, a retreat from mass public housing provision means reduced local ownership of public housing. Similarly, the private delivery of public services diminishes the need for the ownership of equipment, service vehicles, and other materiel. Of course, following Buchanan's service club theorem, municipal ownership reflects local political as well as local public preferences in public services and goods provision (Buchanan 1965). Thus, there is great diversity in the composition and size of assets among local governments. It is also important to note how assets can influence the financial situation of individual local governments in terms of ratings of solvency and access to borrowing.

In many countries, thanks to their long existence, local governments often own a range of significant assets. The largest part is often real-estate property (land, buildings, infrastructure), but also important are movable property (vehicles and equipment), financial assets (various kinds of funds and bonds), assets in companies, and increasingly intangible assets (such as software and copyrights). Usually, a substantial part of these assets is directly owned and administered by local governments. Nevertheless, a significant proportion of assets are often transferred to legal entities, various organisations, or companies wholly or jointly owned by local governments. Frequently, in fields that require more private sector experience, or activities suitable for approaches typical of the non-governmental and non-profit sector, local governments are co-owner of entities with other partners.

Municipal assets can be categorised in several ways. We can distinguish short-term, medium-term, and long-term assets. Long-term assets often attract the highest interest. Municipal asset evaluation plays an important role in this process. One group of assets directly serves the administrative and control functions of local government. This is especially the case for town halls, other office buildings, related facilities, and equipment. Another group of assets supports the provision of public goods and services such as schools, social centres, libraries, police station, fire stations, and municipal infrastructure. These types of property are characterised by the absence of frequent ownership changes. Even in small local government units, a minimum amount of municipal assets is needed to undertake and maintain basic functions. Unneeded parts of municipal property can be used by local governments for business-based activities to generate additional income or provide a contribution or incentive to encourage needed development projects. Such entrepreneurial activities fall outside the main roles of local governments and must be managed with a full awareness of related risks. They are circumscribed by many prudential rules. Municipal entrepreneurial activities can be seen in municipally owned companies or in shareholdings in business entities. Some municipal assets such as property are rented out to serve local activities and businesses. Entrepreneurial activities can also involve financial operations, especially brief use of free financial resources such as short-term loans or loans at reduced interest rates. These kinds of activities are common, particularly among large urban governments.

The ownership rights of local governments are conferred in through legislation. These influence the composition of municipal assets and limit freedom in decision-making over their exploitation, including buying, selling, and renting. Management approaches or transactions regarding municipal assets are, in principle, based on decision-making by elected local bodies but are not free from external interests and public disputes. New municipal assets are gained through purchase, new construction, intergovernmental transfers, and, in exceptional cases, donations from citizens, private organisations, or non-governmental entities. Municipal assets are dynamic, and their structure and value is continually changing. Larger cities often have a large stock of diversified assets. Their efficient exploitation requires experienced management, competent operation, and well-working maintenance. One particular requirement is a precise registration and evaluation of local government assets. Local governments with more assets need specialised departments, well-paid professionals, and access to specific software including, for example, GIS applications.

For local governments with a developed asset base, besides the acquisition or disposal of property, the key issue is efficient administration of those assets. This is not an easy task. As Jolicoeur and Barrett (2004) have noted, public interest and the interest of elected officials often focus on new construction and major renovations and less on the sustainability of current infrastructure. The shortage of funding for acceptable maintenance leads to physical deterioration. Due to their prevalence and visibility, real-estate properties often receive the most attention. Kaganova and Nayyar-Stone (2000) summarised development in this sector as entailing a shift from public ownership of real estate, increased recognition of public real estate as a productive asset, and the adoption of private sector practices. However, local governments should be interested in the efficient financial and operational management of their whole portfolio of assets, applying strategic asset management practices such as those described in Fernholz and Fernholz (2006). Local governments also face market turbulence and must recognise differences between the accounting value and market value of their assets, particularly real estate. Well-administered municipal assets can substantially influence the competitive position of cities and can foster a positive perception among residents and the business community.

Municipal Assets and Their Legal Framework During the Transition Period

Background to Subnational Governance in the Slovak Republic

For local governments in transitional countries, asset-related challenges are different. They need property to function but also need to develop experience in its management and growth. In many transition states, local governments played an important role in providing social and human services. Local government assets frequently played an important role in the economy, and their reform and restructuring was an integral part of economic transition. For example, as Sun (2002) notes, township and village enterprises in China played an important role in the transition economy and the adaptation to a market economy in particular territories. In transition countries, the use of municipal assets provoked debates on service provision, decentralisation, and the democratisation of societies. Assets were an important part of decentralisation in most transition countries. Transfers of property accompanied reforms in local public service provision and the transfer of branches of the national economy such as utilities to local government and the private sector (Péteri and Horváth 2001). Many services were quickly transformed into more contemporary forms of service provision 'outside' the public sector. Problems typically associated with the efficient administration of assets in transition countries were outlined by Kaganova et al. (2001) referring to the case of Kyrgyzstan. The essential feature is that the management, transfer, and sale of municipal assets were among the less transparent issues in local government agendas during the transition, as compared to basic local fiscal policy issues like the formation and adoption of local budgets. This is understandable. Local governments only slowly built asset registers. They managed assets without any well-elaborated strategy, and they were not aware of the real value of assets, particularly since those values were often rapidly changing as land markets and the private sector developed (see, e.g. Kaganova 2008). Integrating assets into the standard fiscal practices of local government and making this process easily understandable and transparent to all citizens has been a long-term task.

The system of administrative divisions and subdivisions in Slovakia has been reshaped during the last two decades. Until the early 2000s, Slovakia was one of the most centralised countries in Central and Southeastern Europe, but over the next decade, the character of country administration changed. Initially, after independence in 1993, subnational public administrative divisions had few powers and resources and less freedom. They relied mainly on transfers from national government. Other sources of revenue included local taxes regulated by national government, fees, and other charges. The powers and resources of subnational governments slowly increased until the onset of the financial crisis in 2007–2009.

Until 1990, the country had three levels of administration below the Republic government: the kraj or region, the okres or district, and local governments. However, there was no local government in the Western European sense of the term.

All the different levels were part of the central state administration. They were mainly administrative and planning organs, used for statistical and accounting purposes. Local governments were also mainly administrative organs which managed local housing, infrastructure, the provision of shop space, and basic local services (see, e.g. Ryder 1990, 1992). Due to the strong identification of regional government with previous regime, the ruling communist party, and the system of central planning, the kraj level of government was abolished in 1990.

After 1990, the construction of a new system of public administration started. Its main and constant feature is its dual character, with separate lines of state administration and local government. Local government was introduced immediately in 1990. Regional government was introduced in 2001. Both levels of government are managed independently from national government by directly elected mayors/chairpersons and local/regional councils. Initially, they had extremely limited competences and few financial powers and relied mainly on transfers from central government. Their scope for action increased as a result of reforms to decentralise government between 2002 and 2005. This was due partly to the need to met European Union accession criteria but also a result of the ideological orientation of the coalition government in power at the time, led by Mikulas Dzurinda. Fiscal decentralisation came into force in 2005. Since then, local and regional governments have been relatively financially independent thanks to more stable and guaranteed sources of income. According to legislation, 23.5 % of income taxes collected by the national government were allocated to the regions and 70.3 % to local governments. The rest went to the central state budget. Since then, the national government has relied mainly on VAT and excise taxes for revenue. Municipal authorities were also given the right to set real-estate taxes and to set other local taxes and fees.

Despite the growth in autonomy, the situation of the state administration has been unstable. The district (okres) state administration continued to exist after 1990, but there were frequent reorganisations and transfers of powers. Regional state administration (the kraj) was reintroduced in 1996, accompanied by a redrawing of territorial boundaries at district and regional (okres and kraj) levels. During the course of decentralisation reforms, the okres were increasingly seen as being a redundant layer of state administration which created unnecessary administrative costs. In 2004 they were abolished as administrative units, although they remain as statistical regions. Similarly, the regional offices of the state administration were abolished in 2007. However, numerous field and circuit offices of the state administration are still organised by okres and kraj, subordinated to various ministries. In addition, in 2013, it has been rumoured that the okres maybe brought back as an administrative district, a move favoured by the government then in power.

Thus, the role of state administration has diminished due to decentralisation. Today, the country has two levels of government below the national level: miestnu samosprávu (local government) and samospravne kraje (self-governing regions). The creation of new layers of government, the creation of new local and regional administrations, and changing and unclear legislation complicated urban and regional management. The extensive and drawn-out process of restructuring inhibited the creation of planning documents and local policies, as well as the creation and implementation of development and management programmes. Uncertainty about the future form of local government and delay in its reorganisation led to a loss of expertise among government employees. It created a vacuum in governance which, in many cases, allowed relatively sophisticated developers and investors to step in and exploit gaps in regulations to promote new projects.

Municipal Assets of Local Governments

The role, size, and exploitation of municipal assets strongly depend on the legal framework and its implementation. Although the legal framework's basic elements were formulated in the early 1990s, they were subject to many amendments during the next two decades. Such changes reflected varying approaches, experiences, and reforms, as well as the growing range of aspirations of local governments in managing their assets. As a result, asset ownership, its autonomous administration, and efforts for its expansion and diversification are features in the evolution of local government in Slovakia. It supports their basic functioning and service provision and has a symbolic value for citizens, such as the case of town halls. Municipal assets are one of the main local government financial resources. Legislation has supported the enlargement, wider use, diversification, and better management practices of municipal assets. These changes reflect the strengthened position of local government in society. In a narrow sense, this position can be seen in the growing role of municipal governments within the public sector. Decision-making over municipal assets is now a typical part of local policy. The current framework allows a more 'businesslike' and sophisticated form of asset use, for example, in the field of local economic development or by allowing participation in financial markets. Nevertheless, long-term experience has led to the introduction of more precise rules and limits, e.g. in favour of more efficient and transparent asset administration.

The position of municipal assets and local government is guaranteed by the Constitution of the Slovak Republic. In legal terms, asset ownership is equal to other forms of ownership. However, local governments must follow current legislation in the administration of assets. The first time that municipal assets were referred to in national legislation was in the initial legislation concerning local government which appeared in 1990 (Act No. 369/1990 Coll.). It defined municipal assets as all properties, as well as property rights, under municipal ownership. These assets were to serve local governments in fulfilling their basic tasks. According to this legislation, municipal assets could be used for public purposes, to serve the needs of local government or to serve its entrepreneurial activities. This introductory and brief mention of municipal assets was not sufficient. It did not substantially influence the practices of asset administration or the size and value of municipal property.

A more detailed framework of asset ownership and administration was defined by the Municipal Asset Act adopted in 1991 (Act No. 138/1991 Coll., later amended on various occasions). The almost exclusive state ownership practised during the previous regime needed specific legislation to enable the distribution of public assets to newly formed lower levels of government and other public-sector bodies. The Act led to the transfer of a large amount of state assets to local governments. It also allowed local governments to transfer property to local government organisations or companies they created. The Act emphasised that local government bodies had to manage municipal assets to support local development, to the benefit of citizens, respecting the limits of local environmental protection. They were obliged to look after their property (prevent damage, destruction, loss) and increase its value. Creating an asset register was among the mandatory tasks. Municipal asset legislation defined which issues had to be regulated by local by-laws and which decisions were only valid if approved by the local (city) council, for example, the sale of real estate, the sale of other assets above a threshold price, the leasing or sale of property rights, and auctions. The legislation also stated that all contracts concerning municipal assets must be in written form. Basic information on municipal assets must be included in the basic accounting documents of local governments. Local governments must follow rules of asset depreciation to list the real value of assets. Nevertheless, some of these tasks were only gradually implemented by local governments.

An important aspect of local government is the level of autonomy that local government bodies have in managing their assets. Thanks to the dual model of public administration adopted in Slovakia with separate lines of government and state administration, intervention by higher levels of government is possible only through legislation. No other direct intrusion by the central state government or other state institutions is permitted. Although ownership rights give local governments a relatively free hand in managing assets, each local government has its own main auditor, and they are subject to internal and external audit supervision. All parts of the Final Account must be approved by the auditor. The Supreme Audit Office of the Slovak Republic also has the right to supervise municipal asset operations, thanks to a constitutional amendment adopted at the beginning of previous decade. Local governments must also follow additional rules defined in legislation concerning their assets, particularly the Budgetary Rules Act (concerning financial resources, Act No. 523/2004 Coll.) and the Business Code (concerning participation in business companies, Act No. 513/1991 Coll. as amended), as well as legislation concerning bonds, securities, investment services, and funds.

In 2009, the latest large-scale changes related to asset management were adopted. They focused on strengthening transparency in municipal asset management and transfers. This was a reaction to widely perceived problems of conflicts of interest, clientelism, and nepotism. More precise rules were introduced concerning transfers of municipal property. Transfers must be made on the basis of public tendering, auctions, or a generalised market value determined by set rules in the case of direct sales to particular persons or legal entities, which have often been disputed. Local governments can no longer make a direct sale if the value exceeds €40 thousand. Local governments also cannot directly sell property to their mayor, councillors, the statutory representatives of local public companies (established by local government), chairs of local offices, local government employees, the internal auditor, or their relatives.

Decision-making is a key element in municipal asset management. As local elected bodies, councils and mayors have a central role. Nevertheless, a significant role is also held by professionals employed in city departments responsible for the administration of all kinds of assets. Decisions in this area require transparency. They require unambiguous rules regulating the use of assets and strategies of acquiring and selling them. There should be a regular assessment of property and property use, focusing on efficiency. In transition countries, the implantation and enforcement of transparent management approaches have been even more important: the rules concerning municipal asset administration include special by-laws adopted by city (local) councils.

The Slovak legal framework allows several forms of municipal assets, discussed in Žárska et al. (2008) and Staroňová and Sičáková-Beblavá (2006). The most complicated issues arise in the use of real-estate property. Most properties are directly used by local governments. In such cases, local government as a legal entity uses property for administrative purposes or the provision of local public services. There are also cases where local government places property under the administration of a subject organisation established by local government without having the status of a legal entity. Two basic forms are used – the budgetary organisation (with its budget and resources determined by local government) and the contributory organisation (functioning thanks to a financial contribution provided from the local government budget). Other forms often involve a formal change of ownership. The most usual is the nonfinancial transfer of municipal assets to a business (usually a limited or a joint stock company). Local governments often establish a wholly owned business in which they are the only owner/shareholder, giving them full control. They can also use their own assets to establish a non-profit organisation or foundation or can share their assets in associations of local governments. Among other cases of municipal asset use, we can find property used as collateral for loans or to issue municipal bonds. However, local governments may not become a guarantor for loans made to a private entrepreneur or legal entity not created by the local governments themselves. Financial assets appear in many forms, including reserve funds, investments in bonds, minority stakes in corporations, etc.

Property rental is another frequent form of municipal property exploitation. Sometimes, the renters of municipal property are established by local governments, but they can be other legal entities, including citizens. Praxis in Slovakia reflects two possible principles in setting rent – market or nonmarket. The local market rent level applies in the case of market-rate renting. Various circumstances are taken into account in the case of nonmarket-rate renting. Nonmarket conditions can be obtained by businesses which provide services lacking in the locality or which are not very accessible. They can also be provided to increase competition in the local market, to compensate firms for their investment in municipal property (maintenance), or to foster employment or entrepreneurship in the locality – that is, to support local economic growth. Another example is the local provision of public-sector rental housing, allowing nonmarket rent for social reasons. Nonmarket conditions are often short term, negotiated for a precisely defined period of time. Related agreements provide opportunities to abrogate such conditions if contractual obligations are not fulfilled.

The most serious decision concerning municipal assets is their sale. Sale means that local governments lose ownership rights to property they sell. It is generally expected that local governments should sell only property they do not need, that is, 'surplus' property, with no special functional or development potential. However, in many cases during the transition, local governments were forced to sell assets despite an expected increase in value and despite their importance to city operations. This is because they more urgently needed resources to rehabilitate or modernise selected property or they faced financial scarcity and needed resources to cover the costs. In such urgent cases, local governments sold very useful assets, often because there was little of no demand for other assets. The highest risk of selling is the risk of a poor price due to the need to sell, as well as the loss of property which could have significant future impacts, such as the sale of land in a good location. In the case of property sales, land and buildings can be sold at less than the market price, but in such circumstances, the future use of property has to be guaranteed. It should be an elementary rule to use the revenue from asset sales to acquire new assets. Local governments should have a well-elaborated policy in 'both directions' – regarding both sales and acquisition, especially in sensitive cases such as the sale of land and buildings.

Information on municipal assets in Slovakia comes mainly from government accounting documents. Slovak accountancy practices concerning assets have improved and currently approximate international standards. Despite numerous changes, especially since the early 2000s, it is possible to use balance sheets published in the annual Final Account of the Slovak Republic to evaluate developments in the field of municipal assets. The situation was different in the 1990s. Reliable data concerning municipal assets were generally unavailable during the early transition period. The main causes were the incomplete registration of property and the large amount of property under litigation. Accounting practices were unstable and non-standard. Changes in the allocation of assets among different categories were frequent. A specific problem was the unwillingness to provide data. In the field of municipal assets, this was characteristic of the transition from a centrally planned economy. Papcúnova and Balážová (2006) argue that this was why such data are missing from Statistical Yearbooks and other documents. It is also why there has not been more extensive publication in this field in Slovakia. Now (2013–2014) the situation has improved. We can view the structure of municipal assets, divided into fixed and current assets, based on the balance sheet, consisting of intangible assets, tangible fixed assets, and financial assets which compose fixed (noncurrent) assets. Current assets consist mainly of financial accounts, inventory, and receivables. Public finance databases, compiled under the supervision of the Ministry of Finance, publish information on assets including depreciation. Separate information is provided concerning organisations to which local governments contribute. Data on individual local governments is still missing from centralised databases because access to such data is legally protected. However, access to such data has improved thanks to the penetration of the Internet, improvement in the content of local government web pages, and the growth of e-government at the local level.

Formation of the Municipal Asset Base During the Transition Period in Slovakia

The formation of the municipal asset base has been one of the most important aspects of the introduction of local government and its full-scale functioning during the transition period. The municipal asset base has developed in two basic ways. First, local governments obtained property located on their territory through transfers defined by legislation as part of the process of wide-scale societal transformation, the introduction of local self-government, and public administration reforms. Second, in parallel, local governments built up their asset base through their own initiative through purchases or other forms of acquisition. These two methods have been applied in varying degrees throughout the 20 years since the start of the transition (described, e.g. in Papcúnová and Balážová 2006). Restructuring and reform-related transfers prevailed in some periods, while in others, local government use of its own resources played the leading role, characterised by the mobilisation of local resources, taking credits for property-related projects, and applying for EU funds. Despite the continued development of the municipal asset base, future waves of property transfers based on legislation and intergovernmental transfers are unlikely. This suggests a move to a post-transformation period – the main transitional processes are almost complete, and more standard ways of asset formations will dominate in the future.

The development of municipal asset bases has gone through several distinct phases: two transfer periods and two consolidation periods. The first years (1991–1993) can be characterised as the *restitution-de-étatisation period*. These two important transformation processes substantially influenced this initial period. The main tool was the Municipal Assets Act and its amendments, adopted in 1992 and 1993. The restitution process meant local governments took over the 'historical' property they owned until December 31, 1949. This included such assets as town halls, other local public buildings, land, agricultural land, and forests. Although the transfer process often took some time, most of this historical property was transferred to local governments during these years. This 'return' of property substantially strengthened the position of local governments, especially in the case of cities which had traditionally owned property. Some assets had been city-owned for centuries. However, it should be noted that local governments faced restitution claims from citizens and other legal entities who asked for the return of property expropriated by the communist regime which had initially been included in local government assets. This affected the extent of municipal assets and led to uncertainty in the exploitation and administration of some of the assets. In turn, this resulted in reduced maintenance and neglect. In many cases, lack of clarity in property rights was a long-term problem, thanks to slow-working courts and the complicated verification of competing ownership rights.

The second main source of property obtained by local governments in the initial transition period reflected their participation in the *de-étatisation* of the economy. The local level was crucial in reducing the role and scope of state ownership in society. Local governments participated in a shift towards a more balanced ownership structure alongside the privatisation process. Local governments received a significant part of the assets (the most valuable was real-estate property) located on their territory, previously managed by their communist-era institutional predecessors (known as 'local/city national committees'). They also obtained property located on their territory that had been part of various state-owned enterprises and other state-owned organisations. Mass housing projects were also transferred to local governments, as well as other state investments on the territory of local government as defined in legislation. Thus local governments became the interim owners of housing stock previously owned by the state. The process of privatising housing stock started at the end of this period. Citizens could buy their apartments, mostly in large-scale housing estates, at favourable prices. Local government served as a 'privatisation agency', obtaining income from selling housing. This process was long and administratively demanding. One result is that there is only a limited number of public housing units in Slovak villages and cities, serving mostly as social housing.

In many cases, the state transferred other properties located on their territory to local governments, based on agreements with the concerned governments. Local governments also gained buildings and facilities built during the communist period by the joint action of local citizens on the principle of community self-help. They also received small firms and service providers not included in so-called small privatisation and small firms which were unsold, even at auction, and not privatised. In addition, during the communist regime, large state enterprises often owned various elements of social infrastructure, such as housing and nurseries. The transformation process led to local governments obtaining such property. They also became owners of many local sport facilities. As a result, within a few years, local governments became the leading owner of public-sector assets on their territory. Due to long-term disinvestment, most of the transferred property was in a poor state, needed reconstruction, and was often no longer suitable for its original use.

The second period (1994–2001) can be considered a *consolidation period*, as far as the asset base is concerned. On the one hand, local governments attempted to administer newly obtained property, while on the other, they tried to formulate approaches to the expansion of their asset base. They developed experience in property management and exploitation. Many of them systematically attempted to harmonise their powers and attempted to develop priorities towards their property. They had to decide on how to the use the existing asset base, particularly real estate, which was often in poor condition. The sale of property was frequent. For financial reasons, local governments were often unable to take care of or develop property. For example, they lacked the resources to rehabilitate all their real-estate property. They were often forced to sell or rent their property, because their incomes, based on transfers from the state budget, were insufficient. Income from property sales was needed to subsidise the costs of providing local services (Tichý and Žárska 2005). Some property was sold following the simple rational calculation that it was not urgently needed or did not fall within the standard remit of local governments. Many local councils decided to concentrate on a limited number of services and priorities whose adequate provision could be secured only with the help of property sales. Resources obtained from selling one part of civic assets were often used for the reconstruction of another, allowing local governments to concentrate on what they thought was the more important part of their assets. Such an approach caused large-scale changes within the structure and use of assets. The large number of property sales was also due to limited borrowing opportunities caused by undeveloped domestic financial markets, accompanied by high borrowing costs. The sale of property peaked in the late 1990s, reaching a maximum in 1997–1998, while the financial situation eased. However, the sale of property reduced the opportunities for raising funds from outside the state budget by renting or leasing, and loss of assets deterred access to credits. Instead, reliance on external resources such as EU pre-accession funds increased.

From 2002 to 2004, Slovakia underwent a wide-scale reform of public administration. This led to the next phase of local government asset formation. The national government decentralised important powers to the local level in stages, in fields such as education, social care, transport, health, and environmental management (Act No. 416/2001). The Municipal Assets Act was amended twice, in 2001 and 2003. This third period, *a decentralisation period* of asset formation (2002–2004), caused another large increase in property owned by local governments. It mainly concerned assets directly related to the transferred powers. These typically included primary schools and other school facilities, social care centres, and health centres. Local governments became the sole owner of those assets located on their territory. Some property of this nature was also transferred to the newly introduced regional (Kraje) governments, such as secondary schools, specialised social care centres, and some cultural facilities. Although some of the transferred property was in a poor state, local governments quickly incorporated these assets into their administration.

The subsequent era, from 2005 onwards, can be described as the *modernisation period*. The local government assets increased but needed extensive upgrading and restructuring for appropriate and efficient use. The improvement of the physical and functional state of property, and the refurbishment and modernisation of real-estate property obtained during the preceding decentralisation period, has been widespread. Local governments concentrated on basic renovations to roofs, windows, and heating systems and the reduction of energy use in buildings to lower operating and maintenance costs. This was possible thanks to increased yields from taxes, better access to EU funds, and more extensive borrowing options under increasingly favourable conditions. Greater reliance on their own resources also played an important role, thanks to setting real-estate taxes. During this time, some local governments gained assets related to reduced activities at border crossings (redundant facilities when Slovakia acceded to the Schengen Zone, leaving customs borders only between Slovakia and Ukraine) and through reductions in the size and number of military bases, including facilities and training areas previously owned by the Army of the Slovak Republic. The restructuring and rehabilitation of these newly acquired assets have been slowed by the impact of the global financial crisis from 2008 onwards. In addition, as the growth of assets has slowed in many cities and villages, pressure to sell property and spend financial reserves has grown.

The transfer of powers and related assets is subject to conditions imposed by the national government. Among the most important has been an obligation to maintain the existing use of transferred property (e.g. schools). A change of use requires a complicated administrative procedure, supervised by the relevant state administrative bodies. Some changes have occurred, for example, in many school buildings which are no longer needed, thanks to a decrease in population growth leading to a drop in the number of school-age children. The legislation defined a fixed time period before the use of newly transferred property could be changed, although in some cases this was later shortened. After the termination of such limits, local councils could freely decide how to use such assets, depending on their preference and local needs. Local governments took over such property not just as passive actors as in connection with the transfer of powers but also because of positive expectations. They viewed the transfer as strengthening their asset base. One minor complication was the fact that in many cases local governments obtained property with significant liabilities. Nevertheless, the functional use of property which was transferred was often changed within a few years.

The latest process of property transfers differed from that at the start of the 1990s. The first transfer was part of a wider post-socialist transformation. At the beginning of this decade, the scope of the transfer was narrower, closely linked to transferred powers, and needed for their execution. In both cases, property was in a poor state. The 'state' was 'pleased' to transfer property to local governments because it also transferred the need for investments in reconstruction and transferred operating costs to the local level. The two consolidation periods, characterised by a more active role for government, was different as well. In the first one, during the 1990s, local governments faced a large-scale scarcity of resources. They remained part of a relatively unreformed economy, with limited access to external resources. This influenced the approach of local governments towards their property, causing frequent sales and reduced investments. Local governments were in a much better situation during the period since 2005. Modernisation expenditures were possible, thanks to increasing income from property, and access to EU funds after 2004, and to increased tax income related to booming economic growth in Slovak economy until 2008.

The Value and Structure of Municipal Assets in Slovakia

Data regarding municipal assets show continual growth in value and only slight changes in structure. In recent years, the most important impact on the structure and size of assets has been the period of economic growth through 2008, replaced by the financial and economic crisis.

Aggregate data for all Slovak local governments clearly documents a continuous growth in asset value, but with visible periods of acceleration or deceleration of growth. Figure 14.1 shows slow asset value growth from 2000, showing the increase in growth since 2002. The total value of municipal property was SKK 394 billion (net of liabilities, €13.1 billion) in 2009, or €16.7 billion (when liabilities are excluded). The highest year-to-year growth rate in the last decade was in 2002,

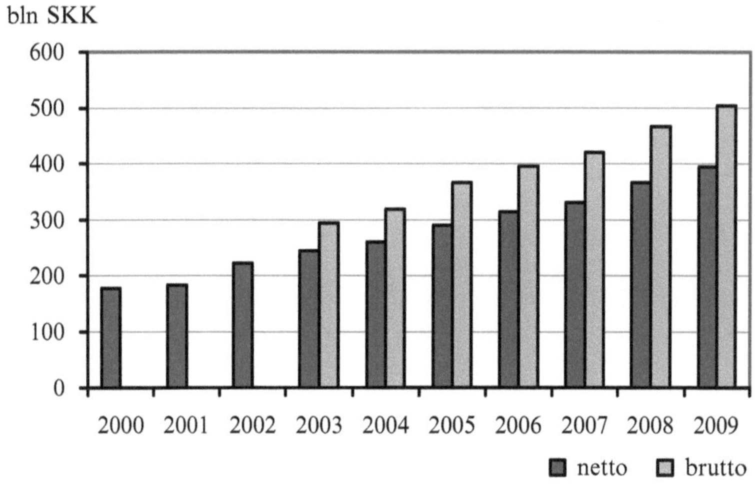

Fig. 14.1 The development of the total value of municipal property in Slovakia. Note: Brutto values not available for years 2000–2002 (Source: Ministry of Finance of the Slovak Republic (2002–2009)

Table 14.1 Basic structure of municipal asset value in 2003–2009 (in per cent)

	2003	2005	2007	2009[a]
Intangible assets	0.11	0.10	0.14	0.2
Tangible assets	81.15	75.66	76.38	77.71
Financial assets	9.20	14.80	15.03	13.75
Inventory	0.29	0.12	0.09	0.07
Receivables	5.51	4.70	3.70	3.28
Financial accounts	3.63	4.45	4.57	4.84
Other	0.13	0.18	0.10	0.15
Total assets	100.00	100.00	100.00	100.00
Liabilities	2.57	10.38	11.35	13.16
Net asset value	97.43	89.62	88.65	86.84

Source: Adopted from Ministry of Finance of the Slovak Republic (2002–2009)
[a]Without accounting relations among public administration entities, introduced in 2008

reflecting the start of the decentralisation of powers, accompanied by the transfer of related assets. A higher rate of asset growth can also be observed in 2005. This is the consequence of two effects. This was the first year of fiscal decentralisation and reflects different flows of resources and reflects the expansion of investment and funding in the year before local elections in 2006. The next upswing in the rate of asset growth 2008 and 2009 reflects the positive effect of the economic cycle and increased access to EU funds, before the effects of the financial crisis effects prevailed. Nevertheless, a precise valuation of assets value must take into account the liabilities of local governments resulting in a lower net value (see Table 14.1). From this point of view, development appears less positive. Total local government liabilities reach the equivalent of 11 % of total asset value in 2007, increasing to

13.6 % in 2009 (or 16.4 % if relations with other public-sector entities are included). By contrast, local government liabilities were less than 3 % of assets in 2003. The level of depreciation of tangible assets also increased, reaching 27.2 % in 2009, compared to 22.4 % in 2004.

The structure of assets has not changed substantially in recent years (see Table 14.1). Throughout the decade just past, noncurrent assets have composed about 90 % of total local government assets in Slovakia. Over the longer term, there has been a reduction of tangible assets in favour of financial assets. This trend stopped due to the financial and economic crisis in 2009, when the value of financial assets decreased. Local governments were forced to use financial assets to meet financial needs unmet by regular income sources, particularly due to a drop in tax revenues.

Tangible assets account for the dominant share of assets, about three quarters. Papcúnova and Balážová (2006), in a detailed analysis of assets, mentioned two groups of tangible assets that are crucial. These are buildings and land, each of which composes about 40 % of total tangible assets. Equipment, vehicles, and similar assets are less important, although assets under construction are often significant. Financial assets, equalling about €1.7 billion in 2009, are composed mostly of securities or consist of participation in companies but have been growing rapidly. The value of intangible assets is small (about €20 million in 2009 after depreciation). However, this class of assets, composed mostly of software and other licences, is growing. The greatest growth is in the field of receivables, consisting mainly of tax receivables, penalties, and transfers from state institutions. The management of receivables has improved over recent years as local governments have become more efficient in this field. Their share in income and assets is decreasing, but their total volume, about €400 million, remains high. Local governments also have a large amount of reserves in bank accounts, mostly in various specialised funds, but despite growing in absolute terms, the relative share of these resources decreased in 2009 compared to 2008.

Despite this more or less optimistic overview, any discussion of municipal assets in Slovakia must recognise that the situation of individual local governments is extremely variable (Buček et al. 2010). Some are quite 'rich' (or their assets are growing quickly), but many have a minimum asset base, with no financial investments, and no real property outside the 'seat' of local offices. This is strongly influenced by a very fragmented administrative system of settlements almost 2,900, which is dominated by rural settlements with very small populations. In such units, the total value of assets is usually low.

Municipal Asset Formation in Large-City Local Government: The Case of Bratislava

The Institutional and Regulatory Framework

As a large capital city, Bratislava provides a good opportunity to observe the development of the portfolio of municipal assets in Slovakia. It is the capital and the largest city with 430 thousand inhabitants and is by far Slovakia's leading economic centre. It has a two-tier model of local government, found only in one other city in Slovakia, the second largest city, Košice. The Bratislava citywide and city district governments together generate approximately about 15 % of the total incomes/ expenditures of local governments in Slovakia. The citywide government owns 11.7 % of local government assets in Slovakia. Alongside these assets, it owns about 30 % of total assets in so-called contributory organisations, which are organisations, often non-governmental, which receive some of their funding from municipal authorities. Assets owned by city districts are outside the scope of this review. However, city districts, particularly larger ones, have a large stock of assets, including those transferred earlier from the citywide level of government.

The local regulatory framework concerning municipal assets must conform to national legislation. Besides the legislation already discussed, the *Act on Bratislava* (Act. No. 377/1990 Coll.) and amendments, valid since 2009, provide a more detailed framework for the sensitive issues of powers and resources, as well as assets of both levels of Bratislava's government. The most important aspect of this Act contains precise rules regarding the distribution of income obtained from real estate sold by citywide as well as city district governments. Ninety percent of income from the sale of real estate by citywide government goes to the citywide government, and 10 % goes to the district government in which the property is located. In the case of real estate owned by city districts, the reverse holds true: 90 % goes to the city district and 10 % to the citywide government. There are also specific rules for real estate owned by the citywide level but administered by city districts. In the case of city districts with more than 40 thousand inhabitants, the income from selling such real property is distributed 50:50 between citywide government and district government. Smaller city districts have more favourable conditions. They obtain 60 % of income, while the citywide government gets only 40 %. Specific powers are directly linked to particular types of assets. City district governments are responsible for primary education – school and nurseries, the maintenance of local third- and fourth-class roads, and local public spaces. The legislation also defines what has to be specified by City Statute, including issues that must be declared in by-laws concerning principles of asset administration.

Asset administration rules are set out in detail by City Statute and other local regulations and by-laws dealing with property (part six of the City Statute). The relation between the citywide government and city district government is very precisely spelled out. For example, any proposed changes from the original function of assets transferred from the city government, or the use of such assets as collateral, must be approved by the citywide government. Further rules concerning municipal assets are stated in the City Council legislative document, *Principles of Assets Administration* (Slov. Zásady hospodárenia s majetkom hlavného mesta Slovenskej republiky Bratislavy 1993), and its amendments. It defines operational rules of asset administration and registration. The section focusing on transfers of property to city district governments and city organisations is particularly important. Rules are specified concerning obsolete or surplus property. It also includes rules regarding the commercial and other uses of municipal assets, including renting. All rental agreements exceeding 10 years must be approved by the City Council. The paragraphs dealing with liabilities are also extremely important. Although their main focus is on real-estate property, there are also rules concerning financial resources, cash, and securities. There are more detailed rules concerning the operation of the city executive rules of the city and its organisations. Additionally the document contains rules regarding the withdrawal of property from city district control and from the control of city-owned organisations. The document very precisely specifies which operations and uses must be approved by the City Council, as well as specifying the powers of the citywide lord mayor. Last, but not least, there is a section defining sanctions for breaking rules set by these local regulatory documents. The management of city assets is regularly supervised by the city's internal main auditor, as well as an external auditor. The *Supreme Audit Office of the Slovak Republic* performed audits focusing on organisations, assets, and liabilities of the Bratislava government during spring 2010, recommending minor adjustments and corrections. These rules should also foster increased transparency in all asset-related operations.

The ownership of such an extensive range of assets requires adequate administration. The City Council has the main decision-making rights over city assets. It must approve any important asset-related decision, including all the creation of any regulations concerning the management and use of each asset. The special Permanent Commission of the City Council has a special role, overseeing all city assets and their utilisation. The everyday administration of municipal assets split among several departments within the City Magistracy is part of the local government executive. Due to the large stock of real-estate property, there are two main specialised departments. The first deals with property administration including divisions focused on legal affairs, information and documentation, registration of property, and the collection of liabilities. The second department focuses on the administration of rented property, related contracts, and litigation. The department responsible for the technical administration and maintenance of buildings, including housing, also has a significant role. These three real-estate management-orientated departments of the City Magistracy employ over 60 officials (2010). Management of the road network is under a separate administration, due to its specialist tasks and needs. A special department focused on financial resources oversees financial assets and related activities such as strategies, analysis, transactions, and debt management, which fall outside standard budgetary issues.

Basic Structure and Main Categories of Assets

Bratislava's citywide government enjoys an extensive asset base. Land and buildings are by far the biggest part. As early as the end of 1993, the City Magistracy reported that the nonmarket value of real property was SKK 90 bln (SKK 74.5 bln in land, the rest in buildings and buildings under construction). The citywide level of government directly administered 60 % of this property by value (Buček 1995). This estimated property value reflected the accounting practices of the time. More realistic information is available for 1997 when reliable accounting rules had been established for property valuation. According to Kresánek et al. (1998), the value of land was app. SKK 17 bln and the value of buildings and buildings under construction was SKK 15.5 bln. They also stated that between 1991 and 1997, citywide government sold only 1 % of its land. At that time, the city owned 47.6 km^2 of land, 13 % of the city's total area, and 6,201 buildings. However, the process of identifying and registering all of the city's real estate was not finished, even as late as 1997.

In 2009–2010, the value of all the municipal assets of Bratislava's citywide government was €1.89 billion, about €20 million less than in 2008. The structure of assets is strongly biased towards noncurrent assets which compose slightly below 80 % of total assets (79.6 % in 2009). Almost 60 % of total assets are tangible assets, mainly land and buildings. The major part of this category serves public service provision. Nevertheless, the citywide government is an important player in property development, thanks to the large stock of real estate suitable for development purposes. Financial assets, comprising 20 % of assets, have stagnated since 2008. Non-tangible assets are marginal, although their share is increasing (0.25 % in 2009). Current assets are dominated by assets jointly owned with other governments and governmental agencies. They accounted for 11.1 % of total assets in 2009. The city successfully reduced receivables which constituted just 6.7 % of assets, or €127 million in 2009. The value of resources in financial accounts fell from €37.6 million in 2008 to €19.2 million in 2009. This is linked to fallout from the financial crisis. In contrast, liabilities did not change substantially in recent years. In 2009, they amounted to €344 million. Figure 14.2 provides an overview of assets owned by Bratislava citywide government and changes over time.

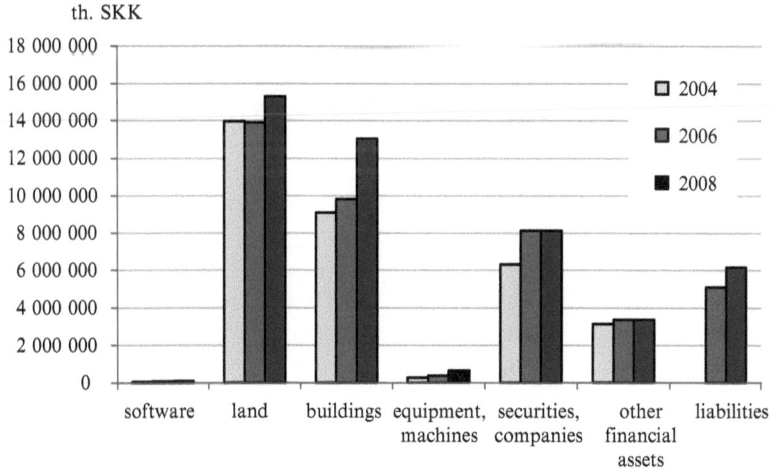

Fig. 14.2 Selected items of Bratislava citywide government assets. Note: date excludes liabilities; liabilities in 2004 not available (Source: Bratislava City Magistracy, Annual Reports (2004–2008) and Final Accounts 2006–2009 (Adopted))

Unsurprisingly, thanks to the large asset stock, the city has been very active in selling, buying, and exchanging property. Selling property is an important source of income for the city, reaching nearly SKK 1.7 billion in 2008 – almost €60 million. Sales of real estate have generally far exceeded purchases. However, one should note that it is difficult to estimate the market value of a large part of the city's real estate, and a significant amount of property consists of service infrastructure, public open space, park land, and similar uses. In recent years, the scope of other assets, for example, in businesses or financial investments, has stagnated. As well, the asset position of the city is reduced by a large volume of liabilities.

City assets are primarily used for city administration and service provision. During the transition processes, assets were an important source of income for the city. Today, 2012–2013, the city obtains an important part of its resources from dividends, rents from real estate, credit yields, borrowing, and bank accounts. However, the sale of assets, predominantly real estate, remains an important source. For example, yields from city assets – mostly sales of property, housing privatisation, and property rental – generated one third of the total income of the citywide government budget as early as 1992 and 1993 (Buček 1995). The erratic nature of this component of the city income is shown by a dramatic 14 % decrease in income from property sales in 2009 compared to the previous year, when this income from this source reached an all-time high. When all asset-based revenues are combined, this revenue source played a substantial role in the city's income, excluding financial operations. Between 2006 and 2008, it provided between 23.4 % and 29.6 % of the total. Through property sales and rental, the citywide government was an active participant in the construction and development boom going on at the time. As a result, the onset of the economic crisis in 2008 had serious consequences for the city revenues. Real-estate income fell from almost a 30 % share of the total city budget income, excluding financial operations in 2008, to just 11 % in 2009. This is by far the worst performance of this type of asset within the last 5 years (in 2005 it was 14 %). As Fig. 14.3 documents, the growth of real-estate revenue in previous years depended on the sale of property, and the drop in sales is the cause of the steep decrease in asset-related income in 2009. As a result, overall spending, from all city resources and reserve funds, declined. Revenues from rents remained more stable, and in recent years they have exceeded €10 million a year. This means that the city has become reliant on sources other than property sales which were vulnerable to the economic crisis.

The financial base of the city also consists of reserves which are mainly loans from commercial banks, as well as transfers from previous budget surpluses. They are an important part of assets. They give the local government flexibility in planning and executing its policies, as well as flexibility if facing unexpected situations or opportunities. Unlike smaller cities, which tend to borrow on a small scale, according to the needs of particular projects or investments, in Bratislava, city representatives prefer to assemble a larger stock of cash reserves in advance. This strategy has been observed since the mid-1990s. The city government has been fiscally conservative, and these reserves have been carefully managed. Although quite high, they remained untapped for a long period. The city government even uses existing

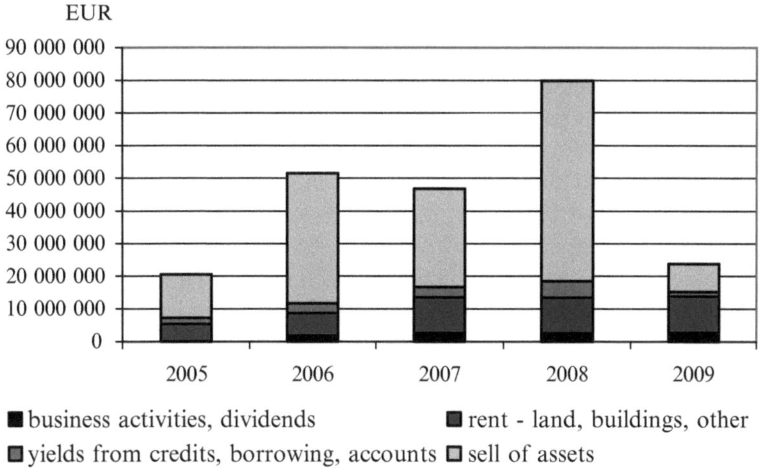

Fig. 14.3 Asset-related revenues in Bratislava citywide government in 2005–2009 (Source: Bratislava City Magistracy, Annual Reports (2004–2008) and Final Accounts 2006–2009 (Adopted))

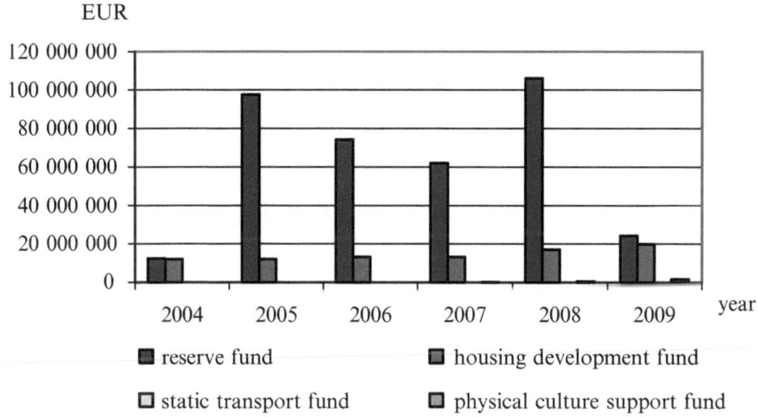

Fig. 14.4 Non-budgetary funds of citywide government (as of December 31) (Source: Bratislava City Magistracy, Annual Reports (2004–2008) and Final Accounts 2006–2009 (Adopted))

reserves to repay borrowing, if suitable projects and investments are not decided on. Frequently, the city has had plans to finance projects, but the preparation was too slow, or they failed to gain political support among the City Council and the general public, and never went ahead. As a result, the city has avoided the extreme debt burden which has appeared in recent years in some other larger Slovak cities (Košice, Banská Bystrica). Nevertheless, in 2009, the main *reserve fund* decreased from more than €100 million to less than €25 million (Fig. 14.4). Besides the impact of the crisis, another reason for such a cut in reserves relates to the end of the electoral term (local elections took place in autumn 2010). The then lord mayor was elected as an MP in the June 2010 parliamentary elections and was moving to the national level of policy-making, but there was still an expansion of investment typical of the end of an electoral cycle. The second main reason for the decline is because Bratislava was preparing for the biggest sport event in Slovakia for years – the 2011 World Ice Hockey Championship. It was involved in reconstruction and other construction activities at the main site of this event – the ice hockey arena, as well as additional investment in infrastructure and mass transport. An important part of the money spent on the construction of the ice hockey stadium was covered by transfers from the state budget, but construction costs far exceeded initial estimates (€96 million versus €40 to 45 million). Construction led to the depletion of the city's financial reserves, leaving it vulnerable to future macroeconomic shocks. These could force the city to borrow more and worsen its financial situation.

In passing, it is worth noting that during the transition period in the 1990s, the city lost cash reserves through bank losses. The City of Bratislava government lost about €12 million (SKK 360 million) in a small bank, Slovenská Kreditná Banka, a.s., that went into bankruptcy. Attracted predominantly by the high rates on deposits offered by the bank, the city transferred part of its reserves, mainly from municipal bonds, during the latter half of the 1990s. Unfortunately, Bratislava failed to respond quickly to indications that the bank was in trouble. Later attempts to obtain back part of money were unsuccessful, and only SKK 13.7 million was recovered. Later investigation proved that as well as acting in a financially imprudent manner, Bratislava had also been the subject of fraud (Hospodárske noviny, June 20, 2005).

Municipal Assets in Companies and Organisations

Local government involvement in businesses constitutes a significant part of municipal assets. Bratislava's citywide government has been involved in company ownership since 1990. The numerous changes in ownership reflect the approaches typical of different periods. The current number of companies was influenced by the existence of 'inherited' entities, their nature of their transformation, as well as the business-oriented activities of local government according to changing priorities. At present, the city has holdings in 15 companies, with a value which exceeded €193 million in 2007. Many of the companies with city involvement have large assets of their own. For example, the most valuable company BVS, a.s. (a regional water and sewage company), owns a total of €340 million assets. A city mass transport company (DPB, a.s.) has total assets of €10.5 million. The waste management and disposal company (OLO, a.s.) has assets of its own totalling €40 million. The city has frequently used real-estate property as part of its basic capital in establishing new companies. Budgetary and contributory organisations established by the city government are particularly dependent on the use of city property.

The city government pays particular attention in overseeing and monitoring those companies with which it is involved. A special department within the Office of the Mayor, the City Magistracy, is responsible for managing city interests in these companies. The City Council regularly obtains information on their operation in several ways. It must approve their main financial documents, especially their final accounts. Besides this, the City Council obtains reports dealing with the situation in these companies and adopts strategic decisions concerning their future (e.g. Bratislava City Magistracy 2006, 2008). A strategic document submitted to the City Council in March 2009 (Bratislava City Magistracy 2009) is among the most recent documents concerning the future of the city's involvement in such companies. It summarised the situation of individual companies and proposed basic decisions concerning their future from the point of view of the city government in terms of 'hold', 'expand', or 'sell', shown in Table 14.3. The City Council also regularly makes important management decisions concerning individual companies, such as 'sell' or 'increase the capital', as well as personnel and other issues, and regularly reviews land and buildings allocated to its organisations, particularly with regard to extensions, renovations, and modernisation.

During the communist-era centrally planned regime, the role of local public administration in managing companies was extensive. It went far beyond providing more or less typical public services and local administration. Before 1990, the centrally planned economy had no private ownership. This meant that the predecessors of local government controlled a range of enterprises in a vast number of economic sectors in which local government is usually not active. They included industrial enterprises, construction companies, restaurants, and other activities. It led to the early post-socialist city government era 'inheriting' ten large and internally diversified companies from the previous era (Kresánek et al. 1998). These companies could only be marginally successful under new market conditions and urgently needed restructuring as a first step. From July 1, 1990, this resulted in their breakup, along sectoral or branch lines, into 31 new companies. That allowed a selective approach to deciding their future during 1991–1992. Bratislava was unique in Slovakia, since no other city had such a large stock of companies under its control.

Local government in Bratislava adopted four main forms in transforming its municipal companies (Buček 2002). Some firms remained under full or partial control of the citywide government. Most started to operate under the new Business Code, with a changed legal status, as joint stock companies (Slov. a.s.) or limited liability companies (Slov. s.r.o.). The initial fragmentation of large, internally diverse companies made it possible for the city to select those firms needed for the city's operation. A second, quite large group of companies were privatised. This group consisted mainly of smaller companies active in retailing, restaurants, accommodation facilities, and personal services (like hairdressers and taxi firms). Local government was not interested in those fields which are more typically private businesses. They were sold under the so-called small privatisation which took place in the early 1990s. Under market conditions, a relatively small number of enterprises underwent formal liquidation. They were not competitive and quickly became deeply indebted. However, this did not negatively influence the provision of local public services. The products and services offered were usually of a purely non-public character (e.g. the 'ZARES' garden centre). Local industrial enterprises, such as those manufacturing furniture or clothing, were also liquidated. In some cases, new private businesses started to operate on their premises. A final group of companies were transferred to state ownership, in expectation of their future transformation and privatisation. This included, for example, smaller companies in the construction industry. Previously, construction companies directly served the development needs of city governments. After the change of regime, local government development projects were low in number for several years, so there was no need to own construction companies. Instead, local governments relied on tendering and contracting out.

After this post-communist 'cleaning' of the old portfolio of city companies, city government shaped its companies according to its own needs (see Table 14.2). These include companies providing essential public services – *DPB, a.s.* (mass transport), *OLO, a.s* (waste collection and disposal) – which have been fully owned by the city government since the early 1990s. The city government holds the controlling share of stock – 59.29 % – of the water and sewage company, *BVS, a.s.*,

Table 14.2 Main companies with city government involvement 1990–2010

Company name	Function	Share on basic capital (in %) In 1998	In 2008
Dopravný podnik Bratislava a.s.	Public transport	100	100
Odvoz a likvidácia odpadu a.s.	Waste collection, disposal, incineration	100	100
Bratislavská vodárenská spoločnosť a.s.	Regional water and sewage company	0	59.28
Mestský parkovací systém s.r.o.	Parking system and services	100	100
Bratislavská integrovaná doprava s.r.o.	Integrated transport company	0	35
KSP s.r.o.	Rent of office/business area	100	100
Metro Bratislava a.s.	City developer company	100	66
Slovenská plavba a prístavy a.s.	River port and transport	0	7
Národné tenisové centrum a.s.	Tennis arena	0	20.52
Incheba a.s.	Exhibition centre	1.58	11.24
Tehelné pole a.s.	Revitalisation of area	0	40
České aerolínie a.s.	Air transport (Czech Airlines)	0.98	0.98
Zámocká spoločnosť a.s.	Revitalisation of under the castle area	10	10
Hasičská poisťovňa a.s.	Insurance company	17.3	17.43
Obchodno spoločenské centrum a.s.	Culture and leisure centre	100	0
Slovakia bus a.s.	Bus maintenance and production	25	0
Inprokon s.r.o.	Clothes cleaning	70	0
Matador – Obnova a.s.	Tyres renewal, car service	49	0
Istrobanka a.s.	Commercial bank	18	0
Intertour a.s.	Tourism	1.52	0

Sources: Kresánek et al. (1998), Bratislava City Magistracy (2006, 2009)

owned in common with other 125 local governments in the region where the company operates. This company resulted from the nationwide transformation of the water and sewage sector and has operated in its current form only since 2003. The city also established another company with full ownership, focusing on providing parking places and managing the parking navigation system, *MPS, s.r.o*. The long-term aim of this company is to develop and manage a more sophisticated city parking system integrated with the mass transport network. The last company under the full control of the city government is *KSP, a.s.* Its main activity is renting real-estate property that was originally used by an old socialist industrial enterprise. These corporate assets have generated a small but stable income stream for the city budget. The position of *Incheba, a.s.*, in which the city owns a minor stake, is similar. This company is a leader in the exhibition and fair market in Slovakia and has its own exhibition area and facilities in Bratislava. It plays a prestigious role for the city and gives the city an opportunity to enhance its competitive position. The city government also wants to retain shares in this company thanks to the income it generates for the city budget.

The city government also uses separate 'project' companies to manage rare physical urban development activities within its sphere of responsibility. This is the case of *Zámocká spoločnosť a.s.*, a new development under Bratislava castle in the area of Zámocká street, and *Tehelné Pole a.s.*, an area close to the ice hockey arena and Tehelné Pole football stadium. The operation of Zámocká spoločnosť has been particularly controversial, ending in long-term litigation (the city holds only 10 % of shares). *METRO a.s.* was created especially for the construction of the Apollo bridge over the Danube with a state contribution, and the state is a partner in this company. After the completion of this large public investment, the city government intends to use this company for the execution of other construction and development projects. Some of these companies illustrate the slow progress in completing their main objectives within a complicated multi-partnership environment. From this point of view, the Bratislava integrated transport company, *BID, a.s*, offers a good example. Creating a well-integrated urban-suburban transport system requires long-term negotiation with many partners.

Besides the large group of companies liquidated at the beginning of the transformation period, over the years, some companies with municipal capital participation went bankrupt, were sold, or were never even active. In most cases, the city government was unsatisfied with their operations or felt it lacked a voice in management. Decisions to end participation in such companies were also influenced by the fact that in most cases they generated no income for the city. In several cases, the city government had to make a considerable effort to regain at least part of its assets. *Slovakia Bus, a.s.*, is a typical case of a later example of a company which was shut down. It was established with the idea of providing services and producing buses for a city-owned mass transport company. This never happened, and the company was wound up after many years of being essentially dormant. *Hasičská Poisťovňa* (an insurance company) was similar, but in this case the city secured back part of its assets. City government more successfully liquidated *Matador-Obnova, a.s.*, a tyre renewal firm, regaining real estate previously transferred to the company. It also obtained financial compensation from its sale. The city government sold its shares in *Intertour*, a.s., a travel agency which also owned hotels, regaining some of its capital. The city government decided to sell its shares at a minimum of 30 % of their original value. The city also sold its share of *Inprokom, s.r.o.*, a civil engineering firm, to the joint private partner, being dissatisfied with its operations and management.

The city government holds a very marginal share in *České Aerolinie, a.s.* (Czech Airlines, CSA, the former Czechoslovak Airlines), which was obtained during the initial post-socialist transformation of the firm in the early 1990s. This gives it no real influence over the company and brings in income to the city. The decision of the City Council was to sell the shares. Initially, it aimed to sell them to the main shareholder, the Czech Ministry of Finance. A 2009 attempt to privatise CSA by selling all government-owned shares, including Bratislava's, was cancelled. In 2010, the city again decided to sell its shares in CSA but, due to lack of demand, was unable to do so.

Some companies have not been used for the planned purpose, usually because the city and its partners changed their intentions. The company created to operate the water and sewage system is typical. It existed in legal terms from 2002 to 2005. The city government and other shareholders decided not to create separate owner and operating companies for water and sewage, and BVS, a.s., is now an integrated company, owning and operating the water and sewage systems. The fate of *Obchodno-spoločenské centrum, a.s.,* was similar. It was created to remodel an old, communist-era, cultural, leisure, and exhibition centre known as PKO (Slovak: Park kultúry a oddychu – park of culture and rest). However, this model for renewing a partly obsolete area was not successful. The city government already had a new exhibition centre, and subsequently, most of the area was sold to a private developer.

Over time, the city government developed more strategies in this field. In the 1990s, some public-sector companies involved with the city attempted to create a circular relationship. They aimed to serve each other as providers of goods or services. The attempt to build such an interlinked local public economy was unrealistic, and it became clear that this approach was inefficient. The core companies recognised that they could obtain better services and goods at better prices on the open competitive market. For example, a group of firms aimed to serve the local mass transport company DPB, a.s. They included Slovakia Bus (buses production and their maintenance); Matador-Obnova, a.s. (tyres and other services); Hasičská Poisťovňa, a.s. (insurance services to the city government); and Istrobanka, a.s (a commercial bank which could have provided financial services or even loans). This circumscribed vision seemed attractive only during the first years of transition. Most of these companies went into bankruptcy or were sold. Subsequently, the city government took a more pragmatic approach towards its companies. Strategic firms active in crucial public services such as mass transport, parking, or waste management have central positions. The city is interested in companies which are needed to provide basic services or which generate income for the city budget (see Table 14.3). The city government has reduced the number of its companies identified as having

Table 14.3 The economic condition of the main companies with city government participation (in thousands of SKK)

	Receipts			Total economic result of the company			City Council decision
	2007	2006	2005	2007	2006	2005	(March 2009)
Dopravný podnik Bratislava a.s.	1,036,316	1,029,567	1,023,715	–75,595	–197,409	–181,763	Hold/not for sale
Odvoz a likvidácia odpadu a.s.	741,089	536,462	554,727	86,120	87,062	56,812	Hold/not for sale
Bratislavská vodárenská spoločnosť a.s.	2,260,721	2,242,324	1,990,435	156,551	227,298	315,063	Hold/not for sale
Mestský parkovací system s.r.o.	7,723	6,507	6,353	970	69	1,183	Hold/not for sale
Bratislavská integrovaná doprava s.r.o.	0	0	0	–3	–1,364	–445	Hold/not for sale
Metro Bratislava a.s.	15,182	77,926	499,564	–8,423	73,632	22,450	Increase basic capital
Slovenská plavba a prístavy a.s.	1,016,492	824,348	854,731	118,997	74,644	53,041	Hold/not for sale
Incheba a.s.	538,355	506,365	509,563	49,153	46,140	39,565	Hold/not for sale
KSP s.r.o.	31,312	34,063	29,931	4,419	6,250	4,042	Hold/not for sale
Národné tenisové centrum a.s.	71,889	67,465	89,998	249	3,600	10,225	Hold/not for sale
Tehelné pole a.s.	0	0	0	–96	–39	0	Sell
České aerolinie a.s. (in th. CZK)	25,035,319	24,377,078	22,455,928	206,600	–396,951	–496,057	Sell

Source: Bratislava City Magistracy (2008, 2009) (Adopted)

a marginal role. This includes firms which generated no income, did not have an important role in service delivery, did not fulfil their original purpose, or where the assets that the city had allocated or contributed were not efficiently used.

In addition, a large number of budgetary and publicly owned organisations depend on the donation or use of municipal property. Most have a long history in the city. For example, the city museum started in 1868 and city public library system in 1900. These organisations provide specialist services. Some of them are so-called 'budgetary' organisations and depend completely on the city budget. Others, so-called contributory organisations, receive contributions from the city. 'Contributory' organisations, listed in Table 14.4, have considerably higher autonomy than 'budgetary' ones. In 2009, the total value of assets used by 'contributory' organisations was €184.8 million. The greater part of these was noncurrent assets (tangible) with a value exceeding €181 million. Liabilities of these organisations are less than 1 % of total assets. Among 'contributory' organisations, about 3,000 ha of mainly forest area is administered by *City Forests* (Slov. Mestské Lesy). *City ZOO* has 96 ha of land and numerous buildings. Palaces and historical buildings have a high value, and museums and galleries are administered by the city government. A substantial amount of property is allocated to organisations which manage sport and recreation facilities in the city, including sport halls, swimming pools, sport centres, and recreation areas. Budgetary organisations which use city property include 11 social services centres (mostly serving the elderly), 12 local art schools, and 4 leisure time centres.

Table 14.4 City contributory organisations

Name of contributory organisation	Main function
Mestská knižnica	City library
Generálny investor Bratislavy	City developer – manage city's construction activities
Mestské lesy	Forests administration
Marianum – pohrebníctvo mesta Bratislavy	Funeral services and cemeteries administration
Galéria mesta Bratislavy	Art galleries
Múzeum mesta Bratislavy	City museum
Bratislavské kultúrne a informačné stredisko	Culture and information services
Paming – Mestský investor pamiatkovej obnovy	Heritage maintenance (buildings, fountains, memorials)
Mestský ústav ochrany pamiatok	Heritage protection, research, administration
Zoologická záhrada	Zoo garden
Správa telovýchovných a rekreačných služieb	Administration of sport and recreation facilities

Source: Bratislava City Magistracy (2004–2008). Annual Reports

Conclusion

Slovak local government significantly increased and improved its asset base during the transition period. As the previous pages show, the asset situation of local government strongly reflects general societal processes, the character and scope of reforms of the state, and the general economic situation. Local government used many opportunities to obtain property previously owned by the state. However, independent local government initiatives to expand and modernise their assets required favourable economic conditions. Higher tax incomes, opportunities to sell surplus assets at good prices, and favourable access to loans and credit markets depend on the wider economic situation. Not surprisingly, the asset situation improved during a period of strong economic growth between 2004 and 2008. The availability of EU funds to local governments also had a positive effect. As a result, in Slovakia, local government property is now greater in value and extent, serves more functions, and provides more opportunities for entrepreneurially orientated activities. Since 2008, pressures from the financial and economic crisis have diminished some classes of assets, especially financial reserves, needed to cope with fiscal austerity.

The initial transition period was characterised by expansion of assets and by asset differentiation among local governments. Introductory transfers often concerned restitution of historic property. The value of real estate was determined by location. The social and economic situation was, and is also, spatially differentiated. This has generated differences among local governments and the types and value of their assets. Transfers of all sorts are now coming to an end. The processes of decentralisation and de-étatisation are complete. Future asset formation will be based on more traditional, standard methods. Local governments need to mobilise their powers and initiative to maintain the expansion and modernisation of their asset base. They should concentrate on formulating asset-orientated development strategies, taking advantage of increased freedom in setting local taxes and fees, better access to external resources and EU funds, and the income generated by the assets they already own.

The transfer of property to the local level has had many positive effects. Accountability to citizens has forced local councils and mayors to preserve, enhance, and maintain local government assets, although there are rate cases where the opposite has happened. Local governments do not consist of anonymous state officials in a remote office, but a clearly identifiable group of local representatives and officials. In the case of Bratislava, the general improvement in management practice is well documented, and the many strategic documents, procedural rules, and executive manuals created by the city can be seen. Nevertheless, disputes emerge, especially in the selling and exchange of real-estate property or the use of financial reserves.

In the case of Bratislava, local government is moving closer to asset-related activities typical of other large cities but, until recently, less evident in Slovakia. Its activities are increasingly focused on large public properties and facilities designed to serve the public. After the construction of the new National Theatre, the city complemented its cultural facilities by taking over one building of the former National Theatre, now the *City Theatre*. Now a theatre without a fixed drama ensemble, it hosts invited performances by other theatre ensembles. A similar change has taken place in sports facilities. Previously, Bratislava administered mainly smaller facilities for its citizens, but in recent years it has moved towards large sports facilities, enabling it to host prestigious professional events. The first sign was the entry of the city government into the *National Tennis Arena* project as a place for major international tennis tournaments in Slovakia, including Davis Cup and Federation Cup matches. A substantial move in this direction was confirmed by the role of the city government in managing the substantial reconstruction of the ice hockey stadium for the world championship in Bratislava in 2011. The history of its construction illustrates the pitfalls facing municipal governments when they move into unfamiliar areas. The city government participated by providing land, managing the development, and contributing to a significant part of the related investment. Ideally, in projects of this kind, the state budget should account for a major source of finance. The arena was completed on time, before the championship. The original costs were estimated at €40–45 million, which later increased to €70 million, but when the project was completed, the total came to €96 million. Of this sum, approximately €40 million was provided by the national government, and the rest, €56 million, came from the Bratislava municipal budget. The result was a significant increase of local government debt. After the championships, the ice hockey stadium became the home ground of a local hockey club, Slovan Bratislava, which is now part of the International Continental Ice Hockey League. This league is slowly expanding its member clubs. As a rule, matches are sold out, bringing over 10,000 visitors per match. Slovan Bratislava pays about €750,000 annually in rent. However, the city remains the main owner.

A similar role was to be played by the city government in building a new national football stadium, but with a larger share of investment coming from private investors and the Bratislavský Region government. Unfortunately, this construction never started. This was partly due to the election of a new national government, which was less supportive of the project, and changes in the leadership of the Slovak Football Association. After the initial proposal, questions were also raised about the proposed location – an inner city site – some arguing that a location on the city's edge, closer to the motorway network, would be more appropriate. In the meantime, it has become clear that the Bratislava municipal government can neither extensively nor exclusively fund this project, particularly in light of the overspending on the hockey arena. The current expectation (2013) is that finance will come from the private sector – the owners of Slovan Bratislava Football Club – and the national government, but the city will probably supply the land in exchange for a long-term lease. In early 2013, the national government announced a grant scheme to support Slovak football which would run through 2020, providing funding for new stadiums and other investments for clubs playing in the first and second divisions. Projects in some cities are already prepared and will start construction in 2014. Financial support for the National Football Stadium might come from this source, although it falls outside the programme's remit. Lack of a national stadium which meets UEFA standards means that some matches may need to be played in Vienna or Budapest.

A feature of municipal asset use is the increasing use of partnerships among municipal and national governments and private and non-governmental organisations. It seems that there is no way to meet the city's development needs without such an approach. Nevertheless, for years, this approach was not part of city policy, as shown by the low number of joint companies with private partners. Such joint companies were more a result of the general privatisation process than the city government's desire to rely on public-private partnerships. Despite the success of several individual projects which relied on a joint partner framework, such as the new Bratislava airport terminal financed by the state, further partnerships are needed. One example is the reconstruction and renewal of the area around the main railway station. This is highlighted in the analytical sections of the *Bratislava Economic and Social Development Programme* (Bratislava City Magistracy 2010b). It requires a more sophisticated model of joint partnerships, including control of property, joint operation, and joint management. Neither the private nor the public sectors can prepare and complete such expensive development projects on their own.

Since 2009, the importance of municipal assets and their diversification has been confirmed by the financial and economic crisis. Bratislava's government shows the crucial role of reserve assets, especially financial assets. The short-term impact of the crisis has been strong. For example, during the first half of 2010, the city stopped new construction. Declining tax revenue put pressure on other sources of income. The citywide government was unable to raise money from property sales, due to decreased demand on the real-estate market. It was able to operate and maintain its investments only due to accumulated resources in the reserve fund. Since 2009–2010, the financial situation of the Bratislava municipal government has improved, thanks mostly to increased income provided by shared taxes. However, the city has not been able to meet its projected income, due mainly to stagnating property prices and a weak property market. This has led to a drop in interest in buying local government property. Within Bratislava, construction activities are only slowly recovering. Income from the sale and lease of assets improved in 2010 as against 2009 but worsened in 2011, falling to 40% of the 2010 level. As a result, the city continues to have problems in funding its capital budget, and development activities remain curtailed. Total local debt remains above €200 million. Current large investment projects are related to transport, such as a tram linked to the right side of Danube and new public transport vehicles. However, these are financed by European Union funds, with small contributions from local budgets. Therefore, the future of local government finance remains uncertain. A longer-term crisis may cause an increase in borrowing and a reduction of investments.

Acknowledgements This chapter was supported by the grant project VEGA No. 1/0424/15 and No. 1/0709/11 of the Slovak Scientific Grant Agency of the Ministry of Education of the Slovak Republic and Slovak Academy of Sciences.

References

Act No. 138/1991 Coll. on municipal assets (as amended)
Act No. 369/1990 Coll. on local government (as amended)
Act No. 377/1990 Coll. on Bratislava the Capital of the Slovak Republic (as amended)
Act No. 416/2001 Coll. on transfer of powers from state administration to communities and higher territorial units.
Act No. 513/1991 Coll. Business Code (as amended)
Act No. 523/2004 Coll. on budgetary rules of public administration (as amended)
Bovaird T (2006) Developing new forms of partnership with the "market" in the procurement of public services. Public Adm 84(1):81–102
Bratislava City Magistracy (1993) VZN č. 2/1993 o zásadách hospodárenia s majetkom hlavného mesta Slovenskej republiky Bratislavy (Local regulation No. 2/1993 – Principles of Assets Administration). Bratislava, March 1993
Bratislava City Magistracy (2004–2008) Výročné správy hlavného mesta SR Bratislavy (Annual reports of the capital city of the Slovak Republic Bratislava)
Bratislava City Magistracy (2006) Informácia o vývoji v obchodných spoločnostiach s kapitálovou účasťou hlavného mesta Slovenskej republiky Bratislavy v období od 1.1.2003 do 20.10.2006 (Information on the development of business companies with asset involvement of the capital city of Slovak Republic Bratislava for period 1.1.2003 to 20.10.2006), Bratislava, October 2006. City Magistracy, Bratislava
Bratislava City Magistracy (2006–2009) Záverečné účty hlavného mesta Slovenskej republiky Bratislavy (Final Accounts of Bratislava). City Magistracy, Bratislava
Bratislava City Magistracy (2008) Informácia o stave a ekonomickej situácii v obchodných spoločnostiach s kapitálovou účasťou hlavného mesta Slovenskej republiky Bratislavy v roku 2007 (Information on state and economy of business companies with asset involvement of the capital city of Slovak Republic Bratislava for year 2007). October 2008, City Magistracy, Bratislava
Bratislava City Magistracy (2009) Návrh ďalšieho postupu pri nakladaní s majetkovou účasťou hlavného mesta SR Bratislavy v obchodných spoločnostiach (Proposal of further steps in use of asset involvement of the capital city of the Slovak Republic in business companies), March 2009, 4 p. City Magistracy, Bratislava
Bratislava City Magistracy (2010a) Štatút hlavného mesta Slovenskej Republiky Bratislavy (The Statute of the capital city of the Slovak Republic Bratislava). City Magistracy, Bratislava
Bratislava City Magistracy (2010b) Program hospodárskeho a sociálneho rozvoja hlavného mesta SR Bratislavy na roky 2010–2020 (Program of economic and social development of the Capital City of Slovak Republic Bratislavy for years 2010–2020). http://www.bratislava.sk. Accessed on 14 Aug 2010
Buček J (1995) City, state and territory: Analyses of relations on the case of the capital of the Slovak Republic Bratislava (in Slovak). Unpublished PhD dissertation, Comenius University Bratislava
Buček J (1999) The citywide/district relationship and the question of integrity: a case study of Bratislava. In: Kimball JD (ed) The transfer of power – decentralisation in central and eastern Europe. Local Government and Public Service Reform Initiative, Budapešť, pp 79–96
Buček J (2002) The role of local government in local development during post-socialist period: the case study of Bratislava. Acta Universitatis Carolinae Geographica 37(2):105–119
Buček J (2006) The position of local government after main phase of public administration reform in Slovakia. Acta Geographica Universitatis Comenianae 48:167–178
Buček J, Borárosová Z, Sopkuliak A (2010) Miestne financie a miestny ekonomický rozvoj (Local finance and local economic development). Geo-grafika, Bratislava
Buchanan JM (1965) An economic theory of clubs. Economica 32:1–14, February
Diefenbach T (2009) New public management in public sector organizations: the dark sides of managerialistic "enlightenment". Public Adm 87(4):892–909

Fernholz F, Fernholz R (2006) Strategic municipal asset management. Municipal Finance Task Force, RTI International. http://logincee.org/file/17900/library. Accessed 4 Aug 2010

Hospodárske Noviny (2005) Kauza SKB: Polícia odhalila ďalší podvod (SKB case: the police found another fraud). Hospodárske Noviny, 20 June 2005. http://www.hnonline.sk. Accessed 14 Sept 2010

Jolicoeur PW, Barrett JT (2004) Coming of age: strategic assets management in the municipal sector. J Facil Manage 3(1):41–52

Kaganova O (2008) Integrating public property in the realm of fiscal transparency and anti-corruption efforts. In: Péteri G (ed) Finding the money: public accountability and service efficiency through fiscal transparency. CEU, Budapest, pp 209–232

Kaganova O, Nayyar-Stone R (2000) Municipal real property asset management: an overview of world experiences, trends and financial implications. J Real Estate Portf Manage 6(4):307–326

Kaganova O, Tian V, Undeland C (2001) Learning how to be efficient property owners and accountable governments: the case of Kyrgyzstan cities. Public Adm Dev 21:331–341

Kresánek P, Vajda M, Štefáková A (1998) Ako v Bratislave – Občiansky almanach o živote a službách hlavného mesta (Citizens' almanach on life and services of the capital city). Inštitút pre verejné otázky, Bratislava

Ministry of Finance (2002–2009) Záverečné účty Slovenskej republiky (Final accounts of the Slovak Republic). Available at www.finance.gov.sk

Papcúnová V, Balážová E (2006) Majetok obcí (Municipal property). SAPV, Nitra

Péteri G, Horváth TM (2001) Navigation to the market: regulation and competition in local utilities in central and eastern Europe. In: Péteri G, Horváth TM (eds) Navigation to the market: regulation and competition in local utilities in central and eastern Europe. OSI/LGI, Budapest, pp 13–98

Ryder A (1990) Growth pole policy in Poland and the Lenin steel-works. Geoforum 21(2 Spring)

Ryder A (1992) Urban planning in post war Poland with reference to Cracow. Cities, August

Staroňová K, Sičáková-Beblavá E (2006) Verejná politika a miestna samospráva: štyri princípy spravovania (Public policy and local government: four principles of administration). ÚVP FSEV UK, Bratislava

Sun L (2002) Fading out of local government ownership: recent ownership reform in China's township and village enterprises. Econ Syst 26:249–269

The Supreme Audit Office of the Slovak Republic (2010) Správa o výsledku kontroly hospodárenia s majetkom, majetkovými právami, finančnými prostriedkami, záväzkami, pohľadávkami a kontrola plnenia opatrení z predchádzajúcej kontroly v meste Bratislava, ktoré je sídlom kraja (The report on the results of control of the assets administration, property rights, financial resources, liabilities and control of measures from previous control in Bratislava, the seat of region). http://www.nku.gov.sk. Accessed 14 June 2010

Tichý D, Žárska E (2005) Municipal borrowing – macroeconomic and microeconomic risks. Biatec 13(10):11–15

Žárska E et al (2008) Komunálna ekonomika a politika (Municipal economy and politics). Ekonóm, Bratislava

Chapter 15
Governing the Transformation of the Built Environment in Post-socialist Bratislava

Pavel Šuška

Abstract The paper focuses on the fate of a former industrial area in Bratislava, the former Kablo plant, where a new central business district (CBD) will be built. Particular attention is paid to the conflict over the preservation of some historic structures on the site, and the role the local government played in efforts to preserve them. First, however, structural forces responsible for the creation of particular patterns of rational landscapes are analysed. These were transformed dramatically during the last century as was the relative location and the functional significance of the location of the Kablo factory within the city and beyond. In the case of Kablo, the local government played a threefold role. First, it was the co-creator of a discourse where the locality was given a new identity and steps to redevelop the site legitimised. Second, its (non)action in both town planning and in listing the heritage buildings allowed the investor to demolish the buildings on the site, despite a lack of permission and a belated attempt on the part of the city to preserve some of the most historic structure, and to excuse its later actions. Thirdly, higher institutions at the regional level did not effectively apply sanctions for the demolition. One can argue that the redevelopment of the site was due to the obsolescence of the buildings on it, and the rise in value of the underlying property – in effect, the penetration of global capitalism into the local economy. However, the failure of local government to protect the site also reflects its relative inexperience with developers, divided public opinion on the value of preserving the fabric of the plant, and unequal relations between relatively sophisticated developers and a relatively young city government.

Keywords Urban redevelopment • Management of redevelopment • Public-private projects • Historic preservation

P. Šuška (✉)
Institute of Geography of the Slovak Academy of Sciences,
Štefánikova 49, 814 73 Bratislava, Slovakia
e-mail: Pavel.Suska@savba.sk

© Springer Science+Business Media Dordrecht 2015
J. Buček, A. Ryder (eds.), *Governance in Transition*, Springer Geography,
DOI 10.1007/978-94-007-5503-1_15

Introduction

This paper briefly discusses the failed attempt by the city of Bratislava to preserve a fragment of a former industrial concern, Kablo, located on the edge of the central part of the city of Bratislava. Developed in the nineteenth century, the factory was finally closed after the collapse of the communist rule in 1990. The territory of the factory was purchased by developers who planned to construct new offices on the site, taking advantage of its proximity to the city centre, as well as the increased demand for offices after Slovakia regained its independence in 1993.

With under 500,000 inhabitants and just over 600,000 in the metropolitan area, Bratislava is a relatively small city by European standards. However, it encapsulates a variety of problems which face all cities in Europe and elsewhere. The first is one of governance. Democracy and self-government are relatively recent innovations in Bratislava, as they are throughout Central Europe. The failure of the city to list historic buildings and monuments in a timely manner reflects the relative inexperience of those in the government, contrasted with the relatively sophisticated background of many investors. In addition, like many other cities in Central-Eastern Europe, Bratislava faced a perfect storm after the collapse of the communist rule. Industry, which was favoured under communism, was revealed to be outdated and uncompetitive. Industrial infrastructure was worn out and obsolete, and the capital needed to modernise factories was not available. Western Europe was going through a process of deindustrialisation, and Slovakia followed a similar path, but with much greater speed. The rise of the capitalist economy revealed that many industrial firms were unprofitable, quickening change. Alongside this, new sectors of the economy came into being, particularly those related to finance, insurance, and real estate. The rise of this sector, alongside with Bratislava's transformation into a national capital, generated numerous demands for renewal and redevelopment. And, as has been noted by Bertaud and Reynaud (1995), under these new conditions, locations, particularly central locations, acquired value as the bid rent curve, invisible under socialism, reasserted itself. Many would argue that one of the roles of the government is to stand up for the "public" against the private sector and to defend the public well-being against market forces – stepping in if there is a market failure, to ensure that the benefits of investment are evenly distributed and to put the general good above individual benefits. Local government draws up the local plan and enforces it and, more broadly, sets rules which it then defends. At the same time, local governments face a conflicting alternative: the need to attract investment, jobs, tax revenues, and economic development for their residents. It is difficult to balance these two imperatives, particularly in relatively young democracies, where government decision makers and administrators have relatively little experience. In addition, Bratislava and Slovakia were (and still are) relatively less well off when compared with neighbouring Vienna and Austria. There was a very real fear that foreign investors would remake the city to suit themselves, regardless of the needs, demands, or tastes of local residents. At one point, it was even suggested that Bratislava could become a suburb of Vienna, the location of inexpensive homes on just across the Austrian border. This debate is not unique to Bratislava. In 2012, it was estimated that 60 % of dwelling purchases in central London were made by non-residents from overseas and that, as a result, London dwelling prices were increasing even as prices were falling elsewhere in the United Kingdom (Hammond 2011, 2012; Thomas 2011). The contexts in which decisions are made to favour or restrict development are also important. In 1989, Bratislava's environment was distressed, buildings in disrepair and industry outdated and badly located. The city needed money to repair and improve sewer and water systems; repair buildings, pavements, and street lighting; and rebuild the local transport system. However, central government was reducing, not increasing, grants, forcing the city, like other cities in the region, to fall back on its own limited resources. A pro-development policy encouraged investment from the private sector to make good local government shortfalls, and at the same time, new investment increased tax revenues to the city. The discussion about development was taking place against a drive to "modernise" – to rebuild damaged areas of the city and damaged buildings, to rebuild and replace old industry, and to make the city more "European". The argument was made that the city's image needed to change. It needed to be more modern and European to compete with neighbouring centres like Prague and Budapest and to attract international investors. This drive is reflected in the style of many of the new projects in Bratislava's city centre. Local government failed to protect the city's historic legacy, but this is no different from debates going on elsewhere, for example, at Kings Cross in London, where initial redevelopment proposals made in 1989 proposed demolishing historic industrial buildings on the site. Local public opinion opposed this and gained the support of the local government and also national institutions like English Heritage (e.g. English Heritage 2012), leading to new plans which preserved a significant amount of historic structures.

Kablo and Bratislava

Between 18 November 2007 and 12 January 2008, the buildings and structures of a former industrial area near the centre of Bratislava known as Kablo were torn down. It was the moment when the industrial past retreated from the wider city centre of Bratislava (the capital of the Slovak Republic) and gave way to a truly post-industrial form of development represented by a 600 million euro investment in order to regenerate a part of the future central business district (see Fig. 15.1). However, two of the destroyed buildings, Feigler's heating plant and a brick chimney, were supposed to appear on the cultural heritage list and thus receive legal protection.

Like many other post-socialist cities, Bratislava has experienced a dramatic construction boom during the last decade. Redevelopment and the transformation of the built environment are not usually uncontested. Often, distinctive ideas, proposals, plans, and interests encounter and compete with each other. Thus, the way these building projects are actually realised can shed light on the composition of governance

Fig. 15.1 The past and the future of the Kablo area

and on the actual power relations and positionality of the actors involved. This can help us evaluate the quality of local government and investigate the level of local democracy within the area.

The story of the place is framed by the concepts of the now classical theory of urban transformation under capitalism which, in general, can be applied in conditions of post-socialist transformations. Therefore, another level of contextualisation is provided by the transformations of post-socialist states which embraced both governance and the transformation of the built environment. Post-socialist transformations can be seen as a radical change – "[a] revolution in temporal and spatial relations [that] often entails not only the destruction of ways of life and social practices built around preceding time-space systems, but the 'creative destruction' of a wide range of physical assets embedded in the landscape" (Harvey 1997, 266). The question is: how is the structural logic of creative destruction articulated in the political praxis of post-socialist Bratislava? Although only one case is presented here, it is by no means unique within the contemporary urban development of Bratislava. On the contrary, the buildings of Kablo share the fate of many other relicts of the area's industrial revolution. Many parks, plazas, promenades, vineyards, etc., had to give way to new forms and functions which fit more properly into the "bright future of the city".

Capitalism and the Transformation of the Built Environment

Capitalism represents a form of economic and social organisation which, despite multiple manifestations or the coexistence of various forms, can be defined by the imperative of the accumulation of capital through its circulation. A characteristic of capitalism is its dynamic flexibility and capacity to transform itself. On the one hand, this is a consequence of its response to changing external conditions and, on the other, of a continuous search for profits and expansion. Concepts, such as post-Fordism (see, e.g. Amin 1994), flexible accumulation (Harvey 1989), or disorganised capitalism (Lash and Urry 1987), are among those used in connection with current transformations of capitalism on the global level.

The fundamental inevitability of capitalist transformations, along with the spatial reorganisation of production, directly impacts on a particular urban space. The built environment is subject to continuous change that places it in the centre of a contradiction inherent in the accumulation of capital, when its circulation produces a durable physical structure which simultaneously limits its subsequent circulation, effectively turning a mobile asset into a fixed one (Smith 2000).

To understand the conflicts which stem from the transformation and regeneration of the built environment, the concept of the dual interpretation of value, namely, exchange value and use value, is indispensable. The briefest characterisation of exchange value is that it is the value for which a commodity can be exchanged for other commodities. According to Harvey (2006), there are several features, contradictions, and consequences in the context of the built environment as a spatially fixed commodity: circulation of capital dynamically fluctuates through the built environment. At particular moments of circulation, elements of the built landscape are devalued, abandoned, destroyed, and selectively reconstructed. Property is commodified: investments in real estate create commodities that become scarce in conjunction with a unique location; capital investments are immobilised in real estate, and the long period of turnover constitutes a barrier to further accumulation; real estate is characterised by poor liquidity and sensitivity to ageing.

The capacity of the built environment to bring returns (rent) depends on changes in the values of two different elements: improvements or buildings on the land (or the level of obsolescence) and the location in space. Locational rent is a contextually dependent variable. The value can rise or fall depending on externalities caused by the location of other elements. This means that proximity and/or the remoteness of other activities and amenities are important. The moment a building is completed, the loss of its value due to depreciation begins. The ageing process can be decelerated by intentional decisions to invest in restoration and upgrading, or on the contrary, ageing and decay can be accelerated (Harvey 2006; Weber 2002).

The general process in which waves of investment and devaluation destroy existing values and prepare opportunities for new rounds of investment has been described as a form of creative destruction. In the context of these processes, Harvey (1989, 2006) described the production of what he referred to as a "rational landscape" formed by a series of investments which answer the needs of circulation and accumulation of a particular time, which then remain "materialised" in a given place. The rational landscape of the past represents a barrier to future development. This contradiction is overcome by the process of creative destruction which destroys the old so that it can be replaced by the new. Yet as Weber (2002) asserts, this is not exclusively a process determined purely by market forces; extraction of a place's value on different levels is determined by a wide range of governmental and other (nonmarket) actors.

The interpretation of value in terms of utility touches on the limitations of the market. This relational quality differs depending on the different actors in the transformation of the built environment and is an important attribute of use value. Real estate contains a use value defined in different ways for a developer, an owner, a governmental institution, or for the people of a place with no property relations to it. Logan and Molotch (1987) expand the definition of use value of a place for its inhabitants. They insist on the inseparability of the material use from the psychological use, where "the daily round that makes physical survival possible takes on emotional meanings through that very capacity to fulfil life's crucial goals" (Logan and Molotch 1987, 20).

In practice, the cohesion of specific value systems within which particular social groups pursue their own interest is manifested along a spectrum ranging from unfettered development at one end, favouring unlimited growth, through efforts to regulate it in varying degrees in the middle, and to efforts at total preservation at the opposite end (Kong 2003). As McManus (2000, 107) wrote, "Decisions to conserve, how to conserve, what to conserve, when to conserve, and so on, are political decisions that represent value-judgments". The outcome of those decisions constructs the identity of the place. The role of local policy here seems crucial.

Examining this process should be the subject of empirical research. Not all the transformations of the built environment taking place in Bratislava are perceived as positive or unproblematic (Ira 2003), and there is debate about the positive aspects of most developments and about the benefits and costs. This can be seen in an attempt to assess the role played by local politics in one particular case of "struggle", to conserve the historic heritage represented by the buildings of the former industrial compound Kablo (the Feigler heating plant and the chimney). During the course of this struggle, local policy was represented by a number of specialised institutions on different hierarchic levels with complicated mutual relations. Here, they are sketched in rough outlines in direct relation to the case presented. First, however, a brief outline of the main changes producing the condition for "creative destruction" is presented.

The Century of Transformations

The Staré Mesto urban district of Bratislava is located between Mlynské Nivy Street and the Danube River. This was an important industrial zone in the late nineteenth and early twentieth centuries, a period of extremely rapid industrialisation in the Hungarian part of the former Austro-Hungarian Monarchy. Bratislava of that time was one of the biggest industrial centres within the mostly agricultural kingdom. The area of Kablo located on the periphery of late nineteenth-century Bratislava and adjacent to crucial transport facilities such as a river port and the former main train station was a favourable location for the capital investment which fostered the industrial zone which emerged there.

Yet its original "peripheral" location "changed" its relative position as a result of the immense growth of the city in the course of the twentieth century (Fig. 15.2). In the end, the area found itself on the boundary of modern Bratislava's city centre. The period of state socialism (1948–1989) affected the Kablo area in several ways. First of all (apart from the city growth that occurred mostly in this period), industrial development (especially of heavy industry) became one of the key goals of the economic and defence policy of socialist Czechoslovakia (Londák 1999). Centrally planned new industrialisation on such a scale could have hardly taken place within the industrial facilities of the early industrial revolution. Flows of industrial investments

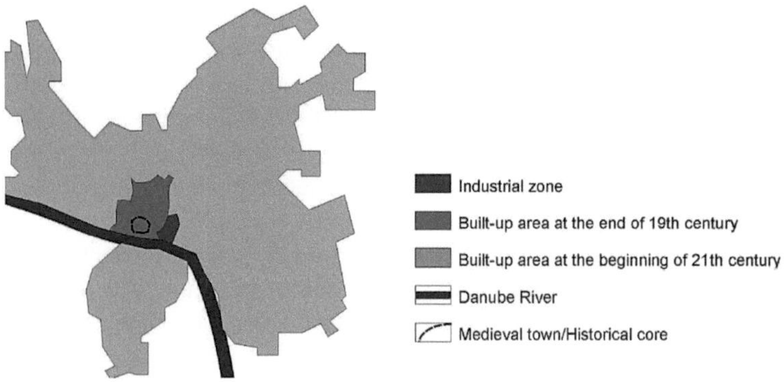

Fig. 15.2 Spatial growth of the city and the location change of the Kablo area within industrial zone during the twentieth century

thus flowed mostly to newly established and much bigger plants on the outskirts of the expanded city. On the other hand, production continued in many viable facilities within the city. Such maintenance of industrial production happened in the case of Kablo, which is why it was not included among listed buildings in the 1980s. In fact, it was in this period that the City Institute of Monument Preservation (MUOP) first attempted to do so. The weak position of institutions advocating a cultural heritage agenda in the wider context of central planning and decision making, however, resulted in an agreement between the Ministries of Economy and Culture not to hinder production in the case of such viable facilities (Obuchová 2009).

It is in the period after the collapse of communism that the contradiction between the then industrial use and the potential value of the area based on potential land use became big enough to enable creative destruction to take place. The material layers of former investments and values were about to be destroyed in order to recreate value in the location. On the other hand, to think of the last 20 years as an unproblematic and unilinear transition towards inevitable ends would be misleading. To the contrary, spatio-temporal variations in development trajectories across the spaces of post-socialism showed the processes of these transformations to be open ended and complicated, with contradictory processes of interplay among many different social actors, operating on different scales and pursuing distinct strategies with path-dependent and path-shaping aspects (Drahokoupil 2007; Smith and Pickles 1998).

Apart from the structural changes analysed below, one particular factor affected the city's development – the formation of the newly independent state of the Slovak Republic, with Bratislava as its capital. New conditions within the national hierarchy of cities and within a new spatial division of labour were set up and shaped (along with other transformations) the future development of Bratislava.

The complexity of the transformation period in Slovakia (and other states in the Visegrád region) is expressed in Drahokoupil's (2008) framework, which consists of three stages, each characterised by distinctive constellations or transformative state projects. The "neoliberal transformational state" was the first stage of post-socialist development in Slovakia. This played a crucial role in introducing the generic conditions of capitalism, represented by the most basic and narrow economic preconditions for a market economy. Emerging in the 1990s, which some might call the heyday of neoliberal hegemony, the transformatory agenda was straightforward. The introduction of market principles through shock therapy would inevitably lead to the development of a range of societal arrangements inherent to all developed societies. Along with this agenda, the processes of fundamental reorientation towards the Western countries and integration into the global political economy started.

This initial stage, however, was soon followed by several distinctive pathways across the Central-Eastern European region during the early and mid-1990s. In Slovakia, this period is often referred to as Mečiarism (after the Prime Minister Vladimír Mečiar, in office 1993–1998). It was characterised by several complementary policies. In the sphere of economic policy, it was "nationalism with economic means" which hindered full integration into the global economy (Drahokoupil 2008, 39). Accompanied by a strongly centralised national administration, this meant anything but the creation of conditions for capital accumulation through local redevelopment. In fact, as Buček and Pitoňák (1997) observed, an initial invasion by the financial sector played an almost exclusive role in the transformation of the built environment in Bratislava's city centre at that time, as banking and financial services moved into the city centre. This was quite a different picture from that of Budapest in the early 1990s, painted by Smith (1996), depicting the vivid atmosphere of a place which was much more "mature" and well integrated into the global political economy.

The current stage of the post-socialist transformation, leading towards a "Porterian workfare regime", is comparable (to a certain extent) with the "competition state" in advanced capitalist countries (Jessop 2002). "The underlying aim of state intervention has become managing the insertion of the local/national economy into the flows of global capitalism. Other social and economic policies were (to different extents) subordinated or integrated into the competitiveness agenda" (Drahokoupil 2008, 46). The logic behind this is to enhance the economic base of a locality by attracting foreign (and domestic) capital with subsequent attempts to embed the investment into the local economy. Along with reforms and the encouragement of foreign direct investments (FDI), a rescaling of the functions of the nation state takes place. Despite the fact that formal decentralisation (downscaling) of the state started immediately after the change of 1989, it was not before the late 1990s that the rhetoric of local autonomy was accompanied with corresponding political and economic powers and responsibilities (Buček 2006; KPMG 2004).

The conditions essential for the process of creative destruction, represented by the introduction of market principles, the resurgence of private property, and the emergence of place-exchange value – that is, land values based on location – were established in the first stage of post-socialist transformation. However, it is the third stage of the "mature" and globally integrated capitalist political economy of both the Slovak state and the city of Bratislava that enabled the process to be realised. Advancing the political and economic transformation of the city, region, and country in conjunction with the growing importance of the city led to an increase in development and an increase in land value. This has resulted in increased pressure on the urban fabric and structures and an unprecedented interest in investment and income generation through urban redevelopment. For example, office floor space within the city doubled during the last decade. Although investors do not necessarily rely on foreign or global capital, the demand for space offered by the planned projects is, to a large extent, a result of integration at the global scale.

Another aspect of post-socialist politics is the actual fulfilment of promises of democracy, transparency, and participation – the emblems of the changes of 1989. Some consider the creation of basic conditions for capital accumulation a sufficient condition, inevitably leading to the development of a mature civil society. However, one of the results of such a one-sided emphasis on the economic aspects of transition is the continuing democratic deficit. Planning is one of the instruments regulating activities within the city and integrating different views and value systems. However, due to the absence of planning skills in the official public boards and a generally shared scepticism towards the planning process, which was entirely compromised during state socialism, regulations regarding private investment have been weak, and the idea of the so-called invisible hand has become the best and only locational arbiter (Altrock et al. 2006). This can be documented, for example, by the long-term absence of a city master plan, and the only recently re-established position of the main city architect, who has only an advisory role, and the lack of legal norms and institutions to enforce adherence to agreed regulations (e.g. building inspectors).

Effective public participation in decision making and planning was very weak, as can be seen in the development of the city of Bratislava master plan, among many other policy and strategy documents (Huba and Trubíniová 2009; Huba 2007; Kvasnicová 2009). Conflicts in other situations in Bratislava (see, e.g. Buček 2006) were characterised by the unrecognised or the under-recognised role of non-statutory actors, to links between local politics, politicians, and businesses and to problems of civic participation in urban development in Slovakia generally (Strussová and Petríková 2009).

It is ironic that urban environmental issues which once played a significant part in civic society and public participation in Bratislava, even during the era of state socialism (see, e.g. Huba *et al.* 2008), have so quickly become regarded as unimportant. The enthusiasm related to expectations from a new "democratic" government (at all levels), its expected sensitivity towards environmental issues, and its acceptance of public nongovernmental voices were soon replaced by disappointment (Podoba 1998; Snajdr 2008).

The End of Kablo

In spite of continuous efforts by the City Institute of Monument Preservation (MUOP) to gain legal protection for the Kablo buildings, a completely new image of the place's future started to emerge. Comments by the then Mayor, Andrej Ďurkovský, emphasised the vision of a "white" Bratislava (as a commercial and administrative centre void of industry) with a new CBD, and industrial production crowded out. This was part of this process of change (for instance, Trend 2004, 20 July). The shape and scope of investment plans also helped to form a new identity for the location. A place once associated with names like Kablo, Gumon, Cvernovka, or BAZ (names referring to industrial production) was gradually been renamed Twin City, Apollo Business Center, City Business Center, Eurovea, or Bratislava Business Center. In the case of Twin City, investments of over one half billion euros "towering high above the present city" dramatically signalled the change. The area was a peripheral zone at the time of its initial development, but since then it has been swallowed up by the growing city and later by its expanding centre. These changes are reflected in the increasing value of the land. However, despite its relative centrality, access to adjacent areas in the city centre, and to the broader city region, is poor. Road access is particularly bad, which has made it less suitable for the kinds of heavy industry which were once located there. As long as the area was dominated by industry, its locational value was obscured. Now, with the collapse of traditional industry and the rise of a capitalist economy, centrality has become more important, and land values have risen, encouraging sweeping land use changes and physical redevelopment of the area.

The owner of the site, Twin City Shareholding Company, acquired the plot of the Kablo area by a purchase contract dated 18 June 2006. According to the architectural projections of the Twin City project (Chudík et al. 2007) produced by the investor HB Reavis, the project worth 570 million euro did not include conservation of the existing buildings. The exchange value of the buildings had dramatically decreased (unlike the land beneath them) and represented a threefold cost burden: their value was not only shrinking, but the developers would have to cope with the costs of conservation and be faced with lower returns because of this conservation. Conservation would be more expensive than new build, and lettable space in the restored buildings would be significantly lower than that in new construction. Hearings and negotiations at the office of construction of the urban district Bratislava – Staré Mesto – about allowing the removal of the buildings from the Kablo area took place from February 2007. By the beginning of June 2007, the investor acquired the permit needed to pull down all structures with the exception of plot No. 9095/7 (the site of the Feigler's heating plant and the chimney). In spite of the survey carried out in the area and in spite of its well-known historic value, the Regional Monuments Board approved the demolition of individual buildings. In such a situation, the municipal office of construction had no choice but to issue the permit. The exception included the brick chimney with adjacent buildings, as they were to be entered into the list of cultural heritage, and because that process had already started, a demolition permit was not given for them. The investor, however, pulled down the chimney and the halls between 18 November 2007 and 12 January 2008. Subsequently, the urban district of Bratislava, Staré Mesto, and its office of construction levied a fine on Twin City amounting to one million Sk (about 30,000 euro). The fine, however, was declared null and void in March 2008 by a resolution of the regional office of construction, based within the larger political region (kraj) within which Bratislava city is located.

Conclusion

Any societal arrangement, not just capitalism, has to produce its own spaces, a rational landscape of its own, responding to the particular needs of that particular society. The inherent contradiction of the production of space based on societal dynamics is overcome by the geographically uneven processes of creative destruction. In the case presented here, during the era of state socialism, both the industrial use and the architectural forms of the Kablo area were preserved, although the latter occurred rather accidentally.

Post-socialist political and economic changes led to the formation of the conditions for creative (and destructive) transformations of the place (Sýkora 2009). Therefore, two intertwined levels of creative destruction affected the Kablo area.

First, the complex post-socialist transformations caused a dramatic restructuring of the economic base. After the collapse of state socialism, industrial production, once the leading sector of the economy, lost its importance generally. Gradually, integration into the global capitalist political economy, along with the changing positionality of both the new Slovak state and the city of Bratislava, favoured new and distinctive ways for the accumulation of capital and the creation of value – in both financial and nonfinancial terms – in places. Thus, various "physical assets embedded in the landscape" came under pressure for redevelopment. Moreover, as can be seen in the case of Kablo, despite high expectations, promises of more inclusive decision-making processes have not been fulfilled.

There were three issues related to the Kablo site. The first was the creation of a new image for the area, through an interactive discourse between developers, politicians, local government administrators, and the general public. The second was the failure of the local government to act quickly to list the historic buildings on the site and to monitor and control the development process. The third was the failure of higher levels of administration to support the local government in its efforts to preserve heritage and manage the built environment. Local politics played a role here because its official responsibilities include acting as an arbiter, representing the public interest and values not associated with direct economic profit. It should help to solve conflicts by creating legal and institutional frameworks.

Several manifestations of local politics and the involvement of government in the politics of built environment change were identified in this case. One was the creation of a genuine discourse between local government and developers, where the new identity of the Kablo site and the surrounding area was discussed and agreed on, and subsequent steps in the development process laid out to the satisfaction of both parties. The second was that the failure to act in both the planning realm and in listing the heritage buildings helped the investor to justify its later actions. A third was that responsible institutions also failed to act effectively in applying sanctions. Although the investigated case illustrates only the rough outlines of these issues, it can serve as a starting point for further investigation. The aim of such investigation should be the detailed detection of the relationships between individual institutions and hierarchies of local politics.

Acknowledgements The study was supported by the project VEGA, No. 2/0035/15 of the Slovak Scientific Grant Agency of the Ministry of Education of the Slovak Republic and Slovak Academy of Sciences.

References

Altrock U, Güntner S, Huning S, Peters D (2006) Spatial planning and urban development in the new EU member states – between adjustment and reinvention. In: Altrock U, Güntner S, Huning S, Peters D (eds) Spatial planning and urban development in the new EU member states: from adjustment to reinvention. Ashgate Publishing, Aldershot/Burlington, pp 1–18

Amin A (1994) Post-Fordism – a reader. Blackwell, Oxford

Bertaud A, Bertrand R (1995) Cities without land markets. World Bank Policy Research Working Paper WPS 1477. Available at: www.worldbank.org. Accessed on 22 June 2012

Buček J (2006) Post-socialist urban development, planning and participation – the case study of Bratislava city centre. In: Enyedi G, Kovács Z (eds) Social changes and social sustainability in historical urban centres – the case of central Europe. Centre for Regional Studies of Hungarian Academy of Sciences, Pécs, pp 65–80

Buček J, Pitoňák D (1997) City-centre transformation in Bratislava – modernisation and financial sector invasion. In: Kovács Z, Wiessner R (eds) Prozesse und Perspektiven der Stadtentwicklung in Ostmitteleuropa. Münchener Geographische Hefte 76:157–168

Chudík M, Klaučo L, Adamczak M (2007) Urbanistická štúdia dostavby územia autobusovej stanice a prestavby areálu Kablo na Mlynských Nivách v Bratislave na mestské polyfunkčné územie (Urbanistic study on bus station area completion and Kablo area rebuilding for polyfunctional area in Bratislava Mlynské Nivy). Aurex, Bratislava

Drahokoupil J (2007) Analyzing the capitalist state in post-socialism. Int J Urban Reg Res 31(2):401–424

Drahokoupil J (2008) Globalization and the state in central and eastern Europe: the politics of foreign direct investment. Routledge, London/New York

English Heritage (2012) Kings cross central. Available at: http://www.english-heritage.org.uk/professional/advice/conservation-principles/constructive-conservation/constructive-conservation-in-practice/kings-cross-central/. Accessed on 26 June 2012

Hammond E (2011) London house prices rise £395 a day. Financial Times, 10 December 2011, p 8

Hammond E (2012) Door to a cloistered world: prime housing. Financial Times, 7 April 2012

Harvey D (1989) The urban experience. Blackwell, Oxford

Harvey D (1997) Between space and time: reflections on the geographical imagination. In: Gregory D, Barnes T (eds) Reading human geography. Edward Arnold, London

Harvey D (2006) Limits to capital. Verso, London

Huba M (ed) (2007) Kapitoly z environmentálnej politiky (Chapters in environmental politics), vol 10. STUŽ SR, Bratislava

Huba M, Trubíniová Ľ (2009) Životné prostredie a trvalo udržateľný rozvoj. In: Kollár M, Mesežnikov G, Bútora M (eds) Slovensko 2008 – Súhrnná správa o stave spoločnosti. IVO, Bratislava, pp 429–455

Huba M, Ira V, Šuška P (eds) (2008) Bratislava/nahlas – ako výzva (Bratislava aloud as a challenge). Geografický Ústav SAV, STUŽ SR, Bratislava

Ira V (2003) The changing intra-urban structure of the Bratislava city and its perception. Geografický časopis 55:91–107

Jessop B (2002) The future of the capitalist state. Polity Press, Cambridge/Malden

Kong LL (2003) Value conflicts, identity construction, and urban change. In: Bridge G, Watson S (eds) A companion to the city. Blackwell, Oxford, pp 354–365

KPMG (2004) Cities in the new EU countries – position, problems, policies. Available at http://www.minbzk.nl/contents/pages/10238/citiessummary0110.pdf. Accessed 6 May 2007

Kvasnicová M (2009) Životné prostredie, občianska spoločnosť a architektúra (The environment, civic society and architecture). Životné prostredie 43(3):164–170

Lash S, Urry J (1987) The end of organised capitalism. Polity Press, Cambridge

Logan J, Molotch H (1987) Urban fortunes: the political economy of place. University of California Press, Berkley

Londák M (1999) Otázky industrializácie Slovenska (1945–1960) (1945–1960 Slovak industrialisation issues). Veda, Bratislava

McManus P (2000) Conservation. In: Johnston RJ, Gregory D, Pratt G, Watts M (eds) The dictionary of human geography, 4th edn. Blackwell, Oxford, pp 106–107

Obuchová V (2009) Priemyslená Bratislava (Industrial Bratislava). PT Marenčin, Bratislava

Podoba J (1998) Rejecting green velvet: transition, environment and nationalism in Slovakia. Environ Polit 7(1):129–144

Smith N (1996) The new urban frontier: gentrification and the revanchist city. Routledge, London

Smith N (2000) Circuit of capital. In: Johnston RJ, Gregory D, Pratt G, Watts M (eds) The dictionary of human geography, 4th edn. Blackwell, Oxford, p 82

Smith A, Pickles J (1998) Introduction: theorising transition and the political economy of transformation. In: Pickles J, Smith A (eds) Theorising transition: the political economy of post-communist transformations. Routledge, London/New York, pp 1–22

Snajdr E (2008) Nature protests: the end of ecology in Slovakia (culture, place, and nature). University of Washington Press, Washington

Strussová M, Petríková D (2009) Možnosti a predpoklady občianskej participácie v udržateľnom sídelnom rozvoji (Opportunities and preconditions of public participation in sustainable settlement development). Sociológia 41(4):354–379

Sýkora L (2009) Post-socialist cities. In: Kitchin R, Thrift N (eds) International encyclopaedia of human geography, vol 8. Elsevier, Oxford, pp 387–395

Thomas D (2011) A world apart. Financial capitals constitute a prime property market that is decoupled from national economies. Financial Times, 4 June 2011

Trend (2004) Bratislava má byť konečne mestom na Dunaji, 20 July 2004. Available at http://reality.etrend.sk/komercne-nehnutelnosti/bratislava-ma-byt-konecne-mestom-na-dunaji.html. Accessed 6 May 2010

Weber R (2002) Extracting value from the city: neoliberalism and urban redevelopment. Antipode 34:519–540

Author Index

© Springer Science+Business Media Dordrecht 2015

J. Buček, A. Ryder (eds.), *Governance in Transition*, Springer Geography,
DOI 10.1007/978-94-007-5503-1

Subject Index

© Springer Science+Business Media Dordrecht 2015

J. Buček, A. Ryder (eds.), *Governance in Transition*, Springer Geography,
DOI 10.1007/978-94-007-5503-1

T

U

MIX
Papier aus verantwortungsvollen Quellen
Paper from responsible sources
FSC® C105338

If you have any concerns about our products,
you can contact us on
ProductSafety@springernature.com

In case Publisher is established outside the EU,
the EU authorized representative is:
Springer Nature Customer Service Center GmbH
Europaplatz 3, 69115 Heidelberg, Germany

Printed by Libri Plureos GmbH
in Hamburg, Germany